Power Electronic Converters: Control and Design

Power Electronic Converters: Control and Design

Edited by Carolina Murray

www.statesacademicpress.com

States Academic Press,
109 South 5th Street,
Brooklyn, NY 11249, USA

Visit us on the World Wide Web at:
www.statesacademicpress.com

ISBN: 978-1-63989-741-4

Cataloging-in-Publication Data

Power electronic converters : control and design / edited by Carolina Murray.
   p. cm.
Includes bibliographical references and index.
ISBN 978-1-63989-741-4
1. Electric current converters. 2. Electric current rectifiers. 3. Electric motors--Electronic control.
4. Power electronics--Design and construction. I. Murray, Carolina.
TK7872.C8 P69 2023
621.381 532 2--dc23

Table of Contents

Preface

This book has been a concerted effort by a group of academicians, researchers and scientists, who have contributed their research works for the realization of the book. This book has materialized in the wake of emerging advancements and innovations in this field. Therefore, the need of the hour was to compile all the required researches and disseminate the knowledge to a broad spectrum of people comprising of students, researchers and specialists of the field.

A power converter is an electromechanical or electrical device that converts electrical energy from one form to a desired form. The power converter regulates the power supplied to the resistor through the use of a parallel DC-DC chopper and an unregulated rectifier. It regulates the power flow from an AC supply to the motor by properly controlling power semiconductor switches. The goal of power converter design is to increase efficiency. Controlling the power converters is crucial for ensuring the standards and requirements of the desired purpose. This book aims to shed light on some of the unexplored aspects of power convertors and the recent researches on them. It consists of contributions made by international experts. Students, researchers, experts and all associated with the design and control of power converters will benefit alike from this book.

At the end of the preface, I would like to thank the authors for their brilliant chapters and the publisher for guiding us all-through the making of the book till its final stage. Also, I would like to thank my family for providing the support and encouragement throughout my academic career and research projects.

Editor

Preface

Direct Digital Design of PIDF Controllers with Complex Zeros for DC-DC Buck Converters

Stefania Cuoghi [1], **Lorenzo Ntogramatzidis** [1], **Fabrizio Padula** [1] and **Gabriele Grandi** [2,*]

[1] School of Electrical Engineering, Computing and Mathematical Sciences, Curtin University, Bentley 6102, Western Australia, Australia; stefania.cuoghi@curtin.edu.au (S.C.); l.ntogramatzidis@curtin.edu.au (L.N.); fabrizio.padula@curtin.edu.au (F.P.)

[2] Department of Electrical, Electronic, and Information Engineering, University of Bologna, 40136 Bologna, Italy

*   Correspondance: gabriele.grandi@unibo.it

Abstract: This paper presents a new direct digital design method for discrete proportional integral derivative PID + filter (PIDF) controllers employed in DC-DC buck converters. The considered controller structure results in a proper transfer function which has the advantage of being directly implementable by a microcontroller algorithm. Secondly, it can be written as an Infinite Impulse Response (IIR) digital filter. Thirdly, the further degree of freedom introduced by the low pass filter of the transfer function can be used to satisfy additional specifications. A new design procedure is proposed, which consists of the conjunction of the pole-zero cancellation method with an analytical design control methodology based on inversion formulae. These two methods are employed to reduce the negative effects introduced by the complex poles in the transfer function of the buck converter while exactly satisfying steady-state specifications on the tracking error and frequency domain requirements on the phase margin and on the gain crossover frequency. The proposed approach allows the designer to assign a closed-loop bandwidth without constraints imposed by the resonance frequency of the buck converter. The response under step variation of the reference value, and the disturbance rejection capability of the proposed control technique under load variations are also evaluated in real-time implementation by using the Arduino DUE board, and compared with other methods.

Keywords: buck converter; inversion formulae; phase margin; gain crossover frequency

1. Introduction

Many industrial applications need the transformation of a constant DC voltage source to a constant value even under load variation, such as photovoltaic systems, mobile power supply equipment, DC supply systems, etc. The buck converter is one of the most widely utilized DC-DC converters, because of its simplicity, high efficiency, and low cost, see e.g., [1], and therefore each improvement has potentially a major economic and commercial impact. However, the presence of its nonlinear characteristics in the switching behavior and the saturation of the duty cycle render the output voltage control a challenging task. Numerous control strategies have been proposed for the voltage regulation of the buck converter. Each of them has advantages and disadvantages, and the selection of the most appropriate one depends mainly on the design task at hand. A brief review of the main digital control techniques can be found in [2]. Among these, the non-linear sliding mode control, which leads to fast transient response under load variation and high robustness, is worth mentioning [3,4]. However, the control performance is reduced by the introduction of high frequency oscillations around the sliding surface, the so-called chattering. Another practical alternative for the voltage regulation of the buck converter is the fuzzy logic control. This non-linear adaptive technique provides a robust performance under parameter variations and load disturbances, and can operate with noise and disturbance of different natures. However, these controllers are traditionally designed by trial-and-error, and this, combined with the rich architecture of the controller, constitutes a major drawback in carrying out stability and performance analysis, as well as transfer function and small signal analysis [5]. By contrast, classical linear proportional integral/proportional integral derivtive (PI/PID) control techniques are widely used by industrial practitioners for their simplicity in the design and implementation, and still by far play a major role. In fact, PID control is often taken as a benchmark for comparison with new strategies since it provides a good compromise among various types of performance indices, including voltage tracking and disturbance rejection, while guaranteeing a satisfactory robustness to small variations of the parameters of the buck converter [1]. Many PID-based strategies have also been combined with non-linear techniques to improve the closed-loop performance [2,3,6–10]. However, in the vast majority of the cases, the PID controller is designed in the continuous-time and then, for its practical implementation, it is converted to the discrete-time, see [10–12].

A common approach to the control feedback design involving PID controllers is to consider the ideal (improper) PID transfer function, which is non-causal. By contrast, the discretization of the ideal PID controller results in a casual, thus feasible, discrete transfer function. This explains why, in this context, the discretization appears to be critical. Indeed, frequency domain specifications assigned in the continuous-time domain can be affected by large undesired variations due to the discretization of the controller. Another critical issue in the design of PID controllers for the buck converter is caused by the presence of a resonance peak in the transfer function of the process. In fact, the resonance usually constrains the assignability of the closed-loop bandwidth, which has to be either well below the resonance frequency, or well above. The former solution is usually discarded since, for obvious reasons, it leads to poor performance. However, the latter is typically associated to a very large bandwidth, which is likely to induce severe saturation in the control variable.

This paper presents a new direct design technique for the discrete PID + filter (PIDF) controllers with complex conjugate zeros. Our method hinges on the classical pole-zero cancellation method [10] combined with the so-called discrete "inversion formulae" [13–18]. In the aforementioned design procedure, two parameters of the PIDF controller are used to achieve pole/zero compensation, as in [10], and the remaining two degrees of freedom are used to exactly meet specifications on the phase margin and gain crossover frequency with the use of the inversion formulae. The design approach based on these formulae was first presented for lead, lag and PID controllers in [13–18]. In this paper we introduce a new set of inversion formulae for the design of the time constant of the discrete PIDF controller.

Thanks to the closed-form design of the filter, which guarantees sufficiently large stability margins even in the presence of uncertainty, our method ensures a satisfactory performance in a neighborhood of the operating point.

The approach based on the inversion formulae results in a proper discrete PIDF controller which is directly implementable on a microcontroller, and which exactly satisfies the design requirements in the discrete domain. Thus, unlike the other techniques described above, the specifications are guaranteed to remain exactly satisfied even when considering the discrete implementation of the controller. Note that we avoid indirect tuning procedures and the inherent trial-and-error nature of graphical tuning techniques based on Bode, Nyquist and Nichols plots.

The procedure presented here is analytical in nature, and can be carried out in finite terms via simple equations which are dependent upon the sampling time of the analog-to-digital converter.

The structure of the discrete PIDF controller is obtained from a continuous-time PIDF [19] transfer function through the matched pole-zero mapping discretization method [20]. In this way, the cancelation results in a discrete transfer function, and therefore the controller can be directly designed in the z-domain.

Simulations and comparisons with other methods show the effectiveness of this new control strategy, which can accommodate plant uncertainties and also, importantly, load variations.

The proposed method has been first simulated in MATLAB Simulink® and then tested in a real-time digital implementation using the Atmel SAM3X8E microcontroller based on the ARM® Cortex® -M3 processor on an Arduino Due board. The experimental results have been analyzed and compared with classical PID control solutions.

The paper is organized as follows. The digital control schemes and discrete buck converter model are described in Section 2. In Section 3, we propose the discrete PIDF controller with complex conjugate zeros. The control problem and the proposed design solution are presented in Section 4. We describe the simulated and experimental results of the proposed DC-DC buck converter control and the performance comparison with other methods in Section 5. Conclusions and remarks will end the paper.

2. Digital Control Schemes and Discrete Buck Converter Model

The DC-DC buck converter is a step-down switching converter extensively described, e.g., in [21]. The block scheme of the digital voltage mode control and the buck converter circuit considered in this paper are shown in Figure 1. It is assumed that the converter operates in continuous-conduction mode (CCM).

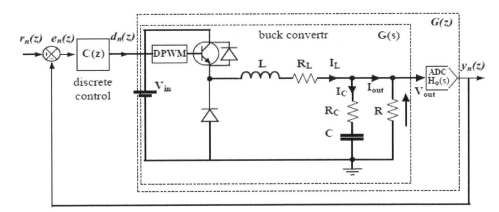

Figure 1. Switching DC-DC converter with digital voltage-mode control.

In the buck converter scheme, V_{out} is the output voltage, V_{in} is the input voltage, L is the filter inductance, C is the filter capacitance, R is the load resistance, R_L and R_C denote, respectively, the parasitic series resistances of the inductor and capacitor. Moreover, $r_n(z)$ represents the reference digital signal, while $y_n(z)$ denotes the sampled output of the process. The sampled signal is obtained by the analog-to-digital converter (ADC) with sampling period T_s. The tracking error signal $e_n(z)$ $= r_n(z) - y_n(z)$ is processed by a discrete-time compensator $C(z)$ to generate the control signal $d_n(z)$. The Digital Pulse Width Modulator (DPWM) converts $d_n(z)$ into the corresponding analog duty cycle with values between 0 and 1 according to the desired ratio of V_{out}/V_{in}, and modulates the PWM signal to drive the buck converter switch.

The transfer function of the discrete plant model $G(z)$ is the Z-transform of the product of the continuous-time converter transfer function $G(s)$ and the transfer function of the zero-order hold:

$$H_0(s) = \frac{1 - e^{-sT_s}}{s}$$

with sampling period T_s:

$$G(z) = Z[H_0(s)G(s)]. \tag{1}$$

Notice that in the hardware device the output voltage of the buck converter is driven into the admissible range of the ADC input voltage by a constant sensor gain H; the resulting output signal of the ADC is then multiplied by the factor *1/H* to be compared with the reference value r_n. In (1) the factors $H \cdot (1/H) = 1$ have been simplified and omitted. According to the buck converter averaged model and Equation (2) of [21], the transfer function of the buck converter:

$$G(s) = \frac{V_{out}(s)}{d(s)} = V_{in} \frac{\left(1 + \frac{s}{\omega_o}\right)}{\left(1 + \frac{2\xi}{\omega_n}s + \frac{s^2}{\omega_n^2}\right)} \tag{2}$$

is a second-order low-pass filter, with a left-half complex plane zero introduced by the equivalent series resistance of the filter capacitance.

The mathematical averaged model is obtained by the following input/state/output equations, where the diode and transistor conduction losses have been neglected [22].

$$\underbrace{\begin{bmatrix} \frac{di_L}{dt} \\ \frac{dV_C}{dt} \end{bmatrix}}_{\dot{x}(t)} = \underbrace{\begin{bmatrix} \frac{-R_L}{L} - \frac{RR_C}{L(R+R_C)} & \frac{-R}{L(R+R_C)} \\ \frac{R}{C(R+R_C)} & \frac{-1}{C(R+R_C)} \end{bmatrix}}_{A} \underbrace{\begin{bmatrix} i_L \\ V_C \end{bmatrix}}_{x(t)} + \underbrace{\begin{bmatrix} \frac{V_{in}}{L} \\ 0 \end{bmatrix}}_{B} d(t),$$

$$V_{out}(t) = \underbrace{\begin{bmatrix} \frac{RR_C}{R+R_C} & \frac{R}{R+R_C} \end{bmatrix}}_{C} \underbrace{\begin{bmatrix} i_L \\ V_C \end{bmatrix}}_{x(t)},$$

where:

$$\omega_n = \frac{1}{\sqrt{LC\frac{R+R_C}{R+R_L}}}, \quad \omega_o = \frac{1}{R_C C}, \quad \xi = \frac{\omega_n}{2}\left(R_C C + \frac{RR_L C + L}{(R + R_L)}\right). \tag{3}$$

The discrete model of the buck converter is:

$$G(z) = V_{in} \frac{(1 - a - bc)z + e^{-2\xi\omega_n T_s} - a + bc}{z^2 - 2az + e^{-2\xi\omega_n T_s}}, \tag{4}$$

where:

$$a = e^{-\xi\omega_n T_s} \cos\left(\omega_n T_s \sqrt{1 - \xi^2}\right), \tag{5}$$

$$b = e^{-\xi\omega_n T_s} \sin\left(\omega_n T_s \sqrt{1 - \xi^2}\right), \tag{6}$$

$$c = \frac{\xi\omega_o - \omega_n}{\omega_o \sqrt{1 - \xi^2}}, \tag{7}$$

and it can be obtained by applying the definition of the Z-transform to the series plants $H_0(s)G(s)H$. Notice that $G(z)$ is characterized the following two complex conjugate poles:

$$z_{1,2} = e^{(-\xi \pm j\sqrt{1-\xi^2})\omega_n T_s}.$$

Indeed, from (1) it follows that:

$$G(z) = Z\left[\frac{1 - e^{-T_s s}}{s}G(s)\right] = \left(1 - z^{-1}\right)Z\left[\frac{G(s)}{s}\right] = \frac{V_{in}\omega_n^2}{\omega_0}\left(1 - z^{-1}\right)Z[R(s)],$$

where:

$$R(s) = \frac{s + \omega_o}{s(s^2 + 2\xi\omega_n s + \omega_n^2)}. \tag{8}$$

Expanding $R(s)$ into partial fractions we have:

$$
\begin{aligned}
R(s) &= \frac{\omega_o}{\omega_n{}^2}\frac{1}{s} - \frac{\omega_o}{\omega_n{}^2}\frac{(s+\xi\omega_n)+\left(\xi\omega_n-\frac{\omega_n{}^2}{\omega_o}\right)}{s^2+2\xi\omega_n s+\omega_n{}^2} \\
&= \frac{\omega_o}{\omega_n{}^2}\frac{1}{s} - \frac{\omega_o}{\omega_n{}^2}\frac{(s+\xi\omega_n)}{(s+\xi\omega_n)^2+\omega_n{}^2(1-\xi^2)} \\
&= -\frac{\xi\omega_o-\omega_n}{\omega_n{}^2\sqrt{1-\xi^2}}\frac{\omega_n\sqrt{1-\xi^2}}{(s+\xi\omega_n)^2+\omega_n{}^2(1-\xi^2)}.
\end{aligned}
$$

Applying the standard manipulation theorems of the Z-transform to $R(s)$ we have:

$$
\begin{aligned}
R(z) &= \frac{\omega_o}{\omega_n{}^2}\frac{z}{z-1} - \frac{\omega_o}{\omega_n{}^2}\frac{z^2-e^{-\xi\omega_n T_s}\cos(\omega_n T_s\sqrt{1-\xi^2})z}{z^2-2e^{-\xi\omega_n T_s}\cos(\omega_n T_s\sqrt{1-\xi^2})z+e^{-2\xi\omega_n T_s}} \\
&\quad -\frac{\xi\omega_o-\omega_n}{\omega_n{}^2\sqrt{1-\xi^2}}\frac{e^{-\xi\omega_n T_s}\sin(\omega_n T_s\sqrt{1-\xi^2})z}{z^2-2e^{-\xi\omega_n T_s}\cos(\omega_n T_s\sqrt{1-\xi^2})z+e^{-2\xi\omega_n T_s}}.
\end{aligned}
$$

It follows that (8) can be written as:

$$
G(z) = V_{in} - V_{in}\frac{(z-e^{-\xi\omega_n T_s}\cos(\omega_n T_s\sqrt{1-\xi^2}))(z-1)}{z^2-2e^{-\xi\omega_n T_s}\cos(\omega_n T_s\sqrt{1-\xi^2})z+e^{-2\xi\omega_n T_s}}
$$

$$
-V_{in}\frac{\xi\omega_o-\omega_n}{\omega_o{}^2\sqrt{1-\xi^2}}\frac{e^{-\xi\omega_n T_s}\sin(\omega_n T_s\sqrt{1-\xi^2})z}{z^2-2e^{-\xi\omega_n T_s}\cos(\omega_n T_s\sqrt{1-\xi^2})z+e^{-2\xi\omega_n T_s}},
$$

which can be rewritten as in (4) using (5)–(7).

3. The Proposed Discrete PIDF Controller with Complex Conjugate Zeros

The controller presented in this paper is a discrete PIDF controller, described by the following transfer function:

$$
C(z) = \widetilde{K}_i\frac{z^2 - 2\delta_d\omega_d z + \omega_d{}^2}{(z-1)\left(z-\frac{\omega_n}{\beta_d}\right)}. \tag{9}
$$

when:

$$
\omega_d = e^{-\frac{\delta}{\tau}T_s},\ \delta_d = \cos\left(\frac{T_s}{\tau}\sqrt{1-\delta^2}\right),\ \beta_d = e^{(\beta-\delta)\frac{T_s}{\tau}},
$$

$$
\widetilde{K}_i = 2k_i\tau\beta\frac{1+e^{-\frac{\beta T_s}{\tau}}}{1+2e^{-\frac{\delta}{\tau}T_s}\cos\left(\frac{T_s}{\tau}\sqrt{1-\delta^2}\right)+e^{-\frac{2\delta}{\tau}T_s}},
$$

the controller (9) represents the discrete pole-zero mapping transformation with the sampling period T_s of the following continuous-time PIDF controller:

$$
C(s) = K_i\frac{1+2\delta\tau s+(\tau s)^2}{s\left(1+\frac{\tau}{\beta}s\right)}. \tag{10}
$$

Here K_i is the integral gain, δ is the damping ratio and $1/\tau$ is the natural frequency of the controller zeros, and:

$$
\beta = \frac{K_\infty}{\tau K_i}
$$

is a parameter that depends on the high frequency controller gain, which is defined as:

$$
K_\infty = \lim_{s\to\infty}C(s).
$$

The PIDF controller (10) is equivalent to the classical parallel PIDF controller:

$$C(s) = K_p\left(1 + \frac{1}{sT_i} + \frac{sT_d}{1 + sT_f}\right). \tag{11}$$

In fact, equivalent parameters for (11) can be obtained from δ, β, K_i, τ by equating (10) and (11): the resulting proportional gain, and the integral, the derivative and the filter time constants are shown in the following Equation (12):

$$K_p = K_i\frac{\tau}{\beta}(2\delta\beta - 1), \; T_i = \frac{\tau}{\beta}(2\delta\beta - 1), \; T_d = \frac{\tau}{\beta}\left(\frac{\beta^2}{2\delta\beta - 1} - 1\right), \; T_f = \frac{\tau}{\beta}. \tag{12}$$

Notice that when $\beta > 1$ and $\delta \geq 1$ the PIDF controller (10) reduces to a series PID controller, when $0 < \beta < 1$ the PIDF controller has complex conjugate zeros, and when $\beta = 1$ and $\delta = 1$ the PIDF controller becomes a PI controller, see [19].

Interestingly, the controller (9) can be written as a digital biquadratic filter:

$$C(z) = \frac{b_0 + b_1 z^{-1} + b_2 z^{-2}}{1 + a_1 z^{-1} + a_2 z^{-2}}, \tag{13}$$

where:

$$b_0 = \widetilde{K}_i, \; b_1 = -2\widetilde{K}_i\delta_d\omega_d, \; b_2 = \widetilde{K}_i\omega_d{}^2,$$
$$a_1 = -\left(\frac{\omega_d}{\beta_d} + 1\right), \; a_2 = \frac{\omega_d}{\beta_d},$$

which has the clear advantage of being directly implementable on a microcontroller by using the difference equation:

$$d[n] = \sum_{i=0}^{2} b_i e[n-i] - \sum_{j=1}^{2} a_j d[n-j]. \tag{14}$$

4. The Design Problem and the Proposed Design Solution

For control design purposes, the control system scheme can be simplified as in Figure 2, where $G(z)$ and $C(z)$ are given by (1) and (9), respectively, while $L(z)$ denotes the loop gain transfer function $L(z) = C(z)G(z)$.

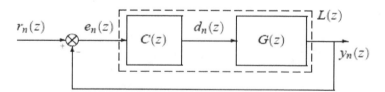

Figure 2. Block scheme of the digital control system.

4.1. Pole/Zero Compensation Method

The PIDF controller (9) introduces a pair of complex conjugate zeros, which can be placed to achieve pole/zero compensation. The PIDF parameters can be selected as follows:

$$\omega_d = e^{-\xi\omega_n T_s}, \; \delta_d = \cos\left(\omega_n T_s\sqrt{1 - \xi^2}\right), \tag{15}$$

where ξ and ω_n are the parameters of the buck converter described in (3). In this way, the transfer function of the controller can be factorized into two parts. The zeros and the integrator:

$$\frac{z^2 - 2\delta_d\omega_d z + \omega_d{}^2}{z - 1} \tag{16}$$

are completely determined by the zero/pole cancellation. The remaining factor is:

$$\widetilde{C}(z) = \frac{\widetilde{K}_i}{z - \frac{\omega_d}{\beta_d}} \tag{17}$$

and it comprises the parameters that are yet to be assigned. Thus, we can consider a new control problem where the controller is $\widetilde{C}(z)$, while the former factor is part of the plant, whose transfer function then becomes:

$$\widetilde{G}(z) = G(z)\frac{z^2 - 2\delta_d\omega_d z + \omega_d^2}{z - 1}.$$

4.2. Discrete Inversion Formulae Method

The design method based on the so-called "inversion formulae" consists in a set of closed-form expressions that deliver the parameters of the controller to exactly satisfy specifications on the gain crossover frequency ω_g, phase margin Φ_m and/or gain margin G_m. In most cases, these specifications are satisfied if the frequency response associated with the loop gain transfer function:

$$L(e^{j\omega T_s}) = C(e^{j\omega T_s})G(e^{j\omega T_s})$$

at frequency ω_g satisfies:

$$\left|L(e^{j\omega_g T_s})\right| = 1, \ \angle L(e^{j\omega_g T_s}) = \Phi_m + \pi. \tag{18}$$

In other words, the design method based on the inversion formulae is a way of constraining the loop gain polar plot to cross a specific point of the complex plane. In practice, in the vast majority of the situations that are interesting in practice this goal alone is sufficient to guarantee that the specifications on the phase (or gain) margin and crossover frequency are met, see [15] for further details.

The classical feedback design problem is to find a controller $C(z)$ that satisfies the steady-state a zero position error specification, and such that the gain crossover frequency and the phase margin of the loop gain transfer function $L(z)$ are, respectively, ω_g and Φ_m.

The first step of the design method consists in guaranteeing that the steady-state requirement is met. In most situations, the pole at $z = 1$ of the controller is sufficient to automatically satisfy the steady-state requirements. However, in some cases, the number of poles at $z = 1$ of the plant and the single pole at $z = 1$ of the controller are not sufficient to meet the desired static requirements, and the factor $\widetilde{K}_i$ in (9) must be chosen accordingly. For example, this is the case of a type-0 plant as the considered buck converter when the steady-state specifications not only require zero position error, but also that the velocity error (i.e., the tracking error in the response of a ramp) be equal to (or smaller than) a given non-zero constant. In the considered case specifications on the steady-state error do not lead to constraints in the value of the integral constant. Let $L(z) = \widetilde{C}(z)\widetilde{G}(z)$ be the loop gain transfer function. We define:

$$M_g \stackrel{def}{=} M(\omega_g) = 1/\left|\widetilde{G}(e^{j\omega_g T_s})\right|, \tag{19}$$

$$\varphi_g \stackrel{def}{=} \varphi(\omega_g) = \Phi_m - \pi - \angle\widetilde{G}(e^{j\omega_g T_s}). \tag{20}$$

The solvability of the feedback design problem amounts to solving the complex equation:

$$L(e^{j\omega_g T_s}) = e^{j(\Phi_m - \pi)}$$

in the unknowns $\widetilde{K}_i > 0$ and $\beta_d > 0$. The closed-form solution to this problem is given in the following theorem:

Theorem 1. *The values of $\widetilde{K}_i$ and β_d that solve the control problem are given by the following expressions:*

$$\beta_d = \frac{\omega_d}{\frac{\sin(\omega_g T_s)}{\tan(\varphi_g)} + \cos(\omega_g T_s)}, \tag{21}$$

$$\widetilde{K}_i = -M_g \sin(\varphi_g) \sin(\omega_g T_s)\left(1 + \frac{1}{\tan^2(\varphi_g)}\right). \tag{22}$$

Proof. From (18), the controller (17) has to be designed in such a way that:

$$\widetilde{C}(e^{j\omega_g T_s}) = M_g e^{j\varphi_g} = M_g(\cos\varphi_g + j\sin\varphi_g), \tag{23}$$

holds. The frequency response of (17) for $\omega = \omega_g$ can be written in Cartesian form as:

$$\widetilde{C}(e^{j\omega_g T_s}) = \frac{\widetilde{K}_i}{e^{j\omega_g T_s} - \frac{\omega_d}{\beta_d}}. \tag{24}$$

Equating (24) and (23) directly leads to (21) and (22). □

Remark 1. *It is easy to verify that the parameters β_d and $\widetilde{K}_i$ in (21) and in (22) are positive if and only if:*

$$\tan(\omega_g T_s) > -\tan(\varphi_g) \tag{25}$$

and one of the following conditions holds:

- $\omega_g \in \left[0, \frac{\pi}{2T_s}\right]$ and $\varphi_g \in \left[\pi, \frac{3}{2}\pi\right]$,
- $\omega_g \in \left[\frac{\pi}{2T_s}, \frac{\pi}{T_s}\right]$ and $\varphi_g \in \left[\frac{3}{2}\pi, 2\pi\right]$,
- $\omega_g \in \left[\frac{\pi}{T_s}, \frac{3\pi}{2T_s}\right]$ and $\varphi_g \in \left[\frac{\pi}{2}, \pi\right]$,
- $\omega_g \in \left[\frac{3\pi}{2T_s}, \frac{2\pi}{T_s}\right]$ and $\varphi_g \in \left[0, \frac{\pi}{2}\right]$,

or:

$$\tan(\omega_g T_s) < -\tan(\varphi_g) \tag{26}$$

and one of the following conditions holds:

- $\omega_g \in \left[0, \frac{\pi}{2T_s}\right]$ and $\varphi_g \in \left[\frac{3}{2}\pi, 2\pi\right]$,
- $\omega_g \in \left[\frac{\pi}{2T_s}, \frac{\pi}{T_s}\right]$ and $\varphi_g \in \left[\pi, \frac{3}{2}\pi\right]$,
- $\omega_g \in \left[\frac{\pi}{T_s}, \frac{3\pi}{2T_s}\right]$ and $\varphi_g \in \left[0, \frac{\pi}{2}\right]$,
- $\omega_g \in \left[\frac{3\pi}{2T_s}, \frac{2\pi}{T_s}\right]$ and $\varphi_g \in \left[\frac{\pi}{2}, \pi\right]$.

Note that, if one of the previous conditions fails, the required frequency-domain constraints are infeasible. In other words, the devised inversion formulae provide a solution whenever a feasible solution exists.

It is worth stressing that the proposed approach is based on closed-form expressions that deliver a discrete-time PIDF controller that satisfies exactly the design specification. This is clearly a major advantage since the imposed stability margin is guaranteed, and it is not subject to variations induced by the discretization method.

5. Design of the Buck Converter

5.1. Design Problem

The aim of this section is to apply the proposed designed procedure to the buck converter circuit with the parameters given in Table 1. The steady state requirement is zero position error, while the phase margin and the gain crossover frequency of the open loop frequency response are required to be equal to $\Phi_m = 85°$ and $\omega_g = 1600$ rad/s, respectively.

Table 1. Circuit parameters of the buck converter.

Parameter	Symbol	Value	Units
Input voltage	V_{in}	20	V
Reference voltage	V_{ref}	12	V
Filter Capacitance	C	100	μF
Filter Inductance	L	680	μH
Load resistance	R	20	Ω
ESR of capacitor	R_C	170	$m\Omega$
ESR of inductor	R_L	173	$m\Omega$

5.2. Proposed Solution Using Discrete-Time PIDF Controller

The discrete plant (4) of the buck converter with the parameters given in Table 1 and with sampling period T_s equal to 5×10^{-5} s is:

$$G(z) = \frac{0.603\,z + 0.1122}{z^2 - 1.916z + 0.9513}. \tag{27}$$

The same result can be obtained using the zero-order-hold discretization method on the transfer function of the continuous time averaged model (2):

$$G(s) = \frac{V_{out}(s)}{d(s)} = \frac{5001s + 2.942 \times 10^8}{s^2 + 998.1s + 1.471 \times 10^7}. \tag{28}$$

The steady-state requirements are automatically satisfied by the pole at $z = 1$ of the discrete PIDF controller. Its zeros can be designed to cancel the complex poles of $G(z)$ at $0.96 \pm j\,0.18$ by selecting $\delta_d = 0.982$ and $\omega_d = 0.97$ rad/s in (9). It follows that:

$$\widetilde{G}(z) = \frac{0.603\,z + 0.1122}{z - 1}. \tag{29}$$

The complex value $\widetilde{G}(e^{\,j\omega_g T_s}) = 8.94\,e^{\,j\,1.54}$ determines the gain $M_g = 1/8.94 = 0.11$ that the controller has to introduce at frequency ω_g, and the phase $\varphi_g = 85° + 180° + 88.4° = 353.4°$ of the controller at ω_g needed to satisfy the design specification on the phase margin. The parameters of the PIDF controller (9) that solves the problem are $\beta_d = 3.22$, $\widetilde{K}_i = 0.078$, and follow directly from (21–22). The resulting PIDF transfer function is:

$$C(z) = \frac{0.0781z^2 - 0.1496z + 0.0743}{z^2 - 1.303z + 0.3033}, \tag{30}$$

which can also be rewritten as:

$$C(z) = \frac{b_0 + b_1 z^{-1} + b_2 z^{-2}}{1 + a_1 z^{-1} + a_2 z^{-2}}, \tag{31}$$

with $a_1 = -1.303$, $a_2 = 0.3033$, $b_0 = 0.0781$, $b_1 = -0.1496$, $b_2 = 0.0743$. Applying this discrete controller, the design requirements are exactly satisfied, as one can observe by the Nyquist and Bode plots of the open loop frequency response $L(e^{\,j\omega T_s})$ shown in red in Figures 3 and 4. The step response of the controlled system is plotted in red in Figure 5 showing the effectiveness of the control in the time domain. Notice that selecting a different value of the sampling time T_s causes the complex conjugate poles of the discrete plant to shift in the complex plane. In this case, new values of δ_d and ω_d can be computed according to (9) as functions of T_s to exactly cancel the shifted poles.

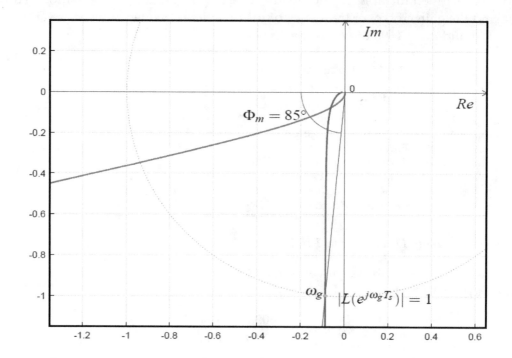

Figure 3. The Nyquist plot of the frequency response of the buck converter (green), and of the open loop frequency response with the discrete-time inversion formulae (red).

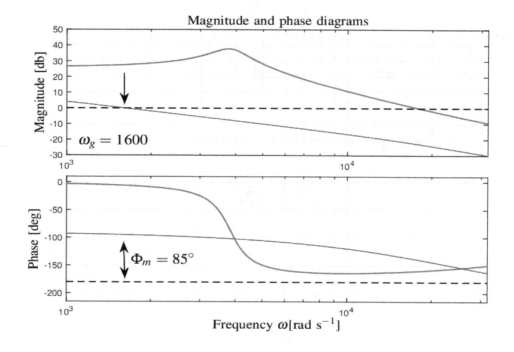

Figure 4. Bode diagrams of the frequency response of the buck converter (green), and of the open loop frequency response with the discrete-time inversion formulae (red).

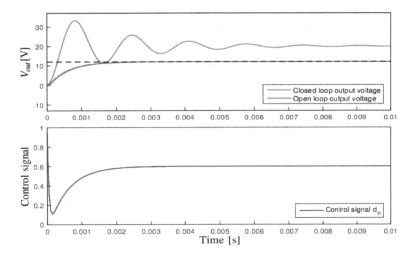

Figure 5. Open-loop step response (green), closed-loop step response (red) and the corresponding control variable (blue).

5.3. Simulation and Experimental Results

The proposed control system for the buck converter regulation has been extensively simulated in MATLAB-Simulink® using the model shown in Figure 6. As a first step, the PIDF controller has been tested introducing the discrete transfer function block contained in the Simulink® library. Then, this block has been substituted with the Infinite Impulse Response (IIR) digital filter shown in Figure 7, which has the advantage to be directly implementable by a microcontroller algorithm.

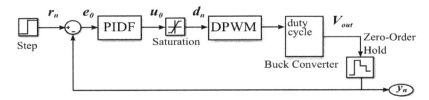

Figure 6. MATLAB-Simulink® model of the buck converter and control system.

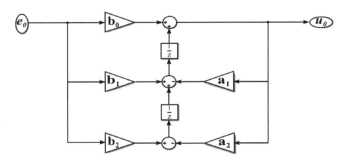

Figure 7. Discrete-time PIDF controller model: biquad cascade IIR filters using a direct form II transposed structure.

The control signal d_n, the inductor current and the output voltage of the converter and inductor under step reference variations from 0 V to 12 V are shown in Figure 8 from which the smoothness and monotonicity of the response achieved with our method can be clearly observed, as well as the notching effect of the complex conjugate zeros, which is well-visible in the first part of the transient response of the control signal. It is also worth noting that the simulated response and the experimental one exhibit a very good matching, demonstrating that our model is effectively descriptive of the real-world buck converter.

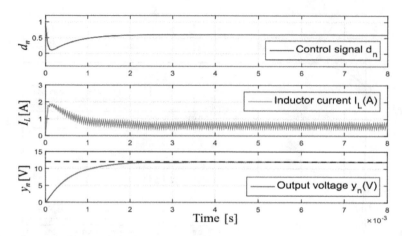

Figure 8. Simulated control signal, inductor current and output voltage under step variation of the reference value from 0 V to 12 V using the proposed control method.

An experimental hardware device has been built to verify the proposed method for the DC-DC buck converter. It is composed by the buck converter and an Arduino Due development board, based on a 32-bit Atmel SAM3X8E ARM® Cortex® M3 CPU, see Figure 9. The main components of the buck converter circuit have been selected as shown in Table 1.

Figure 9. Physical realization of the buck converter.

An interrupt routine is generated every $T_s = 5 \times 10^{-5}$ s. This routine starts the ADC conversion of the output signal of the converter and computes the duty cycle of the PWM control signal. Figure 10 represents the difference Equation (14) of the PIDF controller using the Direct-Form II Transposed structure of a Biquad Cascade IIR Filter shown in Figure 7.

```
// PIDF control algorithm
// implemented in Arduino board
    u0 = b0 * e0 + q1;

    q1 = b1 * e0 - a1 * u0 + q2;
    q2 = b2 * e0 - a2 * u0;
```

Figure 10. The Biquad Cascade IIR Filter algorithm.

The output voltage of the converter and inductor current I_L under step reference variations from 0 V to 12 V are shown in Figure 11. The measure of I_L has been obtained using the analog transducer LEM 6−NP with a 5V supply and a galvanic isolation between the primary and the secondary circuit. The experimental results confirm the behavior already observed in the simulations: the output voltage reaches the desired value in a monotonic fashion, and the inductor current remains always well below the saturation value. A zoom of the inductor current sensor output in steady-state condition is shown in Figure 12 from which the regularity of the PWM duty cycle when the system has reached the new steady-state can be observed. This is a consequence of the selected bandwidth, which is large enough to obtain a fast set point tracking, but narrow enough to avoid the amplification of high frequency noise and discontinuities due to the PWM behavior. Note that assigning such bandwidth without cancelling the complex conjugate poles would result in large oscillations due to the presence of the resonance peak in the closed-loop system.

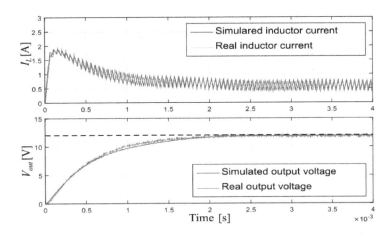

Figure 11. Experimental result using PIDF and the proposed method: inductor current sensor output and converter output signal under reference step variation from 0 V to 12 V.

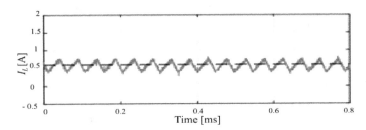

Figure 12. Experimental result using PIDF: zoom of the inductor current sensor output in steady-state condition.

5.4. Comparison with Other Methods

There are various design techniques for determining the parameters of a PID controller when the mathematical model of the plant is explicitly available. For a comparison with the proposed design method, three classical PID tuning techniques have been considered, see Table 2.

Table 2. PID parameters.

Control	K_p	K_i	K_d
IMC-Chien	0.033	958.7	6.519×10^{-5}
Pole placement	0.55	247.1,	7.353×10^{-5}
Pole-zero cancellation	0.02	294.7	2.004×10^{-5}

The first controller has been obtained using the Internal Model Control IMC-Chien method described in [23], and by neglecting the capacitor and inductor resistances in (2). The second has been obtained by placing one zero of the PID controller an octave below the cut-off frequency, approximately at 480 rad/s, while the other zero has been placed at 7×10^3 rad/s, see [12]. The third controller has been obtained by selecting the PID zeros to approximately cancel the complex conjugate poles of the converter at the cut-off frequency and a phase margin equal to $95°$, see [10]. The considered

144

continuous-time PID controllers have been simulated via the Simulink® PID(s) block which implements a PID controller in the form:

$$C_{PID}(s) = K_p + \frac{K_i}{s} + K_d \frac{N}{1 + N/s}. \tag{32}$$

The MATLAB-Simulink® PID(s) model uses a lowpass filter in the derivative term to obtain a proper transfer function. The default value of the coefficient N in the filter is set at 100. Using this value in (32), all the considered PID controls generate large oscillations during the step response transient. These oscillations are considerably reduced by setting the coefficient N of the filter to the value $N = 200,000$. It is clear that the time constant of the filter is a critical component in the design of a PID controller and that a systematic design method should be taken into account. Accordingly, in our method, the time constant of the filter is selected to achieve the desired closed-loop system performance, and it is not designed by using trial-and-error, empiric or rule-of-thumb methods.

For the practical implementation of the controller on the Arduino board, the continuous-time PID controllers are converted to the discrete-time by using the backward Euler's integration method, as suggested in [10]. The discrete control algorithm will therefore implement the causal difference equation:

$$d[n] = K_p e(n) + K_i T_s \sum_{i=0}^{n} e(n) + \frac{K_d}{T_s} [e(n) - e(n-1)].$$

Moreover, an anti-windup filter based on the conditional integration method (see [24] for details) has been implemented in order to minimize the detrimental effect of the large saturation resulting from the techniques listed in Table 3.

Table 3. Phase margin variations from continuous to discrete-time control using backward Euler's discretization method.

PID Control	Phase Margin		
	Continuous-Time	Discrete-Time	
		N = 100,000	N = 200,000
IMC-Chien	90°	47.5°	50.5°
Pole placement	98.6°	26.3°	29.5°
Pole-zero cancellation	95.7°	65.2°	67.6°

The simulated step responses of these three methods in the continuous-time are shown in Figure 13. The simulated step response using the IMC-Chien method is very fast, with a settling time of 0.2 ms. However, the peak of the resulting control signal is approximately 80.

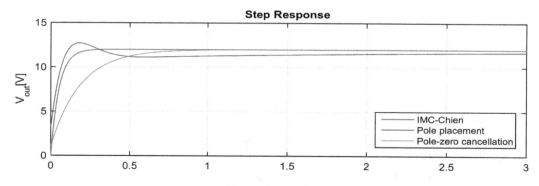

Figure 13. *Cont.*

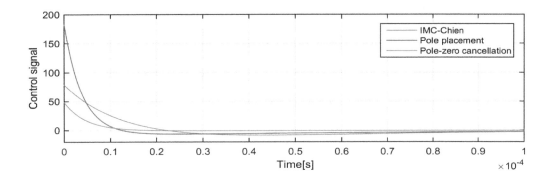

Figure 13. Simulated step responses using IMC-Chien method, the pole placement method and pole-zero control method in the continuous-time case: output voltages and control signals.

On the other hand, the saturation of the duty cycle range [0,1] leads to an oscillatory behavior in the experimental output voltage, see Figure 14b. In fact, the resulting settling time is 20 times greater than the one obtained in the simulated test. Moreover, a steady-state ripple in the output voltage and in the inductor current is present because of the excessively aggressive tuning.

The simulated continuous-time step response using the pole placement method exhibits a rise time of 8.1×10^{-5} s, a settling time of 4.2 ms, and an overshoot equal to 6%. As in the previous case, the control signal reaches a very high value, in this case with a peak of nearly 180. The converter signals obtained using this control method in the experimental hardware device are shown in Figure 14c. The main drawback of this type of control is the large steady-state ripple in the output voltage, see [10]. This is due to an excessively large closed-loop bandwidth that results in an aggressive control action which tries to compensate the high frequency noise. The saturation of the duty cycle in the range [0,1] leads to an ON-OFF behavior in the hardware device and high power dissipation both during the transient response and in the maintenance of the steady-state. Note also that the voltage ripple is unsuitable for most sensitive electronic equipment and the resulting current may cause heating and damage of capacitors over time, see [25].

The simulated continuous-time step response using the pole-zero cancellation method has a rise time of 4.1×10^{-4} s, a settling time of 0.7 ms. The peak of the control signal is 48, which is considerably lower than the ones obtained with the previously described techniques, but still orders of magnitude above the saturation level. The corresponding experimental results are shown in Figure 14d. Notice that the steady-state output ripple is not present using this type of control because of the less aggressive tuning of the parameters, which also results in a lower peak of the control variable. However, the non-linear saturation of the control signal is not considered in (2). As a consequence, the zeros of the controller only partially compensate the oscillatory effects of the buck converter poles in the transient period. It follows that the settling time rises to 4 ms in practice, and the experimental output voltage exhibits an oscillatory behavior with an overshoot of 20–30%.

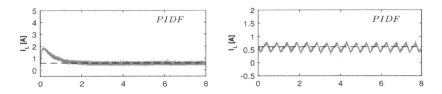

Figure 14. *Cont.*

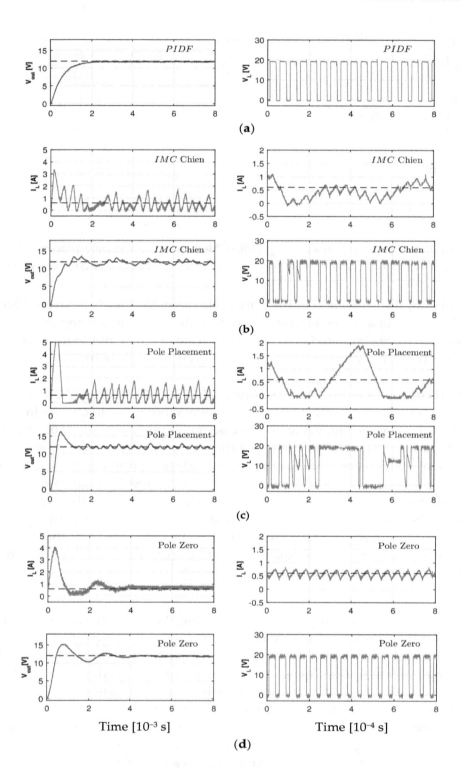

Figure 14. Inductor current and output voltage under step variation of reference value from 0 V to 12 V and zoom of inductor current and inductor current at steady-state using: (**a**) Inversion Formulae, (**b**) IMC-Chien, (**c**) Pole Placement, and (**d**) Pole Zero methods.

Compared to all the considered methods, the proposed control procedure leads to a good matching between simulated and experimental results, due to the design which is carried out directly in discrete-time via closed-form formulae. As such, the phase margin that we obtain with the discrete PIDF controller is exactly the design one. On the contrary, other approaches are based on the design of the controller in the discrete domain, and eventually, on the discretization of the obtained continuous controller. However, this results in a discrete controller that often delivers a phase margin considerably different from the one that would have ideally been obtained in the continuous time, see Table 2, where the phase margins obtained from the considered methods and a discrete PID of the following form are presented:

$$PID = K_p + K_i T_s \frac{z}{z-1} + K_d \frac{N}{1 + NT_s \frac{z}{z-1}}.$$

5.5. The Proposed Control under Output Load and Converter Parameters Variations

For the widespread diffusion of a control technique in practical applications, robustness to parameter variation and model uncertainty is clearly a key feature. For this reason, we study the behavior of the output voltage under different load resistance variations. Experimental results of the step load testing under different output loads (10 Ω, 20 Ω and 30 Ω) are shown in Figure 15a. Notice that the output voltage presents an almost overlapping behavior in the three considered cases, confirming that the control is not affected by load variation in the range ±50% of the nominal value, see Table 1. Other experimental results on load variations from 20 Ω to 10 Ω and from 20 Ω to 30 Ω in steady-state condition are shown in Figure 15b. Notice that the proposed control system promptly stabilizes the voltage output with negligible undershoot and overshoot, thus providing a good performance in the case of load variations. Moreover, the set-point step response remains virtually the same irrespectively of the load resistance.

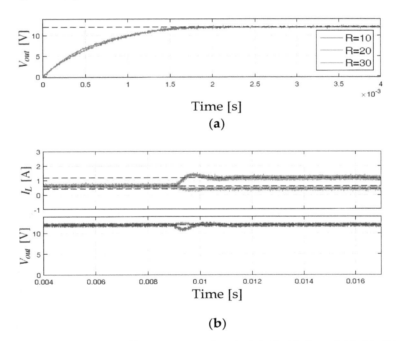

Figure 15. (a) Step responses under different output loads (10 Ω, 20 Ω and 30 Ω), (b) inductor current and output voltage under load variation from 20 Ω to 30 Ω (blue), and from 20 Ω to 10 Ω (red).

While load variations are due to normal operations of the buck converter, other parameters of the circuit of the converter, such as the inductance and capacitance, may vary as well as a result of the uncertainties affecting the production of the electrical components. In particular, the resonance frequency is directly related to the inductance and capacitance. In fact, since $R \gg R_C$ and $R \gg R_L$, in practice we have:

$$\omega_n \cong \frac{1}{\sqrt{LC}}.$$

Therefore, the inductor current and the output voltage under variations of the capacitor and inductor in the buck converter are also studied, and the results are shown in Figures 16 and 17. The proposed system delivers a good robust performance under parameter variations, and a monotonic response is obtained with all the considered combinations.

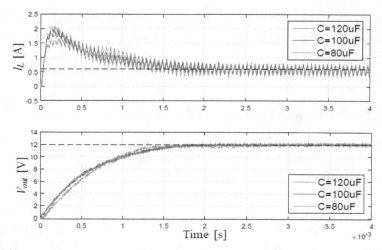

Figure 16. Step responses with the proposed design procedure when the model value of the capacitor is 100 μF−20%, 100 μF+20%.

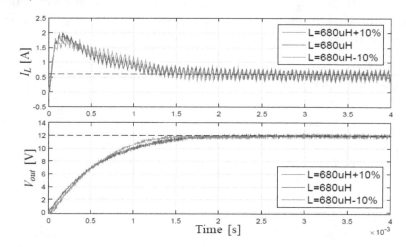

Figure 17. Step responses with the proposed design procedure when the model value of the inductor is set to 680 μH−10% and 680 μH+10%.

6. Conclusions

A new design framework for the control of buck converters has been presented in this paper. The proposed methodology is based on the discrete PIDF controller, and hinges on a direct design procedure that can be easily implemented in any non-specific platform. Indeed, the proposed methodology delivers a closed-form solution to meet suitable phase margin and gain crossover frequency values without a simulation environment. Moreover, the proposed design procedure and the discrete control algorithm are simple, they require small tuning times and they can be implemented by inexpensive microcontrollers.

Numerical and experimental verifications confirm that the proposed method goes well beyond the well-known zero/pole cancellation strategy and other control methods available in the literature. Indeed, the proposed approach enables the designer to assign an arbitrary bandwidth, which is therefore no longer constrained by the resonant peak. This aspect leads to a double benefit. On the one hand, this method avoids an excessively large bandwidth, which would result in noise/ripple amplification and ultimately in an increase in power consumption and a decrease in the component life. On the other hand, this method avoids the discretization problem that derives from discretizing a controller which assigns a bandwidth that is too large with respect to the sampling period. This, in particular, avoids detrimental effects on the stability margin due to the discretization. Moreover, experimental results confirm that the selection of large phase margin with the direct proposed method delivers a good system performance under load variations and plant uncertainties.

Author Contributions: S.C. provided for theoretical developments, simulations and experimental tests, L.N. and F.P. contributed to theoretical developments and manuscript organization, G.G. contributed to experimental results and manuscript finalization.

Abbreviations

The following abbreviations and symbols are used in this manuscript:

ADC	Analog-to-digital converter
CCM	Continuous-conduction mode
DC	Direct current
DC-DC	Direct current to direct current
DPWM	Digital pulse width modulator
ESR	Equivalent series resistance
IIR	Infinite impulse response (digital filter)
IMC	Internal model control
PI	Proportional-integral (controller)
PID	Proportional-integral-derivative (controller)
PIDF	PID + filter
PWM	Pulse width modulation

List of Symbols

a_i, b_i	PIDF coefficients
C	Buck converter capacitance
$C(s)$, $C(z)$	Continuous and discrete-time controller transfer function
D	Control signal
E	Tracking error
$G(s)$, $G(z)$	Continuous and discrete-time plant model
G_m	Gain margin
H	Constant sensor gain
$H_0(s)$	Zero-order hold transfer function
i_L	Buck converter inductor current
K_d	Controller derivative gain
K_i	Controller integral gain
K_∞	High frequency controller gain
K_p	Controller proportional gain
L	Buck converter inductance
$L(s)$, $L(z)$	Continuous and discrete-time loop gain transfer function
N	Filter coefficient
R	Output load resistance
R_C	ESR of buck converter capacitor
R_L	ESR of buck converter inductor
r_n	Reference digital signal
T_d	Derivative time constant of the controller
T_f	Filter time constant of the controller
T_i	Integral time constant of the controller
T_s	Sampling period
u_0	Output signal of IIR filter
V_C	Buck converter capacitor voltage
V_{in}	Buck converter input voltage
V_{out}	Buck converter output voltage
V_{ref}	Reference voltage
y_n	Sampled output of the process
δ	Damping ratio of the controller zeros
Φ_m	Phase margin
ξ	Buck converter damping ratio
ω_g	Gain crossover frequency

ω_n	Buck converter natural frequency
$1/\tau$	Natural frequency of the controller zeros

References

1. Yang, J.; Wu, B.; Li, S.; Yu, X. Design and qualitative robustness analysis of an DOBC approach for DC-DC buck converters with unmatched circuit parameter perturbations. *IEEE Trans. Circuits Syst.* **2016**, *63*, 551–560. [CrossRef]

2. Liu, Y.F.; Meyer, E.; Liu, X. Recent developments in digital control strategies for DC/DC switching power converters. *IEEE Trans. Power Electron.* **2009**, *24*, 2567–2577.

3. Al-Hosani, K.; Malinin, A.; Utkin, V.I. Sliding mode PID control of buck converters. In Proceedings of the 2009 European Control Conference (ECC), Budapest, Hungary, 23–26 August 2009; pp. 2740–2744.

4. Roshan, Y.M.; Moallem, M. Control of nonminimum phase load current in a boost converter using output redefinition. *IEEE Trans. Power Electron.* **2014**, *29*, 5054–5062. [CrossRef]

5. Perry, A.G.; Feng, G.; Liu, Y.F.; Sen, P.C. A design method for PI-like fuzzy logic controllers for DC–DC converter. *IEEE Trans. Ind. Electron.* **2007**, *54*, 2688–2696. [CrossRef]

6. Mantz, R.J.; Tacconi, E.J. Complementary rules to Ziegler and Nichols' rules for a regulating and tracking controller. *Int. J. Control* **1989**, *49*, 1465–1471. [CrossRef]

7. Son, Y.I.; Kim, I.H. Complementary PID controller to passivity-based nonlinear control of boost converters with inductor resistance. *IEEE Trans. Control Syst. Technol.* **2012**, *20*, 826–834. [CrossRef]

8. Kapat, S.; Krein, P.T. Formulation of PID control for DC–DC converters based on capacitor current: A geometric context. *IEEE Trans. Power Electron.* **2012**, *27*, 1424–1432. [CrossRef]

9. Sira-Ramirez, H. Nonlinear PI controller design for switchmode DC-to-DC power converters. *IEEE Trans. Circuits Syst.* **1991**, *38*, 410–417. [CrossRef]

10. Guo, L.; Hung, J.Y.; Nelms, R.M. PID controller modifications to improve steady-state performance of digital controllers for buck and boost converters. In Proceedings of the Seventeenth Annual IEEE Applied Power Electronics Conference and Exposition, Dallas, TX, USA, 10–14 March 2002; Volume 1, pp. 381–388.

11. Guo, L.; Hung, J.Y.; Nelms, R.M. Digital controller design for buck and boost converters using root locus techniques. In Proceedings of the IECON 2003 29th Annual Conference of the IEEE Industrial Electronics Society, Roanoke, VA, USA, 2–6 November 2003; Volume 2, pp. 1864–1869.

12. Guo, L.; Hung, J.Y.; Nelms, R.M. Evaluation of DSP-based PID and fuzzy controllers for DC–DC converters. *IEEE Trans. Ind. Electron.* **2009**, *56*, 2237–2248.

13. Marro, G.; Zanasi, R. New formulae and graphics for compensator design. In Proceedings of the 1998 IEEE International Conference on Control Applications, Trieste, Italy, 4 September 1998; Volume 1, pp. 129–133.

14. Zanasi, R.; Cuoghi, S.; Ntogramatzidis, L. Analytical and graphical design of lead–lag compensators. *Int. J. Control* **2011**, *84*, 1830–1846. [CrossRef]

15. Ntogramatzidis, L.; Ferrante, A. Exact tuning of PID controllers in control feedback design. *IET Control Theory Appl.* **2011**, *5*, 565–578. [CrossRef]

16. Zanasi, R.; Cuoghi, S. Direct methods for the synthesis of PID compensators: Analytical and graphical design. In Proceedings of the IECON 2011 37th Annual Conference of the IEEE Industrial Electronics Society, Melbourne, Australia, 7–10 November 2011; pp. 552–557.

17. Cuoghi, S.; Ntogramatzidis, L. Direct and exact methods for the synthesis of discrete-time proportional–integral–derivative controllers. *IET Control Theory Appl.* **2013**, *7*, 2164–2171. [CrossRef]

18. Cuoghi, S.; Ntogramatzidis, L. New inversion formulae for PIDF controllers with complex zeros for DC-DC buck converter. In Proceedings of the IECON 2011 42th Annual Conference of the IEEE Industrial Electronics Society, Florence, Italy, 23–26 October 2016; pp. 235–240.

19. Kristiansson, B.; Lennartson, B. Robust tuning of PI and PID controllers: Using derivative action despite sensor noise. *IEEE Control Syst. Mag.* **2006**, *26*, 55–69.

20. Dorf, R.C.; Bishop, R.H. *Modern Control Systems*; Pearson: Upper Saddle River, NJ, USA, 2011.

21. Severns, R.P.; Bloom, G. *Modern DC-to-DC Switchmode Power Converter Circuits*; Springer: New York, NY, USA, 1985.

22. Abbas, G.; Abouchi, N.; Pillonnet, G. Optimal state-space controller for power switching converter. In Proceedings of the IEEE Asia Pacific Conference on Circuits and Systems, Kuala Lumpur, Malaysia, 6–9 December 2010; pp. 867–870.

23. Panda, R.C.; Yu, C.C.; Huang, H. PID tuning rules for SOPDT systems: Review and some new results. *ISA Trans.* **2004**, *43*, 283–295. [CrossRef]

24. Visioli, A. *Practical PID Control*; Springer: London, UK, 2006.

25. Mohan, N.; Undeland, T.M. *Power Electronics: Converters, Applications, and Design*; John Wiley and Sons: Hoboken, NJ, USA, 2007.

2

Suppression Research Regarding Low-Frequency Oscillation in the Vehicle-Grid Coupling System using Model-Based Predictive Current Control

Yaqi Wang and Zhigang Liu *

School of Electrical Engineering, Southwest Jiaotong University, Chengdu 610031, China;
wangyaqi@my.swjtu.edu.cn
* Correspondence: liuzg@home.swjtu.edu.cn

Abstract: Recently, low-frequency oscillation (LFO) has occurred many times in high-speed railways and has led to traction blockades. Some of the literature has found that the stability of the vehicle-grid coupling system could be improved by optimizing the control strategy of the traction line-side converter (LSC) to some extent. In this paper, a model-based predictive current control (MBPCC) approach based on continuous control set in the *dq* reference frame for the traction LSC for electric multiple units (EMUs) is proposed. First, the mathematical predictive model of one traction LSC is deduced by discretizing the state equation on the alternating current (AC) side. Then, the optimal control variables are calculated by solving the performance function, which involves the difference between the predicted and reference value of the current, as well as the variations of the control voltage. Finally, combined with bipolar sinusoidal pulse width modulation (SPWM), the whole control algorithm based on MBPCC is formed. The simulation models of EMUs' dual traction LSCs are built in MATLAB/SIMULINK to verify the superior dynamic and static performance, by comparing them with traditional transient direct current control (TDCC). A whole dSPACE semi-physical platform is established to demonstrate the feasibility and effectiveness of MBPCC in real applications. In addition, the simulations of multi-EMUs accessed in the vehicle-grid coupling system are carried out to verify the suppressing effect on LFO. Finally, to find the impact of external parameters (the equivalent leakage inductance of vehicle transformer, the distance to the power supply, and load resistance) on MBPCC's performance, the sensitivity analysis of these parameters is performed. Results indicate that these three parameters have a tiny impact on the proposed method but a significant influence on the performance of TDCC. Both oscillation pattern and oscillation peak under TDCC can be easily influenced when these parameters change.

Keywords: vehicle-grid coupling system; low frequency oscillation; traction line-side converter (LSC); model-based predictive current control (MBPCC); dSPACE semi-physical verification

1. Introduction

With the rapid development of high-speed railway, alternating current (AC)–direct current (DC)–AC drive electric multiple units (EMUs) and electric locomotives are increasingly put into operation. Meanwhile, low-frequency oscillation (LFO) accidents have happened in many countries, such as Norway, Germany, Switzerland, the United States, and France [1–4]. Since 2008, the phenomenon has frequently occurred in China's high-speed railway depots. The LFO in railway is characterized by the amplitude fluctuation of grid-side voltage, current, and DC-side voltage, and happens when the multiple vehicles are concentrated in one power supply district and get power from a traction network [5]. With the larger voltage oscillation peak of the traction network, the protection logic operation of the line-side converter (LSC) would be triggered and result in the traction blockade, transformer breakdown, even an arrestor explosion [5,6].

Some testing, modeling, and simulation studies have been conducted to explore the mechanism of LFO. In [7], the phenomenon of LFO reappeared, and the stability of the traction LSC was analyzed by the adjustment of proportional integral (PI) controller parameters. Authors in [8] investigated the LFO and proposed an advanced multivariable control concept to avoid occurring stability problems. Based on the eigenvalues analysis, a detailed modeling plan was presented to investigate the mechanism of LFO [9]. A forbidden region-based criterion was performed to analyze of the critical condition of LFO [10]. In [11], a small-signal frequency domain model of LSC in a dq reference frame was established, and the impact of controller parameters on the stability of the vehicle-grid coupling system was discussed. According to these studies, the vehicle-grid coupling system can be defined as a dynamic stability problem of a large-scale multi-converter system under specific conditions. Many scholars working with renewable energy plants are dedicated to improving the modeling process of large systems to improve their stability and work efficiency, which can provide a good reference for high-speed railway [12–16]. The oscillation damping methods in railways could be categorized into two types [17]. One is the improvement of the traction network—for example, reducing the equivalent impedance of the traction network or adding power oscillation damping link. The other is modification of the vehicle and control, including decreasing of the number of vehicles, adjusting of the control parameters, and optimizing the control strategy.

Transient direct current control (TDCC) is a traditional control strategy in the traction LSC of China Railway high-speed 3 (CRH3) EMUs, which is a linear combination of error proportion and integral. Its integral feedback can inhibit the constant disturbance, while makes the closed-loop system unresponsive, and prone to oscillation and controlling quantity saturation. Though the integral link can reduce the static errors to a large extent, a tracking static error between the reference value and the actual value is inevitable in TDCC. This will adversely affect the control performance, such as the distortion of input electric parameters. In particular, when the number of EMUs accessed in traction network increases, the distortion will be exacerbated, and may directly lead to LFO.

The most common predictive control methods are generally divided into some types, namely deadbeat control, hysteresis-based control, trajectory-based control, continuous-control-set model predictive control (MPC), and finite-control-set MPC [18]. MPC was originally employed in a process industry that could easily handle multivariable cases, system constraints, and nonlinearities [19], and some studies have verified its good performance by applying it in the power electronics converters in renewable energy systems [20], cascaded H-Bridge Inverters [21], and so on. At present, MPC is regarded as one of the most promising control strategies, is starting to be applied to converters. In combination with predictive selection of a voltage-vector's sequence, predictive direct power control (PDPC) of three-phase converters was proposed, and the effective voltage vector sequence was selected for next sampling period [22]. Based on a finite set of controls, the model PDPC was also proposed, in order to directly control the active and reactive power by forecasting possible future behaviors [23–26]. In [27], the generalized predictive control of three-phase rectifiers developed in the dq frame is introduced.

So far, to solve the LFO in a high-speed railway system, the way of adopting the MPC strategy, based on continuous control set in the traction single-phase LSC, has not yet been tried. In addition, the control performance analysis of the proposed method in a vehicle-grid coupling system is also lacking at present. In this paper, which aims to optimize the control performance in EMUs' traction LSCs and handle the LFO in vehicle-grid coupling system, a model-based predictive current control (MBPCC) strategy based on a continuous control set is presented for EMUs' traction LSC control. Simulation results in MATLAB (R2016b, The MathWorks Incorporated, Natick, MA, USA) and experimental data in the dSPACE semi-physical platform demonstrate the effectiveness and superiority of the proposed method. Furthermore, the suppression effect on the LFO is proved in the simulation when multi-vehicles are accessed in the reduced-order model of the traction network.

The paper is organized as follows. In Section 2, the mathematical model of MBPCC based on the topology structure of one traction LSC of the CRH3 EMUs is deduced. In Section 3, the simulation model of EMUs' dual traction LSCs is built, and a comprehensive simulation verification compared with TDCC is performed in order to verify the superiority of the proposed control strategy from several aspects. In Section 4, a whole dSPACE semi-physical platform is established to certify the control strategy's feasibility in real applications. In Section 5, based on a reduced-order model of the traction network, the simulation model of seven vehicles accessed in the traction network is constructed to verify the suppression effect of LFO, and the sensibility analysis of parameters under the condition of seven EMUs accessed in the traction network is discussed, in order to further explore the impact on LFO occurrence and the performance of traction LSCs when system parameters change. The study idea is shown in Figure 1.

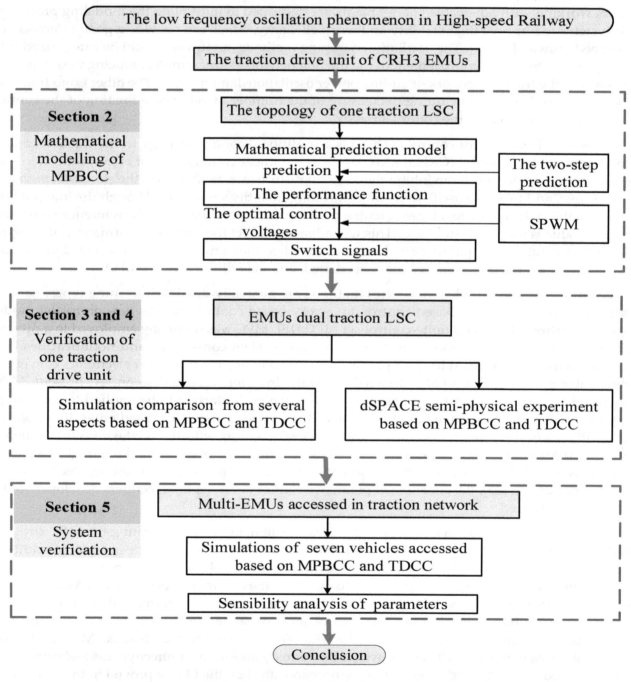

Figure 1. The study idea of the article.

2. Model Predictive Control of One Traction Line-Side Converter of China Railway High-Speed 3 Electric Multiple Units

2.1. Mathematical Model of One Traction Line-Side Converter

The vehicle-grid coupling system of high-speed railway is composed of the traction network and the EMUs traction drive system, as shown in Figure 2. In this paper, a CRH3 EMU is set as the study object; thus, the topology of the traction drive unit of EMUs is a single-phase two-level structure. When the LFO happens, the EMUs just start up and only supply power for the auxiliary facility by DC-side voltage. Therefore, the inverter and motor can be regarded as a pure resistance, and only part of the rectifier is involved in the equivalent circuit of one traction LSC.

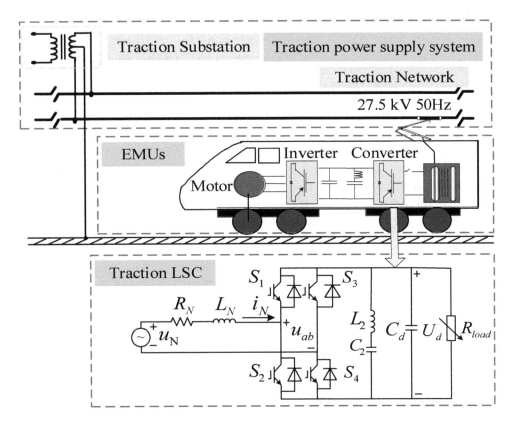

Figure 2. Vehicle-grid coupling system.

In the equivalent circuit of EMUs' traction LSC, L_N and R_N denote the leakage inductance and leakage resistance of the secondary winding of the vehicle transformer in EMUs, respectively. Four IGBTs (S_1, S_2, S_3, and S_4) are used to construct the two-level topology, C_d is the DC-side capacitor, inductance L_2 and capacitor C_2 compose the second-order filter circuit, and the inverter and motor are replaced by a pure resistive load R_{load}.

According to the Kirchhoff voltage law, the relationship between the voltage and current in the AC-side is listed below:

$$u_{ab} = u_N - R_N i_N - L_N \frac{di_N}{dt} \tag{1}$$

where u_N, i_N, and u_{ab} represent the line voltage, line current, and input voltage of the converter, respectively. If the harmonics are neglected, u_N, i_N, and u_{ab} in the dq reference frame are defined as follows:

$$\begin{cases} u_N = u_{Nd} \sin(\omega t) + u_{Nq} \cos(\omega t) \\ i_N = i_{Nd} \sin(\omega t) + i_{Nq} \cos(\omega t) \\ u_{ab} = u_{abd} \sin(\omega t) + u_{abq} \cos(\omega t) \end{cases} \tag{2}$$

where u_{Nd}, i_{Nd}, and u_{abd} are the d-axis components of u_N, i_N and u_{ab}. u_{Nq}, i_{Nq}, and u_{abq} are the q-axis components of u_N, i_N, and u_{ab}, respectively.

Substituting Equation (2) into Equation (1), the mathematical model of one traction LSC can be described as:

$$\begin{cases} u_{abd} = u_{Nd} - R_N i_{Nd} - L_N \frac{di_{Nd}}{dt} + \omega L_N i_{Nq} \\ u_{abq} = u_{Nq} - R_N i_{Nq} - L_N \frac{di_{Nq}}{dt} - \omega L_N i_{Nd} \end{cases} \tag{3}$$

2.2. Model Predictive Control of One Traction Line-Side Converter

2.2.1. Predictive Model of One Traction Line-Side Converter

Applying the first-order discrete approximation to the mathematical model expressed in Equation (3), a discrete dynamic model of one traction LSC in the dq reference frame is depicted by Equation (4):

$$\begin{cases} u_{abd}(k) = u_{Nd}(k) - R_N i_{Nd}(k) - \frac{L_N}{T_s}(i_{Nd}(k+1) - i_{Nd}(k)) + \omega L_N i_{Nq}(k) \\ u_{abq}(k) = u_{Nq}(k) - R_N i_{Nq}(k) - \frac{L_N}{T_s}(i_{Nq}(k+1) - i_{Nq}(k)) - \omega L_N i_{Nd}(k) \end{cases} \tag{4}$$

where T_s is the sampling interval. $i_{Nd}(k)$ and $i_{Nq}(k)$ are the discrete values of i_{Nd} and i_{Nq}, respectively; $i_{Nd}(k+1)$ and $i_{Nq}(k+1)$ are the one-step predictive discrete values of i_{Nd} and i_{Nq}, respectively; and $u_{abd}(k)$ and $u_{abq}(k)$ are the discrete values of the control voltages at the k-th sampling instant.

Thus, $i_{Nd}(k+1)$ and $i_{Nq}(k+1)$ can be predicted at the k-th sampling instant:

$$\begin{cases} i_{Nd}(k+1) = (1 - \frac{T_s R_N}{L_N})i_{Nd}(k) + T_s \omega i_{Nq}(k) - \frac{T_s}{L_N}u_{abd}(k) + \frac{T_s}{L_N}u_{Nd}(k) \\ i_{Nq}(k+1) = (1 - \frac{T_s R_N}{L_N})i_{Nq}(k) - T_s \omega i_{Nd}(k) - \frac{T_s}{L_N}u_{abq}(k) + \frac{T_s}{L_N}u_{Nq}(k) \end{cases} \tag{5}$$

The relationship between $u_{ab}(k)$ and $u_{ab}(k-1)$ is satisfied in the following way:

$$\begin{cases} u_{abd}(k) = u_{abd}(k-1) + \Delta u_{abd}(k) \\ u_{abq}(k) = u_{abq}(k-1) + \Delta u_{abq}(k) \end{cases} \tag{6}$$

where $\Delta u_{abd}(k)$ represents the variation between $u_{abd}(k)$ and $u_{abd}(k-1)$, and $\Delta u_{abq}(k)$ represents the variation between $u_{abq}(k)$ and $u_{abq}(k-1)$.

Substituting Equation (6) into Equation (5), then:

$$\begin{cases} i_{Nd}(k+1) = (1 - \frac{T_s R_N}{L_N})i_{Nd}(k) + T_s \omega i_{Nq}(k) - \frac{T_s}{L_N}u_{abd}(k-1) + \frac{T_s}{L_N}u_{Nd}(k) - \frac{T_s}{L_N}\Delta u_{abd}(k) \\ i_{Nq}(k+1) = (1 - \frac{T_s R_N}{L_N})i_{Nq}(k) - T_s \omega i_{Nd}(k) - \frac{T_s}{L_N}u_{abq}(k-1) + \frac{T_s}{L_N}u_{Nq}(k) - \frac{T_s}{L_N}\Delta u_{abq}(k) \end{cases} \tag{7}$$

Equation (7) is the prediction model. It should be noted that prediction line current values are decided by present currents, voltages, and the LSC's system parameters.

2.2.2. The Two-Step Prediction

In the digital control system, both the computation time delay and sampling time delay surely exist [28]. Because of the restriction of the hardware and the digital control algorithm, the control voltages calculated by the sampling and predictive currents have to be adopted at the $(k + 1)$-th sampling instant. Therefore, there is one sampling period delay (T_d) between the calculated control voltage by the controller and the real adopted control voltage, as shown in Figure 3. To eliminate the error caused by T_d, a two-step current prediction is adopted.

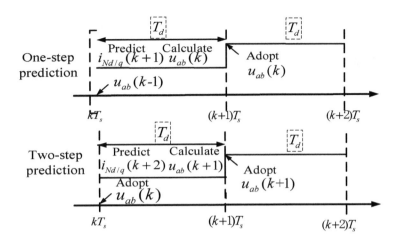

Figure 3. Control voltage in two switching periods.

When two-step prediction is adopted, $i_{Nd}(k+1)$ and $i_{Nq}(k+1)$ are first predicted at the k-th sampling instant according to Equation (8), and then $i_{Nd}(k+2)$ and $i_{Nq}(k+2)$ are predicted according to Equation (9). Thus, the control voltage $u_{abd}(k+1)$ and $u_{abq}(k+1)$ can be calculated at the k-th sampling instant, and then adopted to the rectifier at the $k+1$-th sampling instant. The delay error is compensated by two-step prediction.

It is worth noting that $i_{Nd}(k+1)$ and $i_{Nq}(k+1)$ are also predicted through two-step prediction at the $(k-1)$-th sampling instant. Thus, $i_{Nd}(k+1)$ and $i_{Nq}(k+1)$, which are predicted at the k-th sampling instant, are considered equal to the predicted values at the $k-1$-th sampling instant. Thus, the control voltage variations $\Delta u_{abd}(k)$ and $\Delta u_{abq}(k)$, which exist in the Equation (7), can be set to zero. The prediction Equation (7) can be rewritten as Equation (8). While $i_{Nd}(k+2)$ and $i_{Nq}(k+2)$ are not predicted before the k-th sampling instant, it cannot be affirmed that the prediction current values are equal to the k-th actual current values. The control voltages $\Delta u_{abd}(k+1)$ and $\Delta u_{abq}(k+1)$ that determine $i_{Nd}(k+2)$ and $i_{Nq}(k+2)$, respectively, cannot be omitted in Equation (9):

$$\begin{cases} i_{Nd}(k+1) = (1 - \frac{T_s R_N}{L_N})i_{Nd}(k) + T_s \omega i_{Nq}(k) - \frac{T_s}{L_N}u_{abd}(k-1) + \frac{T_s}{L_N}u_{Nd}(k) \\ i_{Nq}(k+1) = (1 - \frac{T_s R_N}{L_N})i_{Nq}(k) - T_s \omega i_{Nd}(k) - \frac{T_s}{L_N}u_{abq}(k-1) + \frac{T_s}{L_N}u_{Nq}(k) \end{cases} \quad (8)$$

$$\begin{cases} i_{Nd}(k+2) = (1 - \frac{T_s R_N}{L_N})i_{Nd}(k+1) + T_s \omega i_{Nq}(k+1) - \frac{T_s}{L_N}u_{abd}(k) + \frac{T_s}{L_N}u_{Nd}(k+1) - \frac{T_s}{L_N}\Delta u_{abd}(k+1) \\ i_{Nq}(k+2) = (1 - \frac{T_s R_N}{L_N})i_{Nq}(k+1) - T_s \omega i_{Nd}(k+1) - \frac{T_s}{L_N}u_{abq}(k) + \frac{T_s}{L_N}u_{Nq}(k+1) - \frac{T_s}{L_N}\Delta u_{abq}(k+1) \end{cases} \quad (9)$$

2.2.3. The Design of Performance Function

The key to achieving MPC is to obtain the most effective control quantities by solving the performance function optimally. In this paper, the optimal control voltage would be obtained.

The performance function is composed of the predicted current components at the $(k + 2)$-th sampling instant and the control voltage variations, with corresponding weighting coefficients [29]. The function is defined as follows:

$$w(k) = \alpha_1[i_{Nd}{}^*(k) - i_{Nd}(k+2)]^2 + \alpha_2[i_{Nq}{}^*(k) - i_{Nq}(k+2)]^2 + \beta_1\Delta u_{abd}^2(k+1) + \beta_2\Delta u_{abq}^2(k+1) \quad (10)$$

where α_1, α_2, β_1, and β_2 represent the weighting coefficients of the line current and voltage variations.

There are no analytical or numerical methods or control design theories to adjust these parameters; currently, they are determined based on empirical procedures. In [30], an approach based on an empirical procedure is presented to obtain suitable weighting factors. When more objectives are considered, the weighting coefficients are usually obtained using trial and error procedures and running time-consuming simulations [31]. Since the front two components in Equation (10) are both the current variables, weighting factors can be considered as the same, as are the back two components. Thus, if two of weighting coefficients are decided, the other two can also be set, which can be clearly seen from Table A1.

In order to make $i_{Nd}(k+2)$ and $i_{Nq}(k+2)$ track their references $i_{Nd}(k)^*$ and $i_{Nq}(k)^*$, the variations of the control voltage need to be kept as small as possible. To do this, take the derivative of Equation (10) to find the extreme point, as shown in Equation (11):

$$\begin{cases} \frac{\partial w(k)}{\partial \Delta u_{abd}(k+1)} = 0 \\ \frac{\partial w(k)}{\partial \Delta u_{abq}(k+1)} = 0 \end{cases} \tag{11}$$

Substituting $i_{Nd}(k+2)$ and $i_{Nq}(k+2)$ of Equation (9) and $w(k)$ of Equation (10) into Equation (11), the optimal control variables $\Delta u_{abd}(k+1)$ and $\Delta u_{abq}(k+1)$ can be derived as Equation (12). In addition, the output control voltages $u_{abd}(k)^*$ and $u_{abq}(k)^*$ can be depicted by Equation (13):

$$\begin{cases} \Delta u_{abd}(k+1) = \frac{-L_N T_s \alpha_1}{T_s^2 \alpha_1 + L_N^2 \beta_1} \{ i_{Nd}{}^*(k) - [(1 - \frac{T_s R_N}{L_N})i_{Nd}(k+1) + T_s \omega i_{Nq}(k+1) - \frac{T_s}{L_N}u_{abd}(k) + \frac{T_s}{L_N}u_{Nd}(k+1)] \} \\ \Delta u_{abq}(k+1) = \frac{-L_N T_s \alpha_2}{T_s^2 \alpha_2 + L_N^2 \beta_2} \{ i_{Nq}{}^*(k) - [(1 - \frac{T_s R_N}{L_N})i_{Nq}(k+1) - T_s \omega i_{Nd}(k+1) - \frac{T_s}{L_N}u_{abq}(k) + \frac{T_s}{L_N}u_{Nq}(k+1)] \} \end{cases} \tag{12}$$

$$\begin{cases} u_{abd}(k)^* = u_{abd}(k+1) = u_{abd}(k) + \Delta u_{abd}(k+1) \\ u_{abq}(k)^* = u_{abq}(k+1) = u_{abq}(k) + \Delta u_{abq}(k+1) \end{cases} \tag{13}$$

The optimal control voltages that minimize the performance function are calculated by Equation (13), and then fed to a modulator stage to generate the PWM drive signal g at the k-th instant. An MBPCC controller of LSCs can be designed based on the deducing process of prediction model, the delay compensation, and the solution of performance function. Figure 4 shows the block diagram of the MBPCC. A PI controller is adopted in the voltage loop to regulate the DC-side voltage, and its output is set as the current d-axis reference $i_{Nd}(k)^*$. In addition, the q-axis current reference $i_{Nq}(k)^*$ generally needs to equal zero to achieve the unit power factor.

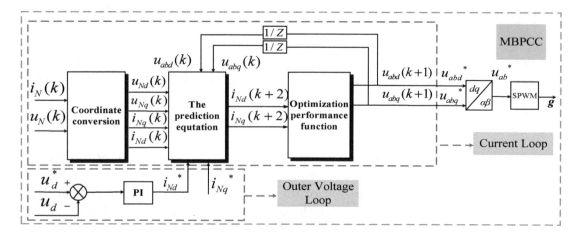

Figure 4. Block diagram of model-based predictive current control (MBPCC).

When LFO happens in high-speed railways, the fluctuation of the traction network's voltage will disturb the performance of the EMUs' LSC. Currents and voltages of the controller cannot trace well with the references, and lead to greater deterioration of fluctuation. MBPCC is a current-controlled strategy constituted by a prediction model and the performance function. By making the performance function minimal, the optimal control variables can be calculated to ensure the prediction current value tracking the references. Therefore, it is possible to adopt MBPCC to optimize the performance of the converter controller and suppress the LFO.

3. Simulations of One Traction Drive Unit of Electric Multiple Units

To validate the feasibility and effectiveness of the proposed method, simulation verifications based on the model of EMUs' dual traction LSCs, namely the traction drive unit, were carried out from several aspects, by comparing with the control performance of traditional TDCC in MATLAB /SIMULINK, respectively. The simulation model of EMUs' dual traction LSCs was built as shown in Figure 5. The adjustable parameters in the TDCC and MBPCC controllers were regulated into the most appropriate values, as listed in Tables A1 and A2.

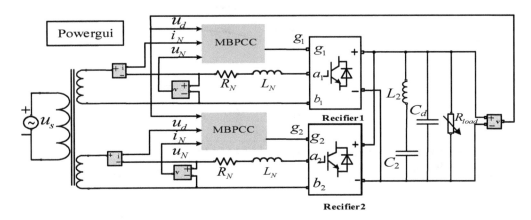

Figure 5. The circuit of electric multiple units' (EMUs') dual traction line-side converters (LSCs), based on MBPCC.

The block diagram of TDCC is shown in Figure 6. The outer voltage control loop uses a PI controller to keep the DC-side voltage equal to its reference value, and the PI output provides the reference of input current. The proportional controller of the inner current control loop makes the input current track its reference value.

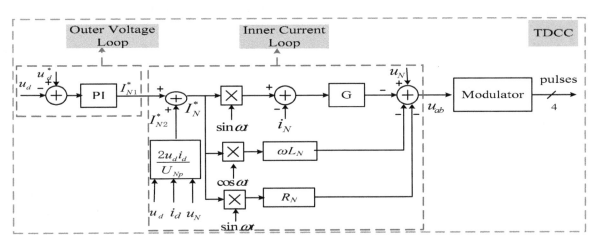

Figure 6. Block diagram of transient direct current control (TDCC).

3.1. Start-Operation Process

The start-operation process of EMUs' dual traction LSCs is divided into three periods—namely, the pre-charge period, the uncontrolled rectifier period, and the nominal load period. In the simulation, the initial time of the three stages was set to 0 s, 0.2 s, and 0.4 s, respectively. To compare MBPCC with

TDCC, the waveforms of u_N and i_N, as well as u_d are depicted in Figure 7. The performance indexes of the u_d of EMUs' dual traction LSCs are shown in Table 1. Figure 8 shows the fast Fourier transform (FFT) analysis results of the i_N with MBPCC and TDCC.

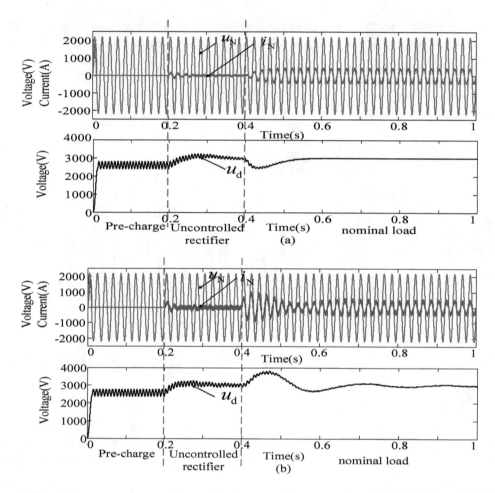

Figure 7. Simulation waveforms of u_N with i_N, as well as u_d, using (**a**) MBPCC and (**b**) TDCC.

Table 1. Comparison of performance indexes.

Control	Overshoot	Peak Time	Adjustment Time	Voltage Fluctuation
MBPCC	3.33%	0.10 s	0.25 s	±10 V
TDCC	26.70%	0.12 s	0.40 s	±40 V

In Figure 7, the pre-charge and uncontrolled rectifier periods of TDCC are almost the same as MBPCC's. When the nominal load is accessed, the u_d based on TDCC achieves the reference value after 0.4 s, with a large overshoot. However, the u_d based on MBPCC achieves stability after 0.25 s, and the overshoot is about 100 V, far less than that of the PI control. The voltage fluctuation range of MBPCC is only ±10 V, while the voltage fluctuation based on TDCC is larger. In Figure 8, the total harmonic distortion (THD) of i_N based on MBPCC is 4.76%, apparently lower than that of TDCC. In the control system based on MBPCC, there are more high-order harmonics around the odd switching frequency, which is considered as a drawback of the control strategy. Overall, MBPCC presents a better performance and dynamic response, due to the smaller overshoot, shorter adjustment time, and tinier voltage fluctuation.

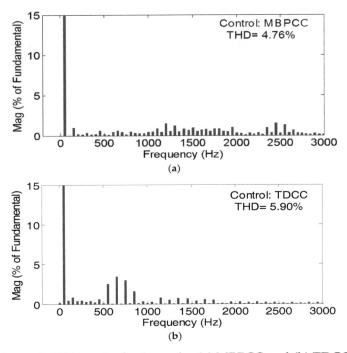

Figure 8. FFT results for i_N under (**a**) MBPCC and (**b**) TDCC.

3.2. Sudden-Load-Change Process

To further validate the dynamic response performance, the simulations under the sudden-load-change condition were carried out. Figure 9 shows the simulation waveforms of the dynamic response of the u_d under the condition of the load changing suddenly, based on MBPCC and TDCC, respectively. The load changes at 1 s, from 10 Ω to 0.01 Ω. Based on MBPCC, the u_d has gone through a period of voltage decline from 3000 V to about 2500 V, and returns to 3000 V after 0.5 s, with a ±20 V voltage fluctuation later. In contrast, the u_d based on TDCC drops by nearly 800 V, and fluctuates in the range of ±50 V after becoming stable again. Therefore, it can be concluded that the MBPCC has a better capacity for resisting disturbance than the transient PI control, because of smaller voltage decline and fluctuation when the load changes suddenly.

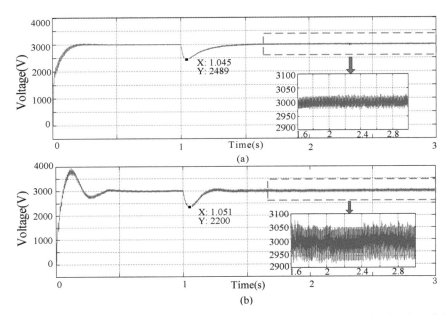

Figure 9. Simulation waveforms of the dynamic response of the u_d when the load suddenly changes, with (**a**) MBPCC and (**b**) TDCC.

3.3. Track Performance

At first, we measured i_{Nd}^* when the outer voltage loop plays a role in obtaining the reference current value. i_{Nd}^* was about equal to 830 A, and i_{Nq}^* was set to zero. Based on the measured values, we replaced the outer voltage loop by a step signal, to test the track performance while the reference current steps up or down similarly, a step signal replaces the constant module, in order to detect the track performance while the reference current i_{Nq}^* steps up or down.

As shown in Figure 10a, i_{Nd} decreased with i_{Nd}^*, varying from 830 A to 600 A at 0.3 s, and i_{Nq} returned to zero after a prompt downward fluctuation. In Figure 10b, i_{Nd} increases, with i_{Nd}^* varying from 830 A to 1000 A, and i_{Nq} rapidly returns the setting value i_{Nq}^* after an upward shock. Similar conditions can be seen in Figure 11, when i_{Nq}^* varies. In two cases, i_{Nd} and i_{Nq} can quickly track the reference current values of i_{Nd}^* and i_{Nq}^* in a very short time, regardless of whether i_{Nd}^* or i_{Nq}^* steps up or down. Thus, MBPCC could guarantee that the system has good track performance when the reference current value changes.

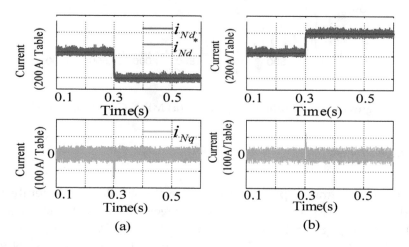

Figure 10. Simulation waveforms of i_{Nd} and i_{Nq} when (**a**) i_{Nd}^* varies from 830 A to 600 A and when (**b**) i_{Nd}^* varies from 830 A to 1000 A.

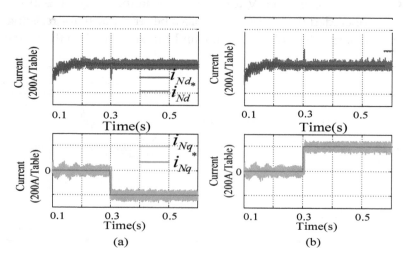

Figure 11. Simulation waveforms of i_{Nd} and i_{Nq} when (**a**) i_{Nq}^* varies from 0 A to −200 A and (**b**) when i_{Nq}^* varies from 0 A to 200 A.

In conclusion, MBPCC presents a better control performance for EMUs' dual traction LSCs, due to the smaller overshoot, shorter adjustment time, and tinier voltage fluctuation in the start-operation process, as well as its greater capacity for resisting disturbance and better track performance between the actual current and the reference current.

4. Semi-Physical Test of Electric Multiple Units' Dual Traction Line-Side Converter

To reflect the real application condition, a whole dSPACE semi-physical experimental platform was established. The dSPACE semi-physical experimental platform included the dSPACE simulator, the physical control circuit chassis, a power supply, and an external PC, as shown in Figure 12. The dSPACE simulator was used to simulate the circuit topology of EMUs' dual traction LSCs, and connected with the PC through a network line, so as to import the circuit simulation model from the PC to the dSPACE simulator. The physical control circuit chassis was connected to the PC through the data cable, in order to achieve the control algorithm program import and debugging. The power supply was supplied to the physical control circuit chassis.

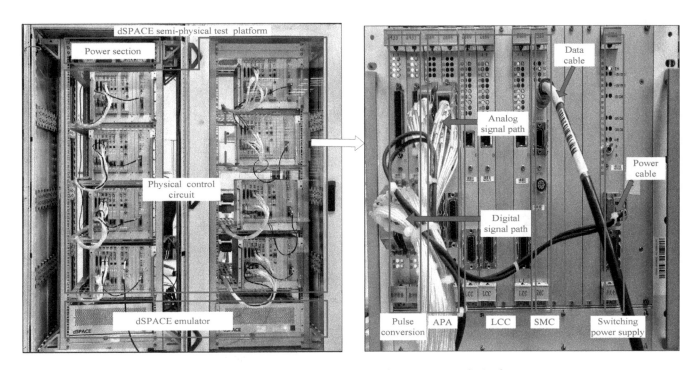

Figure 12. The dSPACE semi-physical experimental platform.

The most important part of the dSPACE semi-physical experimental platform was the physical control circuit chassis, which was composed of the main five modules, as shown in right-hand section of Figure 12:

(1) Switching power supply module: the switching power supply module has the function of a power supply for the entire physical control circuit chassis.

(2) SMC module: the SMC module realizes the data transmission of the physical control circuit chassis and the outside. After control program is compiled on the computer, the SMC module achieves the connection with the computer through the data cable, and control strategy is imported into the physical control circuit chassis.

(3) LCC module: an LCC module contains four internal DSPs, each of which controls a single converter. A physical control circuit chassis contains three LCC modules, so it can control six dual traction LSCs.

(4) APA module: the dSPACE simulation voltage and current signals are transmitted to the APA module through the analog signal channel, and the module achieves the signal acquisition of the physical control circuit chassis.

(5) Pulse conversion module: the pulse conversion module outputs the digital control signal of EMUs; thus, it can realize the control for EMUs directly.

Figures 13 and 14 show the simulation waveforms of u_N and i_N, as well as u_d when the system of dual traction LSCs are tested on the dSPACE semi-physical platform using MBPCC and TDCC, respectively. The numerical value on the abscissa multiplied by $8e^{-5}$ represents the simulation time. The process is performed from startup to normal working conditions.

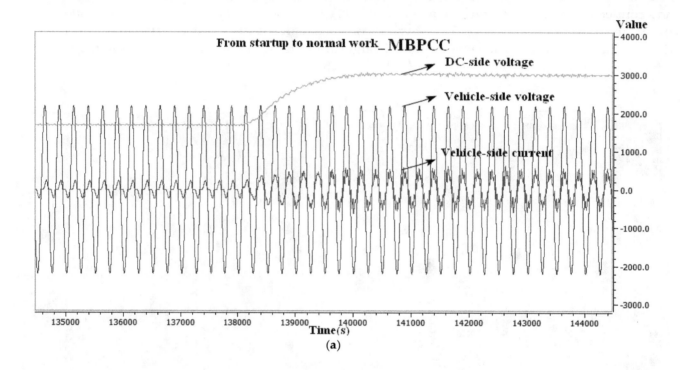

(a)

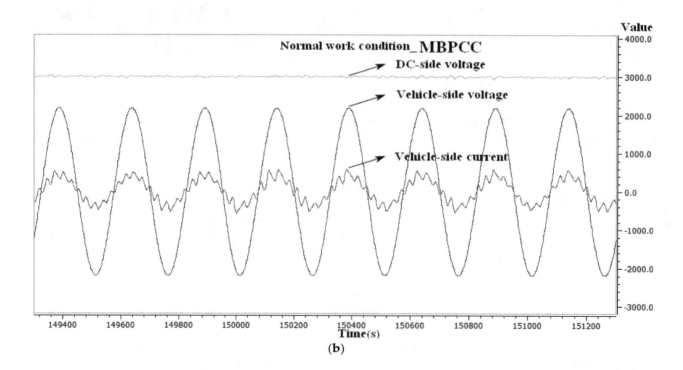

(b)

Figure 13. Waveforms of u_N and i_N, as well as u_d using MBPCC: (**a**) the process from startup to normal work; and (**b**) normal working conditions.

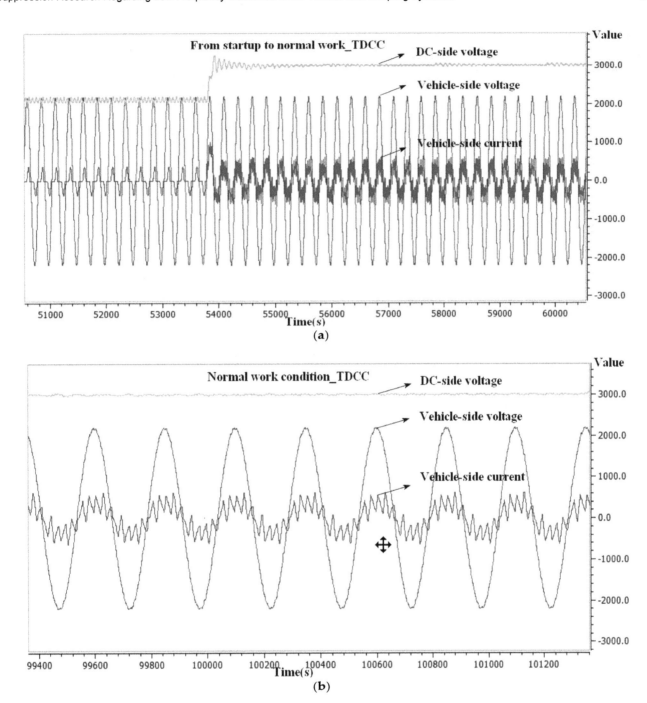

Figure 14. Waveforms of u_N and i_N, as well as u_d, using TDCC (**a**) the process from startup to normal work and (**b**) under normal working conditions.

The experimental results on the dSPACE semi-physical platform are almost in accordance with the simulation results in MATLAB. In Figure 13a, the u_d based on MPBCC achieves the reference value without the overshoot. As seen in Figure 13b, i_N can remain in phase with u_N, which means that the dual traction LSCs operate with a unified power factor. Compared with TDCC, as shown in Figure 14, i_N based on MBPCC has the smaller harmonic distortion.

5. System Verification

The dSPACE semi-physical experiment when multi-EMUs are accessed in the power supply system is still a challenge. To verify the suppression effect of LFO and perform further sensibility analyses, the simulations of seven EMUs accessed in traction network were performed in MATLAB/SIMULINK. The autotransformer power supply system is mostly adopted in high-speed railways of China. The structure of a traction network is very complicated because of a mass of multi-conductor transmission lines that are distributed, as well as the mutual coupling effect [5]. Therefore, considering the practical factors, such as the skin effect of lines and external disturbances, it is more reasonable to adopt a reduced-order method to model the traction network [5,32] than the Thevenin-equivalent method used in [33]. Avoiding a duplication of effort, the modeling process of the vehicle-grid coupling system is no longer described.

5.1. The Effect of Suppressing Low-Frequency Oscillation

Based on the TDCC controller, when seven vehicles are accessed in the reduced-order model of traction network, LFO occurs in both voltage and current, as shown in Figure 15. The u_d fluctuates between 2600 V and 3600 V. It can also be found that the traction network voltage arrives at the wave crest while the grid-side current reaches the wave trough. The amplitude fluctuation of voltages and currents will influence the performance of the EMUs' LSC, and even lead to a traction blockade.

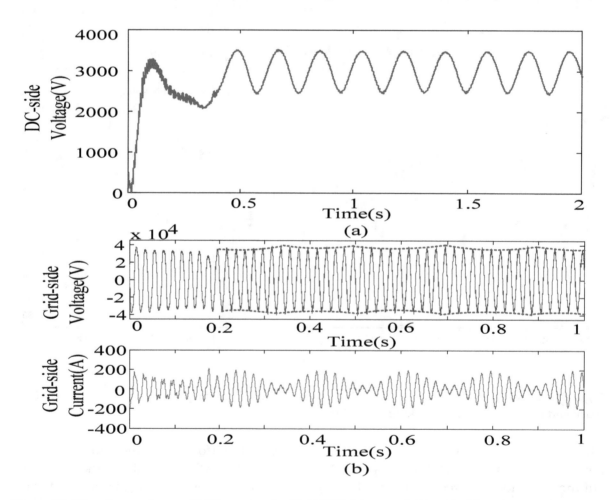

Figure 15. Waveforms of seven EMUs accessed with TDCC: (a) u_d; and (b) grid-side voltage and current.

By adopting the proposed MBPCC strategy, the system can still achieve stability when the number of multi-EMUs accessed in the traction network reaches seven. Figure 16 shows the simulation results when seven EMUs are accessed. It can be observed that the LFO does not appear even though vehicle numbers have reached their critical value. In a stable state, the u_d remains stable at about 3000 V, and the voltage deviation is ±10 V. The fluctuations of grid-side voltage and current are small.

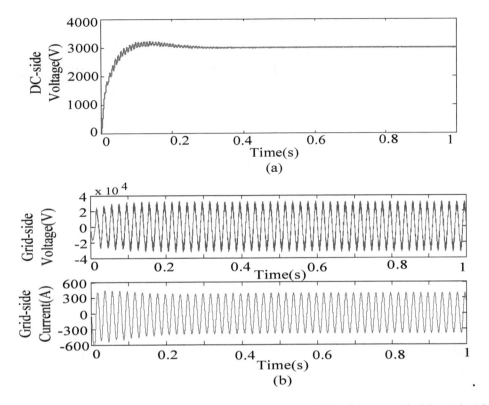

Figure 16. Waveforms of seven EMUs accessed with MBPCC: **(a)** u_d and **(b)** grid-side voltage and current.

5.2. Analysis of System Parameters

The analysis of parameter sensitivity is necessary to find out their influence on the stability of the vehicle-grid coupling system. In this paper, in order to find out the effects of different external parameters on the LFO and the performance of traction LSC, three parameters (the load resistance R_{load}, equivalent leakage inductance, L_N and the distance to power supply D) are discussed when seven vehicles are accessed in the traction network.

5.2.1. Load Resistance R_{load}

Figure 17 shows the waveforms of the u_d under TDCC and MBPCC, when the load resistance R_{load} is set as 20, 50, 75, or 100 Ω, respectively. As seen in Table 2 and Figure 17, when the load resistance R_{load} varies, both the overshoot and adjustment time of the u_d under MBPCC have only some tiny variations.

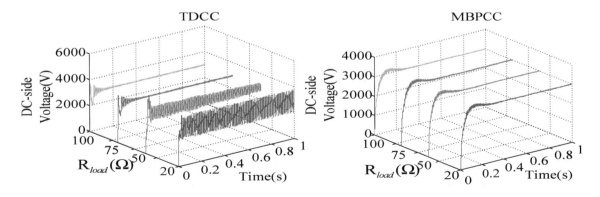

Figure 17. Waveforms of u_d when R_{load} = 20, 50, 75, or 100 Ω.

Table 2. Performance of the system.

Item	Value	TDCC		MBPCC	
		Oscillation Pattern	**Oscillation Peak (V)**	**Overshoot (V)**	**Regulation Time (s)**
R_{load} (Ω)	20	Stable	3600	3200	0.25
	50	Damping	[3500, 3350]	3200	0.25
	75	No	-	3200	0.25
	100	No	-	3200	0.25
L_N (H)	0.001	No	-	3200	0.25
	0.002	Damping	[3500, 3350]	3200	0.25
	0.004	Stable	3600	3200	0.25
	0.006	Stable	3700	3200	0.25
D (km)	10	Damping	[3240, 3530]	3200	0.20
	20	Damping	[3240, 3520]	3200	0.20
	30	Damping	[3300, 3520]	3200	0.30
	40	Stable	3500	3300	0.50

(1) Damping oscillation: the oscillation peak diminishes gradually and returns final to a stable state; (2) Stable oscillation: the oscillation peak abidingly maintains a value [9].

However, LFO happens under TDCC when the load resistance is small. The smaller R_{load} is, the severer the oscillation is. LFO would not occur once R_{load} exceeds a value about 75 Ω. Therefore, it can be concluded that the load resistance has fewer effects on the proposed method than TDCC.

5.2.2. Equivalent Leakage Inductance L_N

In Figure 18, when the equivalent leakage inductance of vehicle transformer L_N varies from 0.001 H to 0.006 H, the system under MBPCC maintains an initially good dynamic performance, with little change in the u_d's overshoot and regulation time. Hence, MBPCC can keep good performance and be unaffected by the change of L_N, while TDCC is very sensitive. When L_N is set as 0.001 H, the LFO does not happen. With the increase of L_N, oscillation appears, and the oscillation peak enlarges gradually.

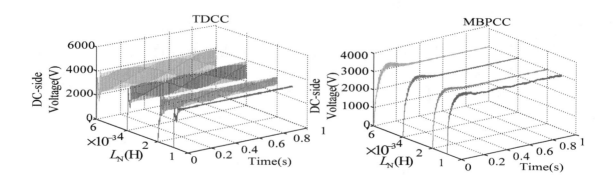

Figure 18. Waveforms of u_d when L_N = 0.001, 0.002, 0.004, and 0.006 H.

5.2.3. Distance to Power Supply D

D represents the distance to the power supply in the traction network. The larger D is, the larger the equivalent impedance on the line side will be. As seen in Figure 19 and Table 2, with the change of D, the influence on MBPCC's control performance is slight, because the overshoot and adjustment time change only slightly more than before, and the LFO does not happen.

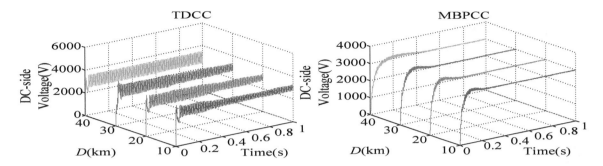

Figure 19. Waveforms of u_d when $D = 10, 20, 30,$ and 40 km.

As for TDCC, the oscillation phenomenon can be observed when D was chosen as 10, 20, 30, and 40 km. The oscillation peak value is about 3500 V at first, and then gradually diminishes in the former three cases; when D is 40 km, the oscillation peak will be kept at 3500 V. The details are listed in Table 2. It was found that the oscillation peak varies from 3530 V to 3240 V within 1 s when D is 10 km, and from 3520 V to 3300 V when D is 30 km. The damping capability of LFO decreases with the increase of D under TDCC.

Contrasting results show that MBPCC can be almost unaffected by the three parameters, and LFO does not happen when these parameters change. However, the performance of TDCC is sensitive to the three parameters. Both the oscillation pattern and oscillation peak can be easily influenced when the parameters change.

6. Conclusions

To optimize the control performance of EMUs' traction drive units, namely dual traction LSCs, and suppress LFO in vehicle-grid coupling system of high-speed railways, MBPCC is proposed in order to apply the traction LSC of CRH3 EMUs. After theoretical analysis, simulation verifications, and semi-physical verifications, the performance of MBPCC is demonstrated in comparison to TDCC, and the advantages of each listed below:

(1) Simulation verifications in MATLAB of EMUs' dual traction LSCs based on MBPCC and TDCC were implemented from three aspects. The results prove that MBPCC can obtain better dynamic and static performances, such as a smaller THD, tinier overshoot in start-operation process, greater capacity for resisting disturbance under load changing suddenly, and a better track performance.

(2) Semi-physical verifications in the dSPACE semi-physical experimental platform were realized. The results certified the effectiveness of MBPCC and its superiority in real applications, when compared with TDCC.

(3) When the multi-EMUs were assessed in the reduced-order model of traction network, the results showed that MBPCC can ensure the system's stability and suppress LFO more efficiently compared with TDCC.

(4) The influences of different external parameters R_{load}, L_N, and D in the vehicle-grid coupling system under MBPCC and TDCC have been discussed in detail. It could be concluded that these three parameters have a tiny impact on MBPCC, while they greatly influence the performance of TDCC. Both the oscillation pattern and oscillation peak under TDCC can be easily influenced when parameters change.

The proposed method can be applied for the control of EMUs' traction LSCs, and provide a good effect on the suppression of LFO. Moreover, MBPCC is insensitive to system parameters, which provides greater possibilities for its application. There is some advanced work regarding MBPCC that can be studied in the future. In the aspect of physical verification, multi-EMUs accessed in the model of the traction network should be realized in order to increase the reliability for suppressing LFO. With respect to algorithm optimization, combining predictive control with the disturbance observer can further advance the performance and robustness of the system, which can deal with the modeling errors and uncertainties, disturbances, and sensor noise.

Author Contributions: The individual contribution of each co-author with regards to the reported research and writing of the paper is as follows. Z.L. and Y.W. conceived the idea, Y.W. performed experiments and data analysis, and all authors wrote the paper. All authors have read and approved the final manuscript.

Acknowledgments: This study is partly supported by the National Nature Science Foundation of China (No. U1734202, U1434203), China Railway (No. 2015J008-A) and the Sichuan Province Youth Science and Technology Innovation Team (No. 2016TD0012).

Appendix

The parameters of the traction LSC of CRH3 EMUs under MBPCC and TDCC are listed in Tables A1 and A2.

Table A1. Parameters of EMUs' dual traction LSCs under MBPCC.

System Parameter	Value	Control Parameter	Value
u_s	25 kV	K_{iv}	0.1
L_N	0.004 H	K_{pv}	9
R_N	0.06 Ω	α_1	1
L_2	0.00084 H	α_2	1
C_2	0.003 F	β_1	0.0002
C_d	0.006 F	β_2	0.0002
U_d	3000 V	–	–
R_{load}	10 Ω	–	–

Table A2. Parameters of EMUs' dual traction LSCs under TDCC.

System Parameter	Value	Control Parameter	Value
u_s	25 kV	K_{iv}	0.1
L_N	0.004 H	K_{pv}	9
R_N	0.06 Ω	G	1
L_2	0.00084 H	–	–
C_2	0.003 F	–	–
C_d	0.006 F	–	–
U_d	3000 V	–	–
R_{load}	10 Ω	–	–

References

1.   Menth, S.; Meyer, M. Low frequency power oscillation in electric railway systems. *EB Elektrische Bahnen* **2006**, *2*, 216–221.

2.   Eitzmann, M.; Paserba, J.; Undrill, J. Model development and stability assessment of the Amtrak 25 Hz traction system from New York to Washington DC. In Proceedings of the 1997 IEEE/ASME Joint Railroad Conference, Boston, MA, USA, 18–20 March 1997; pp. 21–28.

3.   Heising, C.; Fang, J.; Bartelt, R.; Staudt, V.; Steimel, A. Modelling of rotary converter in electrical railway traction power-systems for stability analysis. In Proceedings of the Electrical Systems for Aircraft, Railway and Ship Propulsion, Bologna, Italy, 19–21 October 2010; pp. 1–6.

4.   Liu, Z.; Zhang, G.; Liao, Y. Stability research of high-speed railway EMUs and traction network cascade system considering impedance matching. *IEEE Trans. Ind. Appl.* **2016**, *5*, 4315–4326. [CrossRef]

5.   Zhang, G.; Liu, Z.; Yao, S.; Liao, Y.; Xiang, C. Suppression of low frequency oscillation in traction network of high-speed railway based on auto disturbance rejection control. *IEEE Trans. Transp. Electrif.* **2016**, *2*, 244–245. [CrossRef]

6.   Liu, J.; Zheng, Y. Resonance mechanism between traction drive system of high-speed train and traction network. *Trans. China Electro Tech. Soc.* **2013**, *4*, 221–227.

7.   Wang, H.; Wu, M. Simulation analysis on low-frequency oscillation in traction power supply system and its suppression method. *Power Syst. Technol.* **2015**, *4*, 1088–1095.

8.   Heising, C.; Oettmeier, M.; Staudt, V.; Steimel, A.; Danielsen, S. Improvement of low-frequency railway power system stability using an advanced multivariable control concept. In Proceedings of the IECON, Porto, Portugal, 3–5 November 2009; pp. 560–565.

9.   Danielson, S.; Fosso, O.B.; Molinas, M.; Suul, J.A.; Toftevaag, T. Simplified models of a single-phase power electronic inverter for railway power system stability analysis–development and evaluation. *Electr. Power Syst. Res.* **2010**, *2*, 204–214. [CrossRef]

10.   Liao, Y.; Liu, Z.; Zhang, G.; Xiang, C. Vehicle-grid system modeling and stability analysis with forbidden region-based criterion. *IEEE Trans. Power Electron.* **2017**, *5*, 3499–3512. [CrossRef]

11.   Wang, H.; Wu, M.; San, J. Analysis of low-frequency oscillation in electric railways based on small-signal modeling of vehicle-grid system in d-q frame. *IEEE Trans. Power Electron.* **2015**, *9*, 5318–5328. [CrossRef]

12.   Dabra, V.; Paliwal, K.K.; Sharma, P. Optimization of photovoltaic power system: A comparative study. *Prot. Control Mod. Power Syst.* **2017**, *2*, 3. [CrossRef]

13.   Liu, J.; Yang, D.; Yao, W.; Fang, R.; Zhao, H.; Wang, B. PV-based virtual synchronous generator with variable inertia to enhance power system transient stability utilizing the energy storage system. *Prot. Control Mod. Power Syst.* **2017**, *2*, 429–437. [CrossRef]

14.   Agorreta, J.L.; Borrega, M.; López, J.; Marroyo, L. Modeling and control of N-paralleled grid-connected inverters with LCL filter coupled due to grid impedance in PV plants. *IEEE Trans. Power Electron.* **2011**, *3*, 770–785. [CrossRef]

15.   Hernández, J.C.; De La Cruz, J.; Vidal, P.G.; Ogayar, B. Conflicts in the distribution network protection in the presence of large photovoltaic plants: The case of ENDESA. *Int. Trans. Electr. Energy Syst.* **2013**, *5*, 669–688. [CrossRef]

16.   Enslin, J.H.R.; Hulshorst, W.T.J.; Atmadji, A.M.S.; Heskes, P.J.M.; Kotsopoulos, A.; Cobben, J.F.G.; Van der Sluijs, P. Harmonic interaction between large numbers of photovoltaic inverters and the distribution network. In Proceedings of the IEEE Bologna Power Tech Conference Proceedings, Bologna, Italy, 23–26 June 2003; pp. 75–80.

17.   Wang, H.; Wu, M. Review of low-frequency oscillation in electric railways. *Trans. China Electro Tech. Soc.* **2015**, *17*, 70–78.

18.   Cortés, P.; Kazmierkowski, M.P.; Kennel, R.M.; Quevedo, D.E.; Rodríguez, J. Predictive control in power electronics and drives. *IEEE Trans. Ind. Electron.* **2008**, *12*, 4312–4324. [CrossRef]

19.   Vazquez, S.; Leon, J.I.; Franquelo, L.G.; Rodriguez, J.; Young, H.A.; Marquez, A.; Zanchetta, P. Model predictive control: A review of its applications in power electronics. *IEEE Ind. Electron. Mag.* **2014**, *1*, 16–31. [CrossRef]

20.   Hu, J.; Cheng, K.W.E. Predictive control of power electronics converters in renewable energy systems. *Energies* **2017**, *4*, 515. [CrossRef]

21.   Chan, R.; Kwak, S. Improved Finite-Control-Set Model Predictive Control for Cascaded H-Bridge Inverters. *Energies* **2018**, *2*, 355. [CrossRef]

22.   Vazquez, S.; Marquez, A.; Aguilera, R.; Quevedo, D.; Leon, J.I. Predictive optimal switching sequence direct power control for grid-connected converters. *IEEE Trans. Ind. Electron.* **2015**, *4*, 2010–2020. [CrossRef]

23.   Xia, C.; Liu, T.; Shi, T.; Song, Z. A simplified finite-control-set model-predictive control for power converters. *IEEE Trans. Ind. Inf.* **2014**, *2*, 991–1002.

24.   Aguilera, R.P.; Lezana, P.; Quevedo, D.E. Finite-control-set model predictive control with improved steady-state performance. *IEEE Trans. Ind. Inf.* **2013**, *2*, 658–667. [CrossRef]

25.   Song, W.; Ma, J.; Zhou, L.; Feng, X. Deadbeat predictive power control of single-phase three-level neutral point clamped converters using space-vector modulation for electric railway traction. *IEEE Trans. Power Electron.* **2016**, *1*, 721–732. [CrossRef]

26. Song, W.; Deng, Z.; Feng, X. A simple model predictive power control strategy for single-Phase PWM converters with modulation function optimization. *IEEE Trans. Power Electron.* **2016**, *7*, 5279–5289. [CrossRef]

27. Aguilera, R.P.; Quevedo, D.E.; Vázquez, S.; Franquelo, L.G. Generalized predictive direct power control for ac/dc converters. In Proceedings of the 2013 IEEE ECCE Asia Downunder, Melbourne, Australia, 3–6 June 2013; pp. 1215–1220.

28. Cortés, P.; Rodriguez, J.; Silva, C.; Flores, A. Delay compensation in model predictive current control of a three-phase inverter. *IEEE Trans. Ind. Electron.* **2012**, *2*, 1323–1325. [CrossRef]

29. Ma, H.; Li, Y.; Zheng, Z.; Xu, L.; Wang, K. PWM rectifier using a model predictive control method in the current loop. *Trans. China Electro Tech. Soc.* **2014**, *8*, 136–141.

30. Cortes, P.; Kouro, S.; La Rocca, B.; Vargas, R.; Rodriguez, J.; Leon, J.; Vazquez, S.; Franquelo, L. Guidelines for weighting factors design in model predictive control of power converters and drives. In Proceedings of the IEEE International Conference on Industrial Technology, Gippsland, Australia, 10–13 February 2009; pp. 1–7.

31. Zanchetta, P. Heuristic multi-objective optimization for cost function weights selection in finite states model predictive control. In Proceedings of the 2011 Workshop on Predictive Control of Electrical Drives and Power Electronics, Munich, Germany, 14–15 October 2011; pp. 70–75.

32. Leng, Y.; Yang, H.; Wang, Z. A method of suppressing low-frequency oscillation in traction network based on two-degree-of-freedom internal model control. *Power Syst. Technol.* **2017**, *1*, 258–264.

33. Lee, H.M.; Lee, C.M.; Jiang, G. Harmonic analysis of the Korean high-speed railway using the eight-port representation model. *IEEE Trans. Power Deliv.* **2006**, *2*, 979–986. [CrossRef]

3

Supervisory Control for Wireless Networked Power Converters in Residential Applications

S. M. Rakiul Islam [1,*], Sung-Yeul Park [1], Shaobo Zheng [2], Song Han [2] and Sung-Min Park [3]

[1] Electrical and Computer Engineering Department, University of Connecticut, Storrs, CT 06269, USA; sung_yeul.park@uconn.edu

[2] Computer Science Engineering Department, University of Connecticut, Storrs, CT 06269, USA; shaobo.zheng@uconn.edu (S.Z.); song.han@uconn.edu (S.H.)

[3] Electrical and Electronic Engineering Department, Hongik University, Sejong-si 30016, Korea; smpark@hongik.ac.kr

* Correspondence: s.islam@uconn.edu

† This paper is an extended version of our paper published in the 2017 IEEE Applied Power Electronics Conference and Exposition (APEC), Tampa, FL, USA, 26–30 March 2017.

Abstract: This paper presents a methodology to design and utilize a supervisory controller for networked power converters in residential applications. Wireless networks have been interfaced to multiple power factor correction (PFC) converters which are proposed to support reactive power. Unregulated reactive power support from PFC converters could cause reactive power deficiency and instability. Therefore, a supervisory controller is necessary to govern the operation of PFC converters. WiFi and WirelessHART networks have been used to implement the supervisory controller. Different nodes of the power network are connected by wireless communication links to the supervisory controller. Asynchronous communication links latency and uncertain states affect the control and response of the PFC converters. To overcome these issues, the supervisory controller design method has been proposed based on the system identification and the Ziegler-Nichols rule. The proposed supervisory controller has been validated by using a hardware-in-the-loop (HIL) test bed. The HIL testbed consisted of an OP4510 simulator, a server computer, Texas Instrument-Digital Signal Controllers (TI-DSCs), WiFi and WirelessHART modules. Experimental results show that the proposed supervisory controller can help to support and govern reactive power flow in a residential power network. The proposed method of controller design will be useful for different small-scale power and wireless network integration.

Keywords: networked power converters; PFC converters; reactive power resources; supervisory controller; HIL Testbed

1. Introduction

According to the US Department of Energy, residential loads have consumed 20.44% of the total energy in 2017 [1]. Residential loads consume both active and reactive power. Reactive power demand in a home is usually fulfilled from the grid. Recently, power factor correction converters have been integrated into some of the residential appliance, which minimizes reactive power consumption from the grid [2–5]. A recent study shows that additional reactive power could be supported from renewable resources and power factor correction (PFC) converters in residential applications [6,7]. However, unregulated reactive power support from PFC converters could cause reactive power deficiency and instability. To utilize the additional reactive power resources in a residential power network, a supervisory controller is necessary. A supervisory controller should monitor and control the reactive power flow in the residential power network. This paper presents a methodology to design a supervisory controller which governs reactive power flow in a residential power network. Wireless networks were used to implement the proposed supervisory controller.

The supervisory controller for large power and energy management systems are well established technology. It is known as supervisory control and data acquisition (SCADA) [8]. Because of the scalability, SCADA is not an appropriate tool to control reactive power flow in a small residential power network. Distributed controllers for power converters have been developed using communication network [9,10]. It has potential for wireless network integration but does not provide a centralized solution for reactive power management. A supervisory controller has been validated using an FPGA for power converters in Reference [11]. Reference [11] has only a single controller for multiple converters, which manages all PWM signals without a communication network. Other power and energy management systems have been implemented using fuzzy logic, distributed, model predictive and supervisory controller in References [12–17]. However, these solutions don't give the opportunity to control power flow using a communication network in a residential application. As a result, a solution has been proposed by assuming that there are reactive power resources as well as a communication network available in the residential applications [18,19].

The possible energy sources in residential applications include utility grid, solar panel, wind generator, stored energy, and so forth. These energy sources supply both active and reactive power. In the residential applications, energy is consumed by different kinds of loads , which include high-consumption applications such as HVAC systems, ovens or refrigerators, along with low-power devices such as televisions and light fixtures. Conventional residential loads consume both active power, P and reactive power, Q. However, smart loads which use PFC converters could contribute reactive power, Q, rather than consuming [6]. Examples of smart loads include HVAC systems, electric vehicle (EV) chargers, computers, televisions and other digital appliances [20–22]. Typically, this is accomplished through the use of boost PFC converters [23,24]. As a result, these appliances have the readily available converters to support reactive power. The main role of these converters is to supply specified current and voltage to the appliances. However, these converters can contribute a specified amount of reactive power to the grid without hampering their main functionality.

Recent advances in technology has has trended towards connecting residential appliances to networks. The connectivity to networks and management of appliance features is represented by smart technologies and Internet of Things [21]. The connectivity to appliances has been implemented using WiFi, WirelessHART, Bluetooth, Zigbee and Ethernet [25–28]. By using this connectivity, some specific parameters of power converters can be monitored and controlled [19].

Appliances as reactive power resources and the connection to a wireless network facilitates the opportunity to implement supervisory controller. However, the reactive power management system now becomes more complex because of the integration of power network and wireless network. This system has multiple power nodes which are connected by communication links. Multiple wireless communication links have different delays, asynchronous latency and uncertainty. These factors affects the operation of PFC converters which ultimately affects the reactive power management system. Due to the system complexity and delays, the top-level supervisory controller can't be designed accurately using a conventional state space averaging method. To address this issue, a system identification-based supervisory controller design method has been proposed in this paper. The proposed method considers uncertainty, complexity and delays. It uses the Ziegler-Nichols rule to design a proportional-integral (PI) controller. The proposed supervisory controller has been implemented in a server computer and validated using an hardware in the loop (HIL) test bed set up.

The HIL test bed was built to simulated a residential power network in real-time using OPAL-RT OP4510 [29]. The power converters of have been controlled by Texas Instrument Digital Signal Controllers (TI DSC) [30]. WirelessHART and Wi-Fi networks were used for connectivity [31,32]. Level shifting and scaling circuit were built to make the TI DSC compatible for OP4510.

Multiple tests were conducted using the HIL test bed to validate the feasibility and compatibility of the supervisory controller. The performance of the supervisory controller has been evaluated by analyzing the dynamic response for reactive power support. Test results show that PFC converters can support reactive power to the residential application in different conditions with the help of an optimal supervisory controller. Section 2 of this paper describes the proposed scenario for this power network. Section 3 describes the proposed supervisory controller design method. HIL testbed and experimental results are presented in Sections 4 and 5, respectively.

2. Networked Power Converters

Considering the wireless connectivity to smart residential appliances, a scenario for a residential power network has been proposed in the following.

2.1. Power Network Configuration

The proposed residential power network configuration is shown in Figure 1 which has both conventional lagging load and PFC connected smart loads. A supervisory controller has been implemented in the power network using wireless nodes. Figure 1 shows the active and reactive power supplies from grid. Conventional loads with a lagging power factor are represented by a single block in Figure 1. Three separate PFC connected smart loads are distributed and are assumed to contribute reactive power. Based on the active and reactive power supply and consumption, the balanced power condition could be described by (1) and (2).

$$P_g = P_L + \sum_{n=1}^{N} P_n \tag{1}$$

$$Q_g = Q_L - \sum_{n=1}^{N} Q_n \tag{2}$$

where, P_g is the active power consumption from grid, Q_g is the reactive power consumption from grid, P_L is the active power consumption of conventional residential loads, Q_L is the reactive power consumption of conventional residential loads, P_n is the active power consumption of PFC connected loads, Q_n is the reactive power consumption or contribution of PFC connected loads. Here, n denotes the PFC number and N is the number of available PFC connected resources. The reactive power resources and loads are described pictorially in Figure 1 for $N = 3$.

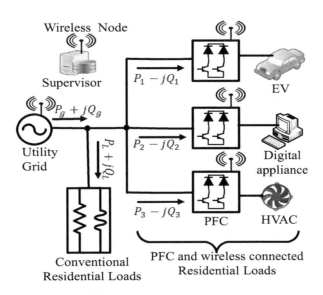

Figure 1. Reactive power resources in wireless networked residential loads.

2.2. PFC Connected Smart Residential Load

PFC converters for the appliance in Figure 1 can support reactive power. The circuit configuration for a PFC converter supporting reactive power is shown in Figure 2. Bridgeless unidirectional single phase boost PFC converter topology has been used in Figure 2. This converter has two diodes, D_1, D_2 and two semiconductor switches, Q_1, Q_2, input inductor, L and DC bus capacitor, C. The detailed analysis of the circuit is available in References [6,7,33]. Each converter is controlled by using a PFC controller. It generates complementary pulse width modulated (PWM) signals for Q_1 and Q_2 [34].

The controller for PFC converter can supply a specified amount of reactive power.

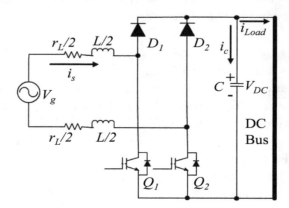

Figure 2. Circuit diagram of a bridgeless ac-dc boost power factor correction (PFC) converter.

This type of PFC controller is shown in Figure 3. The controller takes Q^* as reactive power reference and contributes Q to the power network.

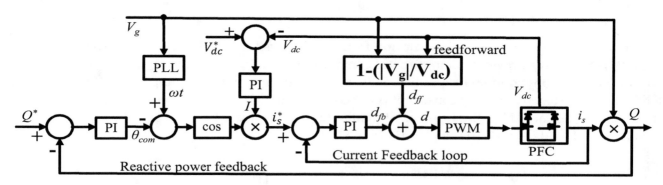

Figure 3. Control diagram of a bridgeless ac-dc boost PFC converter.

The controller initially maintains the proper DC bus voltage and in-phase current. Then, by controlling current, it makes the PFC converter the leading load and contributes reactive power to the network. This controller uses both feedback and feedforward terms to maintain inner current and outer voltage loops. Output current, DC link voltage and reactive power measurements are used as feedback signals for PI controllers. Phase angle, ωt, is determined by phase lock loop (PLL) using grid voltage, V_g .Feedback and feedforward duty cycle values are added before the generation of PWM signals.

The inner current loop is the fastest control loop in this controller and maintains leading-phase current based on the compensation angle, θ_{com} and the output of the voltage control. DC voltage feedback control is maintained by a PI controller, voltage reference, V_{dc}^* and feedback V_{dc}. The θ_{com} is determined by using a PI controller and reactive power feedback. A sinusoidal signal is reconstructed using PLL and a cosine function. The reactive power compensation loop has 2 key Equations (3) and (4).

$$Q_{err}(t) = Q^*(t) - Q(t) \tag{3}$$

$$\theta_{com}(s) = (k_p + \frac{k_i}{s})Q_{err}(s) \tag{4}$$

The DC output voltage is maintained by generating current reference, $I(s)$, using (5) and (6).

$$V_{dc_{err}}(t) = V_{dc}^*(t) - V_{DC}(t) \tag{5}$$

$$I(s) = (k_p + \frac{k_i}{s})V_{dc_{err}}(s) \tag{6}$$

AC current reference, i_s^*, is obtained by I, ωt and θ_{com}.

$$i_s^*(t) = I \times cos(\omega t - \theta_{com}) \tag{7}$$

Error in the inner current loop is calculated by applying the following equation.

$$i_{err}(t) = i_s^*(t) - i_s(t) \tag{8}$$

Current error is compensated by feedback duty, d_{fb}, using (9).

$$d_{fb}(s) = (k_p + \frac{k_i}{s})i_{err}(s) \tag{9}$$

Feedforward duty, d_{fb} is generated from (10).

$$d_{ff}(s) = 1 - \frac{V_g}{V_{dc}} \tag{10}$$

Finally, feedback and feedforward duty is added to generate PWM signal.

$$d(s) = d_{fb}(s) + d_{ff}(s) \tag{11}$$

The detailed description and design of the PFC controller for reactive power support is available in Reference [6]. The waveforms of input current, duty and PWM switching signal are shown Figure 4.

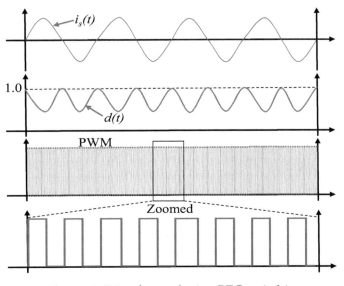

Figure 4. Waveforms during PFC switching.

2.3. Wireless Network and Supervisory Reactive Power Control

To improve the overall power factor in the proposed scenario, a supervisory controller has been included in Figure 1 using the wireless connectivity. The proposed power network has three PFC converters.These PFC converter can contribute a specified amount of reactive power Q^* as discussed in Section 2. However, reactive power demand could change with time. In demand-varying conditions, the supervisory controller identifies the reactive power demand and maintains equal amount of reactive power support from the PFC converters. If the load demand for reactive power increases, the supervisory controller ensures more reactive power. On the other hand, if the load demand decreases, the supervisory controller decreases reactive power supplied. Through this method, the supervisory controller tries to maintain unity power factor for the proposed scenario. It is recommended to maintain the grid's power factor $(p.f.)$ as close to 1.00 possible [35]. From the definition of power factor, the following equation can be written.

$$p.f. = cos\phi = \frac{P_g}{\sqrt{P_g^2 + Q_g^2}} \tag{12}$$

From (12), it is clear that the power factor will be unity if and only if $Q_g = 0$. In that case, the condition of (2) becomes the condition of (13).

$$Q_L = \sum_{n=1}^{N} Q_n \tag{13}$$

In dynamic conditions, (13) can be expressed by (14).

$$Q_g(t) = Q_L(t) - \sum_{n=1}^{N} Q_n(t) \tag{14}$$

where, t represents time. The grid reactive power is minimized by making $Q_g(t)$ close to 0. To minimize $Q_g(t)$, the supervisory control technique was applied as shown in Figure 5. In this proposed technique, a proportional integral (PI) controller is used to compensate $Q_g(t)$ to 0.

The supervisory controller is shown in Figure 5 and named as supervisor. Supervisor is comprised of a PI controller, reference and distributor. Q_{ref} is set to zero, so that Q_g become zero. From the difference of the Q_{ref} and Q_g, the error is generated. Based on the error, the PI controller sets a reference of total reactive power demand, Q_d. Q_d is total reactive power that need to be fulfilled by all available PFC converters. Using a distributor, this demand is distributed and assigned to individual PFC controllers. Individual PFC controllers get command for reactive power references of $Q_1^*, Q_2^*, ..., Q_N^*$. The commands are obtained though wireless network. After performing the control, PFC converters contribute $Q_1, Q_2, ..., Q_N$ amounts of reactive power to the power network. Individual PFC converter has individual local control loops to maintain reactive power support. These local controllers get the reference from supervisory controller based on their capacity.

The response time, wireless communication delay and capacity of the PFC converters are factors of the supervisory control design consideration. The supervisor is a single input multiple output (SIMO) system. The power network of the smart home is considered as a multiple input single output system (MISO). The whole system is a single input single output (SISO) feedback system. A host program maintains the supervisory controller. A varmeter measures reactive power consumption from the grid, Q_g. Q_g is the feedback for the supervisory controller.

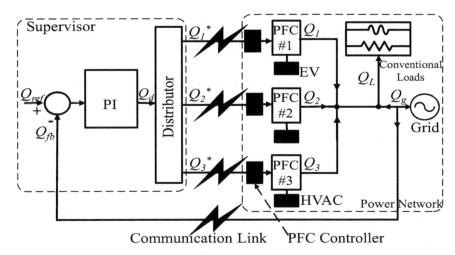

Figure 5. Supervisory control for reactive power resources using wireless connectivity.

3. Supervisory Controller Design

The supervisory controller governs the distributed PFC connected loads using wireless communication link and a server. The response of the supervisory controller is slower than the individual PFC controllers. The control system configuration, design challenges and solution are discussed below.

3.1. Control System Configuration

The functional block diagram of the supervisory reactive power control scheme is shown in Figure 5 for the power network of Figure 1. Figure 5 describes the role of the supervisor which is comprised of a PI controller and a distributor. To design the proper PI controller, the system should be analyzed in the control system perspective. By that manner, supervisory reactive power compensation system can be modeled as in the control system as shown in Figure 6.

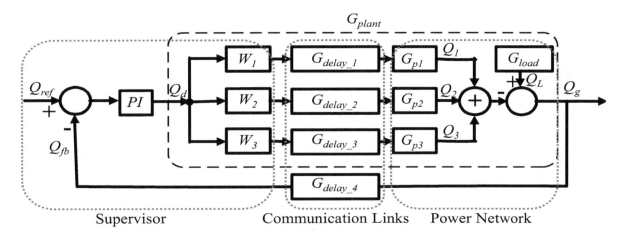

Figure 6. Control block diagram for reactive power compensation with multiple PFCs.

In the control block diagram, W_1, W_2 and W_3 are the weighting factors of the distributor. $G_{delay\_n}(s)$ is the transfer function of the communication delay due to wireless links. G_{load} is the transfer function for conventional loads. $G_{p1}(s)$, $G_{p2}(s)$ and $G_{p3}(s)$ are transfer functions of the PFC converters including their controllers for reactive power compensation. To design the proper PI controller, all PFC converters, delays and loads transfer functions have been put together and considered as plant, $G_{plant}(s)$. The output of the PI controller determines the reactive power demand, Q_d. The transfer function of the overall feedback loop can be written as in (15).

$$H(s) = \frac{G_{PI}(s)G_{delay\_4}(s)G_{plant}(s)}{1 + G_{PI}(s)G_{delay\_4}(s)G_{plant}(s)} \tag{15}$$

For asynchronous unequal delays and dynamic loading conditions the transfer function of the plant can be written as in (16).

$$G_{plant}(s) = G_{load}(s) - tr\left(\begin{pmatrix} W_1 & 0 & 0 \\ 0 & W_2 & 0 \\ 0 & 0 & W_3 \end{pmatrix}\begin{pmatrix} G_{delay\_1}(s) & 0 & 0 \\ 0 & G_{delay\_2}(s) & 0 \\ 0 & 0 & G_{delay\_3}(s) \end{pmatrix}\begin{pmatrix} G_{p1}(s) & 0 & 0 \\ 0 & G_{p2}(s) & 0 \\ 0 & 0 & G_{p3}(s) \end{pmatrix}\right) \tag{16}$$

where, $G_{plant}(s) = \frac{Q_g(s)}{Q_d(s)}$, $G_{p1}(s) = \frac{Q_1(s)}{Q_1^*(s)}$, $G_{p2}(s) = \frac{Q_2(s)}{Q_2^*(s)}$, $G_{p3}(s) = \frac{Q_3(s)}{Q_3^*(s)}$. tr denotes trace of the matrix. Although Equations (15) and (16) are seem like deterministic representation of the feedback system, the dynamic nature of $G_{plant}(s)$ can cause this representation to yield inaccurate results. $G_{plant}(s)$, changes with the change of $G_{load}(s)$, $G_{delay}(s)$, $G_{p1}(s)$, $G_{p2}(s)$, $G_{p3}(s)$, W_1, W_2 and W_3 . $G_{load}(s)$ changes with the change of load demand. The response of $G_{p1}(s)$, $G_{p2}(s)$ and $G_{p3}(s)$ are mutually independent and asynchronous because of the different command schedules for different PFC converters. The transfer functions for $G_{p1}(s)$, $G_{p2}(s)$ and $G_{p3}(s)$) can be determined from Section 2.2. $G_{delay}(s)$ is the reason for multiple phase shifts between commands. W_1, W_2 and W_3 are variable with respect to the capacity of the PFC converters. As a result, $G_{plant}(s)$ has uncertainty, non-linearity and multiple asynchronous phase shifts in dynamic conditions.

3.2. System Identification

Because of the uncertainty, non-linearity and asynchrony, accurate G_{plant} can't be determined from (16) at any instant. So, alternative method is proposed to get the transfer function response characteristics. System identification method is applied to get the characteristic of the system. The system has been identified for G_{plant}, G_{p1}, G_{p2} and G_{p3} by applying step input and measuring settling time. The settling time is the main feature that has been used to design supervisory controller considering it is an asynchronous distributed system. The settling time depends on the transfer function of the system.

3.3. Solution for Multiple Asynchronous Phase Shift

Different settling time, wireless communication delay time and overall response time for the proposed supervisory control system are explained by timing diagram in Figure 7. The timeslots t_1, t_2 and t_3 are communication delay time for PFCs; in other words those are settling time of G_{delay_n}. After these time slots PFCs execute their received command from supervisory controller within their settling times. The received commands are reference amount of reactive power assigned for the individual PFC converters. The necessary times for reactive power compensation of PFC's are t_{1_s} , t_{2_s} and t_{3_s}; in other words these are settling time of G_{p1}, G_{p2} and G_{p3}. t_w is a flexible waiting time before getting reactive power measurement from varmeter.

t_g is the communication time slot assigned for varmeter to send data to supervisor. The optimal value of t_w and t_g are determined by from the experiments. The value of t_{1_s} , t_{2_s} , t_{3_s} , t_w and t_g are determined by system identification method as discussed in Section 3.2. The overall value of t_1, t_2 and t_3 depends on the communication link and should be determined with certainty. The value of t_w can be adjusted to ensure certainty of the response. All the time slots are added in (17) to get the minimum optimal sample time, T_s, for the supervisory controller.

$$T_s = t_w + t_g + t_1 + \sum_{n=1}^{N=3} t_{n\_s} \tag{17}$$

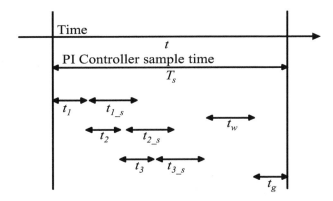

Figure 7. Time slots consideration for supervisory controller design.

3.4. PI Controller Design

The minimum sample time, T_s, is used to design the discrete PI controller. The block diagram for the discrete PI controller is shown in Figure 8a. The integrator for the PI controller has a sample time of T_s. The output of the PI controller is Q_d, which is the reactive power demand from all of the PFC converters. Since the system has certainty of response using the sampling time of T_s, Ziegler-Nichols rule can be used to design PI compensator for this system [36–38]. Using the rule, the values of K_p and K_i are determined by (18) and (19) for the ultimate gain K_u.

$$K_p = 0.5K_u \tag{18}$$

$$K_i = 0.45K_u \tag{19}$$

For the wireless communication, communication time is much higher than response time of PFC converters that is, $t_1 >> t_{1\_s}$. The value of ultimate gain K_u is 1 for such system.

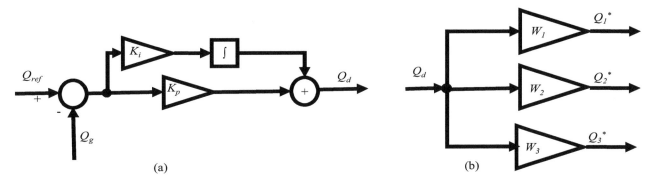

Figure 8. Detail diagram of supervisory control (**a**) proportional-integral (PI) controller (**b**) Distributor.

3.5. Distributor Design

The value of the Q_d is distributed by the weighting factors W_1, W_2 and W_3. The weighting factors are determined from the maximum reactive power support capacities (Q_{1_max}, Q_{2_max} and Q_{3_max}) of the PFC converters. The value of weighting factors can be determined from (20).

$$W_n = \frac{Q_{n\_max}}{Q_{1\_max} + Q_{2\_max} + Q_{3\_max}} \tag{20}$$

4. Testbed Description

The configuration for hardware-in-the-loop (HIL) testbed for the proposed supervisory controller and scenario is shown in Figure 9.

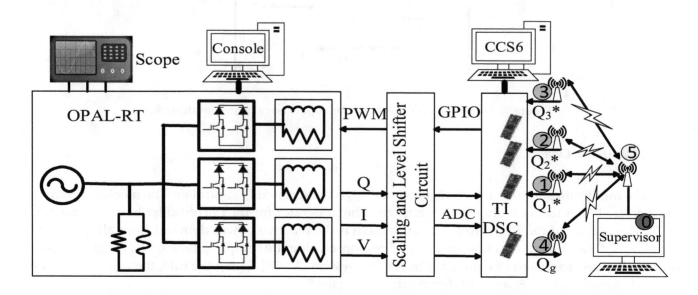

Figure 9. Block diagram of the testbed.

The testbed consists of an Opal-RT real-time simulator OP4510, TI-DSCs, scaling and level shifter circuits, AwiaTech wirelessHART modules [39], ESP8266 Wi-Fi modules [40] and communication links. The proposed power network has been simulated in real-time using OP4510 real-time simulator. The waveforms of voltages, currents and reactive power were observed from both the oscilloscope and the computer console connected to the OP4510. The real time simulation data is also stored in .mat files from the console for further analysis. The measured voltages, currents and reactive powers from OPAL-RT are scaled down using the Op-Amp based circuit to use in the ADC of the TI-DSCs. One of the TI-DSCs transmits the value of reactive power consumption from the grid input power node. The rest of the TI- DSCs are designated to control the PFC converters. The controller of Figure 3 has been implemented to those TI-DSCs. TI-DSCs were programmed using Code Composer Studio v.6 (CCS6). All the TI-DSCs are connected to AwiaTech wirelessHART modules and ESP8266 Wi-Fi modules via serial (UART) connections for the communication links to the supervisor. Communication nodes are numbered from 0 to 5 in Figure 9. The connections of ESP8266 Wi-Fi and AwiaTech wirelessHART modules were switchable in the node 1–4. Node 5 has different configurations based on the communication preference. The description of wireless network configuration will be discussed in the Subsection 4.3.1.

The HIL testbed is validated by three types of physical devices: wireless modules, digital controllers and real time simulator. The actual testbed setup is shown in Figure 10. OP4510 is used to simulate the power stages in real time; TI-DSCs are used as controllers of the PFC converters; AwiaTech wirelessHART or ESP8266 Wi-Fi modules are used as interface of communication link; Op-Amp based circuits have been used for the ADC scaling and buffers have been used for the level shifting of PWM signals; a supervisory controller have been implemented in the computer that is connected to the internet and the master AwiaTechHART module. The supervisory controller ensures zero reactive power consumption from the grid. The components of the test bed are described in the following subsections.

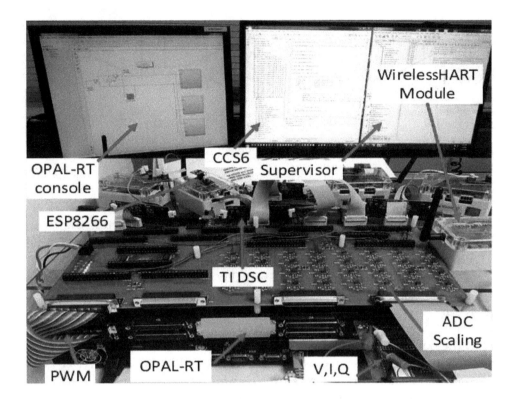

Figure 10. Hardware in the loop testbed set up.

4.1. Hardware Description

4.1.1. Real Time Simulator

The power stages of the PFC converters, loads, measurement of voltage and current and active and reactive power are simulated in real-time (RT) using OPAL-RT real-time simulator, OP4510. It has 128 I/O channels, a quad core INTEL Xeon 3.3 GHz processor as well as a Kintex 7 FPGA for sub-microsecond simulation time steps [41]. Each core of the processor has the capability to simulate up to 20 μs time step. The system is compatible with Simulink and the SimPowerSystems library. The OP4510 is also connected to an oscilloscope to monitor voltage, current and reactive power in real time. For this testbed, analog values of voltages and currents are routed from the FPGA based model to the DAC channels directly. The PWM signals are interfaced with digital input pins which drive the FPGA based modesl of IGBTs. These two steps ensure sub-microsecond RT simulation. The reactive power is calculated using Fourier transform in the CPU core at a step of 25 μs and later routed to DAC channels.

4.1.2. ADC Scaling and Level Shifting Circuit

The analog output signals for V, I and Qs are ±16 V in OP4510. The ADC level of TI-DSCs is limited to 0–3.3 V. To interpret and match the analog signal level properly, Op-Amp (TL082C) based scaling and offsetting circuits have built. The GPIO of TI-DSC has a logic level of 3.3 V and OP4510 takes 5 V PWM input. As a result, a 74HC240D IC has been used as a buffer for level shifting. The level shifting and ADC scaling circuit has been built up in the same PCB as shown in Figure 10. The PCB also has connectivity to Opal-RT by two DB37 connectors and connect TI-DSCs with the 20 pin connectors. This circuit has 32 analog and 32 digital re-routable channels. The ADC scaling and level shifting circuit is shown in Figure 11.

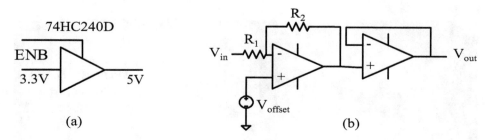

Figure 11. Interface circuit: (**a**) level shifting (**b**) ADC scaling.

4.1.3. Texas Instruments Digital Signal Controller

The TI-DSC, TMS320F28335, was selected to control the PFC converters that were simulated in real-time on the OP4510. In addition, the DSC code includes UART communication which interfaces either the AwiaTech wirelessHART or the ESP8266 Wi-Fi modules. The TMS320F28335 is a 32 bit floating point processor with clock speeds up to 150 MHz and 18 PWM channels [42].

4.1.4. WirelessHART and AwiaTech Wireless 220 Module

To implement the communication system for the testbed, AwiaTech WirelessHART (Wireless Highway Addressable Remote Transducer) modules were chosen. WirelessHART is a simple, secure, reliable, real-time and open-standard networking technology, operating in the 2.4 GHz ISM radio band [43]. It uses a time-synchronized, self-organizing and self-healing mesh network architecture. At the bottom of its communication stack, WirelessHART adopts IEEE 802.15.4-2006 [31] as the physical layer. On top of that, WirelessHART defines its own time-synchronized data link layer. In WirelessHART, communications are precisely scheduled based on Time Division Multiple Access (TDMA) and employ a channel hopping scheme for added system data bandwidth and robustness [43]. AwiaTech Wireless provides a variety of interfaces such as UART, JTAG, SPI/I2C and USB, which provides us the flexibility for interconnecting the TI DSC. The features of this module are shown in Figure 12a.

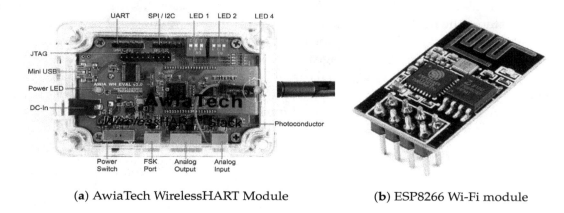

(**a**) AwiaTech WirelessHART Module   (**b**) ESP8266 Wi-Fi module

Figure 12. Wireless communication modules.

4.1.5. WiFi and ESP8266 Module

Four ESP8266 modules have been used in the test bed [40]. ESP8266 has serial communication (UART) features as in Figure 12b. It can give any microcontroller access to your WiFi network and maintain TCP/IP protocol by IEEE standard 802.11/b/g/e [32]. These modules have been programmed using an Arduino programming environment.

4.2. Software Components Description

4.2.1. TI DSC Code Architecture

The PFC controller has a 20 µs sample time. It generates a 50 kHz PWM signal. One interrupt has been used to compute ADC values, scaling, controller operation and the PWM update. For the low speed communication links, a polling method has been used. The communication function is executed in an infinite while loop. All the functions are executed within a 20 µs window.

The ADC interrupt and controller computations consume only 12 µs and the remaining 8 µs is designated for communication through UART. Figure 13 describes the interrupt and while loop structure.

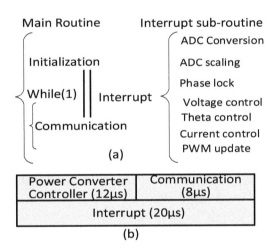

Figure 13. (a) Code structure of TI DSC (b) Execution time of code sections.

4.2.2. Host Program

AwiaTech wireless provides a Java package "Host" which is encapsulated within a series of APIs, enabling users to write programs which process all the data acquired from WirelessHART network. Based on these APIs, "Host" software is customized as an interface between the wireless network and the supervisory controller. This design separates communication and control, which provides scalability to use other wireless communication technologies without modifying the controller code. In addition, this architecture enables the convenient use of other languages. The host program has been modified using TCP/IP link as well. The TCP/IP link has been used to maintain communication links to the ESP8266.

4.2.3. Supervisory Controller Implementation

The supervisory controller gives the control signal to the power converters. It is also capable of collecting data from the input port of the power network. The algorithm for the supervisory controller is written in C++ code and later called from the JAVA platform which maintains communication functions. The supervisory controller has a low speed proportional integral controller with a limiter and distributor as discussed in Section 3.

4.2.4. Dataflow

A complete dataflow is presented in Figure 14. The data flow in the test bed is composed of 8 steps, starting from a TI DSC and ending in a TI DSC. First, a TI-DSC sends a HART-IP command containing data to the AwiaTech wireless (device) through UART. Second, AwiaTech wireless forwards this command to the access point based on WirelessHART standard. Third, the access point sends data to a desktop running Host and Gateway software with UART over USB using a FTDI chip. Fourth, after Host receives the data, it forwards it to the supervisory controller. For the WiFi network, HART-IP has been decoded at ESP8266 and sent to the AP and host through the Wi-Fi. Our Host program will exchange messages with the controller through Stdin and Stdout. The dataflow in the other direction will follow the same steps in reverse.

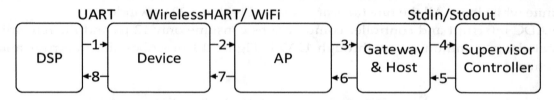

Figure 14. Dataflow between TI DSC and supervisory controller.

4.3. Communication Link Description

4.3.1. Wireless Communication Set Up

The communication network topology for WiFi and WirelessHART has been shown in Figure 15. Either all ESP8266 or all AwaiaTech modules are connected to the TI-DSC through UART. The UART to module connections are switchable for Wi-Fi and wirelessHART. Four AwiaTech modules maintain communication links with a fifth AwiaTech wireless module, which is connected to a host computer, where the supervisory controller is running.

On the other hand ESP8266 is connected to the internet via Wi-Fi and a router. The supervisory controller can use internet to control the PFC converters for Wi-Fi network preference. Multiple unidirectional PFC converters contribute reactive power support by following a reference command by a supervisory controller via wireless modules. Real-time algorithms or strategies for energy and power management systems can validate their performance in this testbed. The network topology in our experiments is shown in Figure 16, the blue circles are devices connected to TI DSCs; the yellow circle is the access point and the red circle is the gateway. A dotted line indicates the wireless communication through WirelessHART or WiFi. The solid line indicates the wired communication through USB or Ethernet. The numbers on the lines show the signal strength of the wireless links. The Figure 16 shows all the components in our network and their unique IDs, types and scan periods.

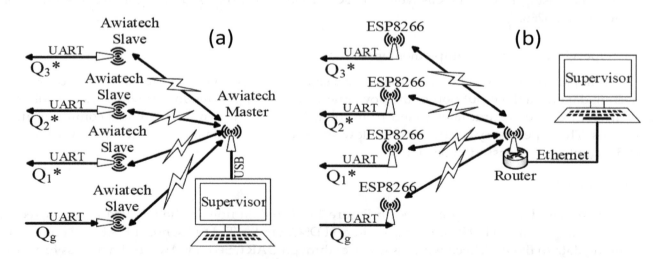

Figure 15. Wireless network topologies for testbed (**a**) WirelessHART (**b**) WiFi.

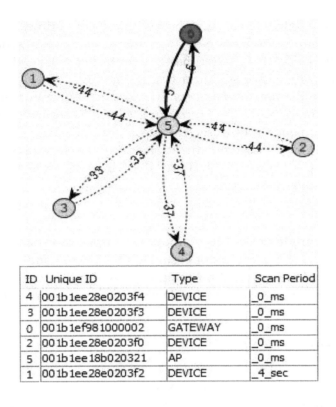

ID	Unique ID	Type	Scan Period
4	00 1b 1ee 28e0 20 3f4	DEVICE	_0_ms
3	00 1b 1ee 28e0 20 3f3	DEVICE	_0_ms
0	00 1b 1ef98 1000002	GATEWAY	_0_ms
2	00 1b 1ee 28e0 20 3f0	DEVICE	_0_ms
5	00 1b 1ee 18b0 20321	AP	_0_ms
1	00 1b 1ee 28e0 20 3f2	DEVICE	_4_sec

Figure 16. Wireless network topology.

4.3.2. Integrity of the Specification

The Figure 17 shows an example of simplified network schedule which is shared with all the devices and the access point in the network. The super-frame size is 500 ms and two up-links and down-links for each device are statically allocated inside one superframe. In Figure 17, a time slot is represented by a small square with a number and an arrow. For example, 2 ↑ means the time slot is scheduled for Device 2 as an up-link. The wireless message exchanging sequence is shown in Figure 17. At the beginning of a super-frame, the TI-DSC sends measurement to Device 1.

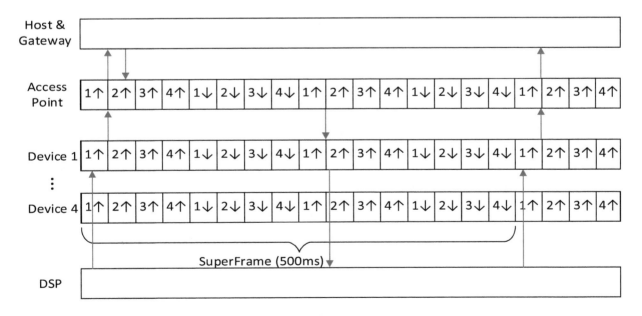

Figure 17. Communication schedule configuration.

Then Device 1 sends the data to an access point within its own timeslot. After receiving the data, the access point sends the data to the host. Next, the Host software will send what it just received to the supervisor controller for calculation. After that, the controller will send data back to control the TI DSC. It will go through the path in reverse. The access point will get the data and wait for the time slot allocated for Device 1. Once the time slot arrives, the access point sends the data down to Device 1 and then Device 1 sends data to the TI-DSC through UART. Theoretically, if we set the super-frame length to be 500 ms and within one super-frame and we allocate two time slots as up-links, the smallest sampling period would be 250 ms. However, due to the noisy wireless environment, we chose a much larger sampling rate and super-frame length. From the wireless link test, we were able to make the super-frame size 1.5 s without any data loss.

5. Experimental Results

The test bed has been implemented to validate the supervisory controller for reactive power support using the proposed power network of Figure 1. OP4510 has simulated the power network in real-time. It takes PWM signal as input and gives analog signals as output. Rest of the components are actual physical device in the HIL test bed. Multiple tests have been conducted to validate the idea of reactive power support and supervisory controller.

5.1. Reactive Power Compensation in a Single PFC Converter

Reactive power has been supported by applying controller of Figure 3 to a PFC converter. The wave form of voltage and current in a PFC converter for different conditions has been captured using an oscilloscope and shown in the following figure. The waveforms are collected from the analog output of the OP4510 real-time simulator. PFC 1 has been built up as in the circuit of Figure 2 using $r_L = 1$ mΩ, $L = 2$ mH, $C = 2000$ μF and $r_C = 2$ mΩ. PFC 1 is rated for $V_{in} = 120$ V(rms), $f = 60$ Hz, $P_{out} = 1.1$ kW, $V_{out} = 250$ V(DC) and $R_{Load} = 56.79753$ Ω. Different power conversion criteria for PFC 1 under conditions of Figure 18 are shown in the Table 1.

Figure 18a shows the input current and voltage waveforms without applying any control that is, current flows through the diode only.

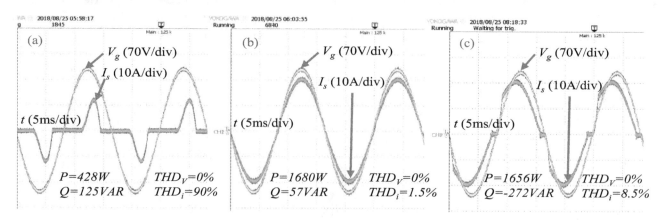

Figure 18. Input voltage and current waveforms in PFC 1: (**a**) no controller, (**b**) in phase controller, $\theta \approx 0$ and (**c**) leading phase controller, $\theta < 0$.

Table 1. Power conversion in PFC 1 for different conditions.

Conditions at Figure 18	P_{in} (W)	Q_{in} (VAR)	THD_{V_g} (%)	THD_{I_s} (%)	V_{out} (DC) (Volt)	P_{out} (W)
a	428	125	0	90	154	396
b	1680	57	0	1.5	250	1100
c	1656	−272	0	8.5	250	1100

This condition has very poor performance and can not fulfil the rated conditions. It has very high total harmonic distortion (THD) for the input current. After applying the reactive power compensation algorithm of Figure 3, voltage and current are almost in phase as in Figure 18b. Reactive power has not been injected for this case that is, reactive power reference is zero, $Q_1^* = 0$. Although Q_1^* is equal to zero but, Q_1 is not exactly zero because of delays, offset and measurement errors. In this case, current distortion is minimum. Leading phase of current has been achieved as in Figure 18c as Q_1^* has been set to negative values. Hence, reactive power has been supported by PFC 1. In this case, the current is not a pure sine wave at zero crossing because of zero crossing distortion. The distortion phenomenon has been considered to determine the maximum capacity of reactive power support (Q_{max1}) for PFC 1. Since, the household load is very small with respect to grid, the voltage supply has been considered to be coming from an infinite bus. As a result, distortion for input voltage, THD_{V_g} is zero in the realtime simulation.

5.2. Reactive Power Support Using Supervisory Controller

After setting up the communication networks successfully, the HIL test was conducted for the proposed residential power network. For the proposed scenario, rated grid voltage is 120 V(rms), 60 Hz. Conventional load is 1500 VA with 0.8 power factor that is, $P_L = 1200$ W, $Q_L = 900$ VAR. Rated output voltage, V_{out}, of all three PFCs is 250 V(DC). Output power, P_{out}, rating of the PFC's are 1.1 kW, 1.5 kW and 1.7 kW.

Weighting factors used in supervisory controller W_1, W_2 and W_3 are 0.25, 0.35 and 0.4 respectively. Reactive power has been supported by the PFC converters to the load as shown in the in Figure 19. In this case, real-time simulation has been conducted for communication and control sample time, $T_s = 4$ s. Supervisory controller engagement time has been set as reference, $T = 0$, on the time axis. Logged data from OP4510 shows that reactive power consumption from grid Q_g becomes 0 VAR within 75 s of applying the controller. The initial and final value of the reactive power for the grid, the load and the PFCs is shown in the Table 2.

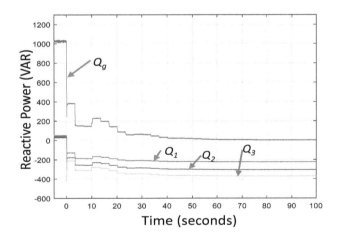

Figure 19. Reactive power support by 3 PFC converters and supervisory controller.

Table 2. Reactive power in different nodes of the proposed power network.

State	Time (s)	Q_g (VAR)	Q_L (VAR)	Q_1 (VAR)	Q_2 (VAR)	Q_3 (VAR)
Initial	0	1025	900	44	33	48
Final	75	0	900	−225	−305	−371

5.3. Reactive Power Support in Different Conditions

The performance of the supervisory controller has been tested for different conditions.

5.3.1. Variation of Load

The designed supervisory controller has been tested for loading conditions of (P_L = 1350 W, Q_L = 654 VAR), (P_L = 1200 W, Q_L = 900 VAR) and (P_L = 1050 W, Q_L = 1072 VAR). For all loading conditions, Q_g become 0 VAR within 75 s.

5.3.2. Variation of Wireless Network

The supervisory controller was tested using WirelessHART and Wi-Fi. It compensated grid reactive power for both networks as shown in Figure 20a.

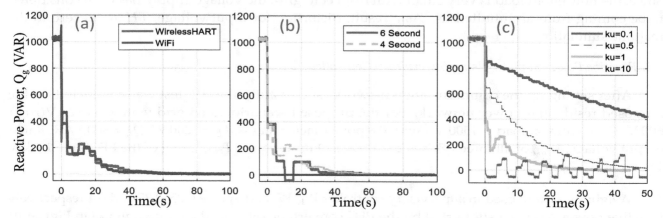

Figure 20. Performance of supervisory controller: (**a**) network variation, (**b**) sampling time variation and (**c**) gain variation.

5.3.3. Sampling Time Variation

The effect of different communication sampling times has been evaluated in Figure 20b. For four and six second sampling times, supervisory controller regulated Q_g to 0 VAR. The higher the sampling time, the slower the response time for overall reactive power compensation. Minimum stable communication and control sampling time for this system has been measured to be 1.5 s. This sampling time has been used for optimal controller.

5.3.4. Gain Variation and Optimal Controller

To verify the optimal supervisory controller, the response of reactive power compensation has been compared for different k_u, k_p and k_i gains. The response has been judged by settling time, overshoot, undershoot and stability. Figure 20c shows performance of supervisory controller with different gains. This controller has used T_s = 1.5 s communication sampling time. The settling time and other factors of performance for the PI compensator are shown in Table 3.

Table 3. Supervisory controllers performance with different PI gains.

K_u	K_p	K_i	Stability	Settling Time	Comments
0.1	0.05	0.045	Stable	250 s	Very slow response
0.5	0.25	0.225	Stable	75 s	slow response
0.8	0.4	0.36	Stable	45 s	Good response
1	0.5	0.45	Stable	40 s	Optimum response
2	1	0.9	Marginally stable	20 s	Undershoots happen, large steady state error, inject Q to grid
5	2.5	2.25	Unstable	–	Undershoots happen, large steady state error, distributor failure
10	5	4.5	Unstable	–	Completely unstable

5.4. Reactive Power Support in Dynamic Load

Different load profiles of Table 4 have been applied to test the supervisory controller. Reactive power demand were changed as the load changed.

Load profile 1 has been applied for the conditions of Figure 21a,c. Load profile 2 has been applied and results have been gathered in Figure 21b. The PFC converters used for Figure 21a,b have the power rating(P_{out}) of P_1 = 1.1 kW, P_2 = 1.5 kW and P_3 = 1.7 kW. The maximum reactive power supply capacity of those PFC converters are Q_{max1} = −300 VAR, Q_{max2} = −400 VAR and Q_{max3} = −600 VAR. Q_{max1}, Q_{max2} and Q_{max3} have determined the weighting factors W_1, W_2 and W_3 by (20). Load profile 1 is based on the power factor change and load profile 2 is based on apparent power demand increment.

Dynamic response of Figure 21c used load profile 1 but it used identical 3 PFC converters. In this case, P_1 = 1.5 kW, P_2 = 1.5 kW, P_3 = 1.5 kW, Q_{max1} = −400 VAR, Q_{max2} = −400 VAR, and Q_{max3} = −400 VAR. As a result, reactive power contribution from PFC converters is equal at different times in the real time simulation.

Load profiles have been implemented by changing the loads using an external switch. As a result there is a transient reactive power spike and an unequal load profile duration. For all of these conditions, the optimal controller has been used, which ensured the stable and quickest response. THD_{V_g} and THD_{I_g} have also been recorded for all the tests. THD_{V_g} for all the tests is 0% as the grid has been considered as an ideal voltage source in the real-time simulation. THD_{I_g} has varied between 4% to 10.65% for all the loading conditions after the settling times. It was observed that an increase in real and reactive power consumption ratio correlated with a decrease in THD_{I_g}.

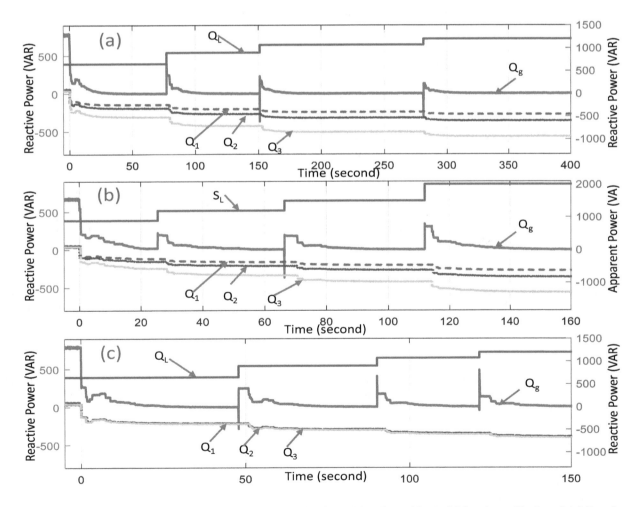

Figure 21. Reactive power support in dynamic loads: (a) load profile 1, (b) load profile 2 and (c) Load profile 1 with equal reactive power support capability.

Table 4. Load profiles for dynamic performance test.

Loading Stage	S_L (VA)	P_L (W)	Q_L (VAR)	Power Factor
Load Profile 1				
1	1500	1350	654	0.9
2	1500	1200	900	0.8
3	1500	1050	1072	0.7
4	1500	900	1200	0.6
Load Profile 2				
1	900	720	540	0.8
2	1200	960	720	0.8
3	1500	1200	900	0.8
4	2000	1600	1200	0.8

6. Conclusions

The PFC converters are considered as reactive power resources in this paper. The supervisory controller is proposed to manage those resources. The WiFi and WirelessHART have provided the interface between the supervisory controller and reactive power resources. The optimal sampling time of the supervisory controller has been determined by various tests. The optimal gain is explained theoretically and validated experimentally. The dynamic performance of the supervisory controller has been validated using different load profiles with the reactive power demand, PFC capacity and load variations. The HIL test results prove the concept and feasibility of additional reactive power support from PFC converters in residential applications. The proposed controller design method will be useful for other small-scale power and wireless network integration.

Author Contributions: All the authors contributed substantially to the manuscript.Contributions of each author are as follows: conceptualization, S.M.R.I. and S.-Y.P.; methodology, S.M.R.I., S.-M.P. and S.Z.; software, S.M.R.I. and S.Z.; validation, S.M.R.I. and S.-Y.P.; formal analysis, S.M.R.I.; investigation, S.M.R.I.; resources, S.M.R.I. and S.-M.P.; data curation, S.M.R.I.; writing—original draft preparation, S.M.R.I.; writing—review and editing, S.M.R.I. and S.-Y.P.; visualization, S.M.R.I.; supervision, S.-Y.P. and S.H.; project administration, S.-Y.P.; funding acquisition, S.-Y.P.

Acknowledgments: This work was supported by the National Science Foundation under grant no. 1446157. However, any opinions, findings, conclusions or recommendations expressed here in are those of the authors and do not necessarily reflect the views of the National Science Foundation.

References

1. The U.S. Energy iNformation Administration Report 2017: Energy Consumption by Sector. Available online: https://www.eia.gov/totalenergy/data/monthly/pdf/sec2_3.pdf (accessed on 17 May 2019).
2. Dixon, J.; Moran, L.; Rodriguez, J.; Domke, R. Reactive Power Compensation Technologies: State-of-the-Art Review. *Proc. IEEE* **2005**, *93*, 2144–2164. [CrossRef]
3. Bing, Z.; Chen, M.; Miller, S.K.T.; Nishida, Y.; Sun, J. Recent Developments in Single-Phase Power Factor Correction. In Proceedings of the 2007 Power Conversion Conference, Nagoya, Japan, 2–5 April 2007; pp. 1520–1526. [CrossRef]
4. Mansouri, M.; Kaboli, S.H.A.; Selvaraj, J.; Rahim, N.A. A review of single phase power factor correction A.C.-D.C. converters. In Proceedings of the 2013 IEEE Conference on Clean Energy and Technology (CEAT), Lankgkawi, Malaysia, 18–20 November 2013; pp. 389–394. [CrossRef]

5.   Nallusamy, S.; Velayutham, D.; Govindarajan, U.; Parvathyshankar, D. Power quality improvement in a low-voltage DC ceiling grid powered system. *IET Power Electron.* **2015**, *8*, 1902–1911. [CrossRef]

6.   Park, S.M.; Park, S. Versatile Control of Unidirectional AC–DC Boost Converters for Power Quality Mitigation. *IEEE Trans. Power Electron.* **2015**, *30*, 4738–4749. [CrossRef]

7.   Ivaldi, J.; Park, S.; Park, S. Integration strategy for bidirectional and unidirectional converters aiming for zero power pollution in residential applications. In Proceedings of the 2015 9th International Conference on Power Electronics and ECCE Asia (ICPE-ECCE Asia), Seoul, Korea, 1–5 June 2015; pp. 1756–1761. [CrossRef]

8.   Radvanovsky, R.; Brodsky, J. *Handbook of SCADA/Control Systems Security*; CRC Press: Boca Raton, FL, USA, 2013.

9.   Nasirian, V.; Yadav, A.P.; Lewis, F.L.; Davoudi, A. Distributed Assistive Control of Power Buffers in DC Microgrids. *IEEE Trans. Energy Convers.* **2017**, *32*, 1396–1406. [CrossRef]

10.   Azidehak, A.; Yousefpoor, N.; Bhattacharya, S. Control and synchronization of distributed controllers in modular converters. In Proceedings of the IECON 2014—40th Annual Conference of the IEEE Industrial Electronics Society, Dallas, TX, USA, 29 October–1 November 2014; pp. 3644–3650. [CrossRef]

11.   Gutierrez, A.; Chamorro, H.R.; Jimenez, J.F. Supervisory control for interleaved boost converters using HiLeS-designer. In Proceedings of the 2014 16th European Conference on Power Electronics and Applications, Lappeenranta, Finland, 26–28 August 2014; pp. 1–6. [CrossRef]

12.   Shen, J.; Khaligh, A. A Supervisory Energy Management Control Strategy in a Battery/Ultracapacitor Hybrid Energy Storage System. *IEEE Trans. Transp. Electrif.* **2015**, *1*, 223–231. [CrossRef]

13.   Babakmehr, M.; Harirchi, F.; Alsaleem, A.; Bubshait, A.; Simões, M.G. Designing an intelligent low power residential PV-based Microgrid. In Proceedings of the 2016 IEEE Industry Applications Society Annual Meeting, Portland, OR, USA, 2–6 October 2016; pp. 1–8. [CrossRef]

14.   Gabbar, H.A.; El-Hendawi, M.; El-Saady, G.; Ibrahim, E.A. Supervisory controller for power management of AC/DC microgrid. In Proceedings of the 2016 IEEE Smart Energy Grid Engineering (SEGE), Oshawa, ON, Canada, 21–24 August 2016; pp. 147–152. [CrossRef]

15.   Hosseinzadeh, M.; Salmasi, F.R. Robust Optimal Power Management System for a Hybrid AC/DC Micro-Grid. *IEEE Trans. Sustain. Energy* **2015**, *6*, 675–687. [CrossRef]

16.   AlBadwawi, R.; Issa, W.; Abusara, M.; Tapas, M. Supervisory Control for Power Management of an Islanded AC Microgrid Using Frequency Signalling-Based Fuzzy Logic Controller. *IEEE Trans. Sustain. Energy* **2018**, *1*. [CrossRef]

17.   Kothari, D.P.; Singh, B.; Pandey, A. Fuzzy supervisory controller for improved voltage dynamics in power factor corrected converter. In Proceedings of the 2002 IEEE Internatinal Symposium on Intelligent Control, Vancouver, BC, Canada, 30–30 October 2002; pp. 93–97.

18.   Islam, S.M.R.; Maxwell, S.; Hossain, M.K.; Park, S.; Park, S. Reactive power distribution strategy using power factor correction converters for smart home application. In Proceedings of the 2016 IEEE Energy Conversion Congress and Exposition (ECCE), Milwaukee, WI, USA, 18–22 September 2016; pp. 1–6. [CrossRef]

19.   Islam, S.M.R.; Maxwell, S.; Park, S.; Zheng, S.; Gong, T.; Han, S. Wireless networked dynamic control testbed for power converters in smart home applications. In Proceedings of the 2017 IEEE Applied Power Electronics Conference and Exposition (APEC), Tampa, FL, USA, 26–30 March 2017; pp. 1196–1202. [CrossRef]

20.   Harper, R. (Ed.) *The Connected Home: The Future of Domestic Life*; Springer: Cambridge, UK, 2011.

21.   Kayas, O. *How to Smart Home: A Step by Step Guide for Smart Homes & Building Automation*, 5th ed.; Key Concept Press: Wyk, Germany, 2017.

22.   White, R.M. *How to Calculate Electrical Loads and Design Power Systems*; CreateSpace Independent Publishing Platform: Scotts Valley, CA, USA, 2011.

23.   Murray, A.; Li, Y. Motion Control Engine Achieves High Efficiency with Digital PFC Integration in Air Conditioner Applications. In Proceedings of the 2006 IEEE International Symposium on Electronics and the Environment, Scottsdale, AZ, USA, 8–11 May 2006; pp. 120–125. [CrossRef]

24.   Fasugba, M.A.; Krein, P.T. Gaining vehicle-to-grid benefits with unidirectional electric and plug-in hybrid vehicle chargers. In Proceedings of the 2011 IEEE Vehicle Power and Propulsion Conference, Chicago, IL, USA, 6–9 September 2011; pp. 1–6. [CrossRef]

25.   Gungor, V.C.; Lu, B.; Hancke, G.P. Opportunities and Challenges of Wireless Sensor Networks in Smart Grid. *IEEE Trans. Ind. Electron.* **2010**, *57*, 3557–3564. [CrossRef]

26.   Gungor, V.C.; Sahin, D.; Kocak, T.; Ergut, S.; Buccella, C.; Cecati, C.; Hancke, G.P. Smart Grid Technologies: Communication Technologies and Standards. *IEEE Trans. Ind. Inform.* **2011**, *7*, 529–539. [CrossRef]

27.   Yarali, A. Wireless Mesh Networking technology for commercial and industrial customers. In Proceedings

of the 2008 Canadian Conference on Electrical and Computer Engineering, Niagara Falls, ON, Canada, 4–7 May 2008; pp. 000047–000052. [CrossRef]

28. Luan, W.; Sharp, D.; Lancashire, S. Smart grid communication network capacity planning for power utilities. In Proceedings of the IEEE PES T D 2010, New Orleans, LA, USA, 19–22 April 2010; pp. 1–4. [CrossRef]

29. Technologies OPAL-RT. OP4510 RT-LAB-RCP/HIL SYSTEMS User Guide. Available online: https://www.opal-rt.com/simulator-platform-op4510/ (accessed on 17 May 2019).

30. Texas Instruments Inc. TMS320F28335, TMS320F28334,TMS320F28332, TMS320F28235, TMS320F28234, TMS320F28232 Digital Signal Controllers (DSCs) Data Manual. Available online: http://www.ti.com/lit/ds/symlink/tms320f28335.pdf (accessed on 17 May 2019).

31. IEEE Standards. IEEE 802.15.4 WPAN TG. Available online: www.ieee802.org/15/pub/TG4.html (accessed on 17 May 2019).

32. IEEE Standards. IEEE 802.11TM WIRELESS LOCAL AREA NETWORKS. Available online: http://www.ieee802.org/11/ (accessed on 17 May 2019).

33. Mohan, N. *Power Electronics: Converters, Applications, and Design*; John Wiley and Sons: New York, NY, USA, 2002.

34. Monmasson, E. *Power Electronic Converters: PWM Strategies and Current Control Techniques*; John Wiley and Sons: New York, NY, USA, 2013.

35. Chang, J.M.; Pedram, M. *Power Optimization and Synthesis at Behavioral and System Levels Using Formal Methods*; Springer: New York, NY, USA, 1999.

36. Ogata, K. *Modern Control Engineering*; Pearson: New York, NY, USA, 2009.

37. Ziegler, J.; Nichols, N. Optimum settings for automatic controllers. *Trans. ASME* **1942**, *64*, 759–768. [CrossRef]

38. Co, T.B. Ziegler-Nichols Closed Loop Tuning. Available online: http://pages.mtu.edu/~tbco/cm416/zn.html (accessed on 17 May 2019).

39. AwiaTech. Advancing Wireless Industrial Automation. Available online: http://www.AwiaTech.com (accessed on 17 May 2019).

40. SparkFun Electronics. ESP8266 WiFi Module, Datasheet. Available online: https://www.sparkfun.com/products/13678 (accessed on 17 May 2019).

41. Xilinx. Kintex 7 FPGA Product Table. Available online: http://www.xilinx.com/support/documentation/selection-guides/kintex7-product-table.pdf (accessed on 17 May 2019).

42. Texas Instruments Inc. TMS320x2833x, 2823x High Resolution Pulse Width Modulator (HRPWM). Available online: http://www.ti.com.cn/cn/lit/ug/sprug02b/sprug02b.pdf (accessed on 17 May 2019).

43. Nixon, M. A Comparison of WirelessHARTTM and ISA100. 11a. Available online: https://www.emerson.com/documents/automation/a-comparison-of-wirelesshart-isa100-11a-en-42598.pdf (accessed on 17 May 2019).

4

Analysis of Intrinsic Switching Losses in Superjunction MOSFETs under Zero Voltage Switching

Maria R. Rogina [1,*], Alberto Rodriguez [1], Diego G. Lamar [1], Jaume Roig [2], German Gomez [2] and Piet Vanmeerbeek [2]

[1] Electrical, Electronic, Computers and Systems Engineering Department, University of Oviedo, 33204 Gijón, Spain; rodriguezalberto@uniovi.es (A.R.); gonzalezdiego@uniovi.es (D.G.L.)
[2] On Semiconductor, 9700 Oudenaarde, Belgium; Jaume.Roig@onsemi.com (J.R.); german.gomez@onsemi.com (G.G.); Piet.Vanmeerbeek@onsemi.com (P.V.)
[*] Correspondence: rodriguezrmaria@uniovi.es;

Abstract: Switching losses of power transistors usually are the most relevant energy losses in high-frequency power converters. Soft-switching techniques allow a reduction of these losses, but even under soft-switching conditions, these losses can be significant, especially at light load and very high switching frequency. In this paper, hysteresis and energy losses are shown during the charge and discharge of the output capacitance (C_{OSS}) of commercial high voltage Superjunction MOSFETs. Moreover, a simple methodology to include information about these two phenomena in datasheets using a commercial system is suggested to manufacturers. Simulation models including C_{OSS} hysteresis and a figure of merit considering these intrinsic energy losses are also proposed. Simulation and experimental measurements using an LLC resonant converter have been performed to validate the proposed mechanism and the usefulness of the proposed simulation models.

Keywords: soft-switching; Superjunction MOSFET; LLC resonant converter; zero voltage switching; C_{OSS} hysteresis; C_{OSS} intrinsic energy losses

1. Introduction

During the last 15 years, the acceptance of resonant converters in the industry application market has been massive, especially regarding adapters, flat panel TVs, electric and hybrid vehicle (EV/HEV), datacenters, and photovoltaic (PV) inverters, among others [1–3] (Figure 1). Besides, new markets and research centers are focusing on moving to higher frequencies to obtain further advantages and gaining power density, taking the present technologies in semiconductors to their physical limit. This is the case of gallium nitride (GaN) and silicon carbide (SiC) technologies, which are thought to be used in the market of resonant converters for low power and high-frequency applications, besides other well-known high-power applications.

A resonant topology operating at a high switching frequency and zero voltage switching (ZVS) provides high power density and is commonly chosen for the previously mentioned applications. The primary side power transistors used in a resonant converter must comply with high-voltage and high-frequency requirements, and need to be properly selected to provide good performance. However, the information given by the manufacturers of these transistors is not usually enough to calculate all the existing energy losses.

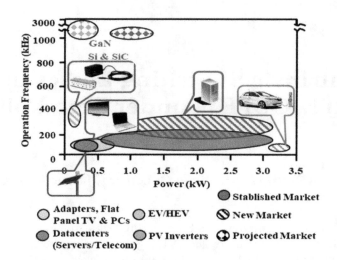

Figure 1. Markets in which resonant converters are used, for different ranges of frequencies (kHz) and power (kW). EV/HEV, electric and hybrid vehicle; PV, photovoltaic.

The parasitic output capacitance (C_{OSS}) of the power transistors has an important role in energy losses, even under ZVS conditions. Traditional switching losses models are not valid for very high switching frequencies. In the work of [4], significant energy dissipation in the process of charging and discharging C_{OSS} of Superjunction MOSFET (SJ-MOSFET) while the gate is shorted to the source is observed. In another paper [5], these intrinsic energy losses (E_i) are measured and compared in different power switches, including Silicon SJ-MOSFETs, GaN cascode, SiC cascode, and SiC MOSFETs. These E_i cannot be eliminated by using ZVS and set an upper limit for the switching frequency of the converters. Similar losses are presented in the work of [6], where energy dissipation during the charging and discharging of the junction capacitance of SiC Schottky diodes is evaluated. In the work of [7], the E_i are included for the determination of soft-switching losses of 10 kV SiC MOSFET modules. Calorimetric measurements are used to evaluate these losses (based on the charge and discharge of the C_{OSS}, and especially of the antiparallel junction barrier Schottky diode). In the work of [8], the variation of E_i with dV/dt is evaluated at very high switching frequency (1–35 MHz) in silicon (Si) and wide-bandgap active devices.

In the work of [9], these E_i are related to a significant hysteresis exhibited by the C_{OSS} of some of the most advanced SJ-MOSFETs. In a further paper [10], the physical mechanism responsible for this C_{OSS} hysteresis is briefly shown by means of mixed-mode simulations. Finally, mixed-mode simulations are also proposed to analyze E_i and the cause of the C_{OSS} hysteresis in different SJ-MOSFETs in the work of [11].

The authors of this paper have previously analyzed the C_{OSS} hysteresis and its related switching losses (including E_i) for different dead-times of three generations of SJ-MOSFETs in an LLC resonant converter in the work of [12]. Moreover, they provide a guideline to select SJ-MOSFETs of different manufacturers to improve the efficiency of this converter in a wide power range in the work of [13].

In this paper, a simple methodology is suggested to manufacturers to include information related to the C_{OSS} hysteresis and E_i of power devices in their datasheets. These data will be useful to select the optimum devices in high-frequency and soft-switching applications. Moreover, a spice model including the C_{OSS} hysteresis and a figure of merit (FoM) including E_i are proposed in Section 2 and 3, respectively. Both proposals are validated using simulation and experimental results of an LLC resonant converter in Section 4.

2. Spice Model Including C_{OSS} Hysteresis Effect

In order to design a resonant converter with low cost and high efficiency and power density, special attention is crucial during the selection of the high-voltage (~600 V) silicon SJ-MOSFET device needed.

However, even if the high-voltage SJ-MOSFETs are selected based on major vendors recommendations for soft-switching applications in resonant converters, they present different values of E_i. E_i might not seem so significant for hard-switching conditions, but it can make the difference under soft-switching operation, especially for low and medium load demands, where conduction power losses are lower and switching losses are relevant because of the high-frequency operation.

E_i is intrinsic to the structure of SJ-MOSFETs, as it is briefly reproduced in Figure 2a–c and explained in detail in the work of [12], where a physical relationship between C_{OSS} hysteresis and E_i was demonstrated, elucidating the existence of energy losses during the charge and discharge of C_{OSS}. Holes and electrons (h+ and e-, respectively, in Figure 2), flowing in parallel to the capacitance, originate stucked charges between the pillars that need to be removed through highly resistive paths that may vary among devices.

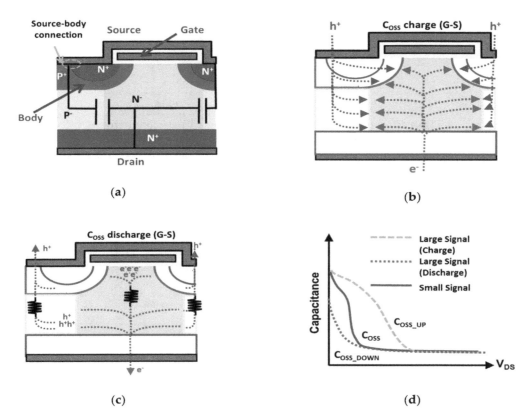

Figure 2. (a) Cross section of a Superjunction (SJ)-MOSFET basic cell. Description of (b) C_{OSS} charge and (c) C_{OSS} discharge. Electron (e-) and hole (h+) currents and charge pockets are indicated (red and blue). (d) Illustrative comparison between C_{OSS} extracted by small-signal (solid blue line) and large-signal (green dashed and red dotted lines).

E_i used to be neglected [14,15], but some simulations models started to consider non-linear C_{OSS} effects, and non-ZVS operation of SJ-MOSFETs [16–18]. However, C_{OSS} hysteresis discoveries are not still considered in those simulation models.

The degree of severity of C_{OSS} hysteresis varies from device to device depending on technological and geometrical features. Up to now, application notes and datasheets do not provide any information regarding this phenomenon. Besides, manufacturers only give small-signal characterization of the transistors, whereas C_{OSS} hysteresis is only detectable during large-signal analysis (Figure 2d). In order to solve this fact, a simple methodology to include in the datasheets enough information to generate simulation models predicting this behavior will be proposed.

In contrast to other reported techniques, a commercial system commonly used by power devices manufactures, an Auriga pulsed I–V system [19], is proposed. This characterization system is able to capture measurements with very high speed and resolution (up to 0.01% of max current), and it is temperature independent. Moreover, voltage/current measurements have emerged as the preferred method of capturing different characteristics of active devices. The simple setup and the voltage and current waveforms obtained using one of the SJ-MOSFETs under test are shown in Figure 3.

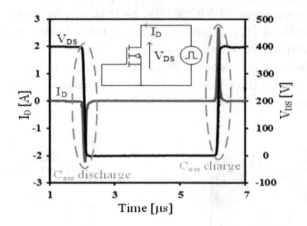

Figure 3. Auriga pulsed I–V tests: I_D and V_{DS} waveforms.

Using these voltage/current measurements, C_{OSS} large-signal curves during its charge and discharge can be inferred using

$$C_{OSS} = \frac{I_D}{\frac{dV_{DS}}{dt}}. \tag{1}$$

Following the presented procedure, C_{OSS} large-signal curves during its charge and discharge of the SJ-MOSFETs under test (Table 1) were estimated (as an example, results of device under test 1 (DUT1) are included in Figure 4).

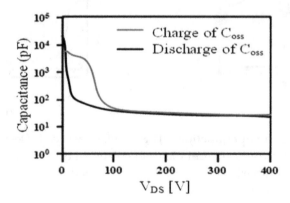

Figure 4. C_{OSS} large-signal curves of device under test 1 (DUT1) (Table 1) during its charge and discharge obtained using (1) and the Auriga pulsed I–V curves.

A detailed simulation model should include this behavior to obtain accurate simulation waveforms of the switching process. The calculated C_{OSS} could be described using a polynomial expression, but in this paper, the use of look-up-tables with pairs of values voltage-capacitance is proposed. Two different look-up-tables, one for the charge and one for the discharge, can be easily included in the spice model of the SJ-MOSFETs. This option is preferred (compared with polynomial expressions) from the point of view of saving computational time and the ease to use, follow, and change data if a different power device is chosen for simulation. The simulations results using the proposed model will be shown and compared with the experimental results in Section 4.

Table 1. List of Superjunction (SJ)-MOSFETs explicitly for LLC primary side with main electrical characteristics and figures of merit (FoMs). DUT, device under test.

DUT	Characteristics from Datasheet								FoM from Datasheet				Proposed FoM
	R_{ON} (mΩ)	BV_{DSS} (V)	R_G (Ω)	Q_G (nC)	Q_{GD} (nC)	Q_{GS} (nC)	E_{OSS} (µJ)[1]	Q_{OSS} (nC)[2]	$R_{ON}·Q_G$ (Ω*nC)	$R_{ON}·Q_{GD}$ (Ω*nC)	$R_{ON}·E_{OSS}$ (Ω*µJ)	$R_{ON}·Q_{OSS}$ (Ω*nC)	$R_{ON}·E_i$ (Ω*µJ)
1	155	600	0.9	24	8	5	2.7	140	3.72	1.24	0.42	21.64	0.049
2	168	650	0.6	60	25	12	6.4	122	10.08	4.20	1.08	20.43	0.194
3	171	600	3.4	37	13	11	4.9	106	6.33	2.22	0.84	18.03	0.116
4	165	650	6	30	13	7.4	3.6	111	4.95	2.15	0.59	18.28	0.793
5	175	600	7	29	12	6	4.6	102	5.08	2.10	0.81	17.85	0.116
6	168	600	7	29	12	6	4.1	103	4.87	2.02	0.69	17.34	0.375

[1] C_{OSS} stored energy at V_{DS} = 400V. [2] Output characteristic charge, result of charging C_{OSS} (time-related effective output capacitance is considered) rising from 0 to 400 V.

3. Simple Methodology to Include E_i in the Datasheets

The cumulative energy (E_{CUM}) of C_{OSS} can be calculated with the previously shown voltage and current waveforms obtained using the Auriga pulsed I–V system.

$$E_{CUM} = \int I_D \cdot V_{DS}\ dt. \tag{2}$$

Using (2), the energy stored during the charge and extracted during the discharge of C_{OSS} can be easily estimated. In Figure 5, an example of the value of E_{CUM} using one of the SJ-MOSFETs under test is shown as an example. As can be seen, the stored energy is higher than the extracted energy, and this difference is the value of E_i. Concretely, E_i is considered as the energy losses after applying a complete cycle of charge–discharge to the device, and consequently is directly related to C_{OSS} hysteresis.

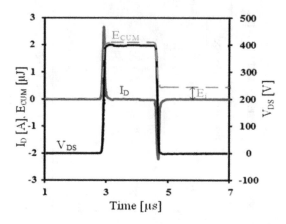

Figure 5. Auriga pulsed I–V tests: I_D and V_{DS} waveforms. Procedure to obtain the cumulative energy and the value of E_i to propose the new figure of merit (FoM).

The proposed FoM, defined as the conduction resistance (R_{ON}) multiplied by E_i, considers both R_{ON} (important for heavy loads) and E_i (crucial for low and medium loads), allowing a proper selection of SJ-MOSFETs in soft-switching applications operating at high frequencies. The lower the FoM value of an SJ-MOSFET, the higher its performance. In Section 4, the prediction of the performance of different SJ-MOSFETs using this FoM is validated with experimental efficiency results.

Besides, it is worth mentioning that, as occurring in other common FoMs, the direct and indirect proportionalities of R_{ON} and E_i with the die area result in an area-independent FoM. This is a preferred FoM approach to facilitate the benchmarking between technologies. In addition, common to other FoMs are the limitations for devices with a small die area, where the termination area could be as relevant as the active area of the transistor (R_{ON} does not perfectly scale with the die area).

4. Validation of the Proposed Simulation Model and FoM

The power supplies used in the applications mentioned in the introduction of this paper must comply with challenging standards, such as 80PLUS Titanium® [20]. An LLC resonant converter is the topology generally selected to develop this kind of power supply, mainly owing to their high efficiency and power density. More information and new models are needed to properly design these power converters operating at a very high switching frequency.

Silicon SJ-MOSFETs are the preferred devices during the design of the primary side of the LLC resonant converter as they meet the requirements regarding voltage, current, and frequency, and an accurate procedure for their proper selection for each specification is important, especially operating at a high switching frequency. The devices under test (DUT) in this work are detailed in Table 1. SJ-MOSFETs with similar voltage blocking capability, R_{ON}, and Q_{OSS} are selected in order to obtain a fair comparison under the same working conditions. In all the tests, ZVS is assured and, consequently, differences in the value of R_G are not relevant because the switching losses were forced to be independent of R_G. Exhaustive experimental tests are carried out using an LLC resonant converter with the DUTs. Waveforms, breakdown losses, and efficiency results are analyzed and compared.

4.1. LLC Resonant Converter Description

The previously described SJ-MOSFETs were tested in a commercial evaluation board of an LLC resonant converter [21] featuring the specifications of Table 2.

Table 2. LLC resonant converter evaluation board specifications.

Parameter	Value	Parameter	Value
Primary side devices	Si SJ-MOSFETs (DUTs)	Input Voltage, V_{IN} (V)	350–410
Secondary side devices	OptiMOS BSC010N04LS	Output voltage, V_{OUT} (V)	12
Gate driver IC	2EDL05N06PF	C_R (nF)	66
Maximum power (W)	600	L_R (uH)	15.5
Resonant frequency, f_{RES} (kHz)	157	L_M (uH)	195
Frequency range (kHz)	90–250	Transformer turns-ratio	16:1

The fundamental requirements related to a fixed resonant tank (C_R, L_R, and L_M) and deadtimes (t_D) are fulfilled, guaranteeing the ZVS inductive mode for the whole power range and for all the transistors under examination [13]. As the devices selected share very close values of R_{ON} and Q_{OSS}, there is no need to redesign different L_M values for each transistor, reassuring ZVS the whole load range. A simplified scheme of the LLC resonant converter with the main components and the sensing methods is shown in Figure 6.

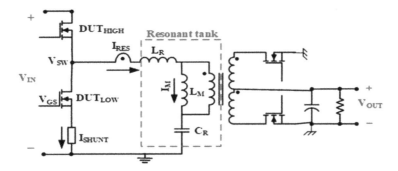

Figure 6. Simplified circuit scheme of the LLC resonant converter and sensing method.

Mixed-mode (MM) simulations were also carried out to analyze the evolution of certain signals that cannot be experimentally measured (as the current through the channel of the MOSFETs). The developed MM simulations consist of spice circuits, where the half-bridge (HB) structure of the primary side of the LLC resonant converter is replaced by TCAD (Technology Computer Aided Design) structures (finite-element structures) simulating the power transistors (DUT$_{HIGH}$ and DUT$_{LOW}$ in Figure 6).

Calibration of TCAD structures was done by means of process simulations in the case of own SJ-MOSFETs technologies, and by means of reverse engineering and reverse calibration technique in the case of other commercial SJ-MOSFETs technologies. Information about the doping profiles is included in the TCAD structures and data regarding voltages, power, magnetics, frequencies, and so on are extracted from the evaluation board datasheet [21].

The accuracy of the MM simulations and its good match with experimental waveforms can be seen in Figure 7. Moreover, the current through the channel of the DUT$_{LOW}$ (I_{CH}) obtained using MM simulation (it cannot be experimentally measured) was included to verify the ZVS operation (I_{CH} falls to zero before V_{SW} is increased). As I_{CH} is zero before V_{SW} rises, the area below P_{INS} waveform represents the energy stored in the output capacitance of the SJ-MOSFET during the turn-off (E_{off}). This energy cannot be considered as losses, because it can be retrieved in the turn-on.

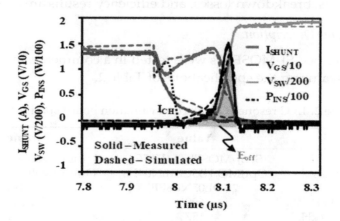

Figure 7. Simulated and measured waveforms for DUT_{LOW} during the turn-off. I_{SHUNT}, V_{GS}, and V_{SW} are referenced in Figure 6. P_{INS} is the instantaneous power (product of V_{SW} and I_{SHUNT}) and I_{CH} is the simulated current through the channel of the SJ-MOSFET.

4.2. Experimental Results, Efficiency, and Power Losses Break-Down

Several experimental measurements and waveforms are analyzed to validate the proper operation of the converter at different loads and with different DUTs. Examples of experimental waveforms measured in the converter are shown in Figure 8.

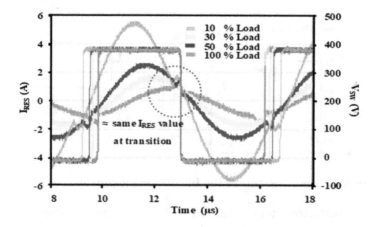

Figure 8. Experimental I_{RES} and V_{SW} measured at different loads for DUT1 as an example of how the resonant current varies with the load. As can be seen, different switching frequencies are also used for different loads.

In Figure 8, the current through the resonant tank (I_{RES} in Figure 6) is shown for different power levels, as well as its corresponding V_{SW} waveform. As can be seen, the I_{RES} value during the transition of both DUTs remains almost the same regardless of the load level, which will be helpful to estimate switching losses (they are calculated by means of the energy dissipated during the turn-on and turn-off) and to understand the behavior of the transistors during these transitions.

An efficiency comparison of the LLC resonant converter using all the DUTs as the primary side transistors is carried out in the whole power range, going from 10% to 100% of full load (600 W), always following the same test protocol and operating conditions. In order to minimize error measurements and its influence on the efficiency comparison, a repetitive protocol was performed using an automatic program based on Java. First, the converter is turned-on at 10% of maximum load, and it remains under this working condition for 15 minutes to achieve a constant working temperature. Then, the efficiency at 10% of maximum load is measured (this measurement is the result of averaging 10 consecutives measurements of V_{IN}, V_{OUT}, I_{IN}, and I_{OUT}). Finally, the load is increased, and new measurements are done after one minute. This procedure is repeated to 20%, 30%, 50%, 70%, and 100% of full load. In Figure 9, the differential efficiencies obtained are shown, taken as reference DUT1, as it is the device that shows best performance for the whole range.

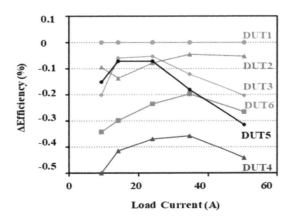

Figure 9. Differential efficiencies of the LLC resonant converter with respect to the best SJ-MOSFET (DUT1).

Using experimental waveforms of V_{GS}, V_{SW}, and I_{SHUNT}, switching (P_{SW}), driving (P_{DR}), and conduction (P_{ON}) power losses contributions from each DUT are calculated for three power loads demands of 60 W (10%), 300 W (50%), and 600 W (100%), and are shown in Figure 10a–c, respectively.

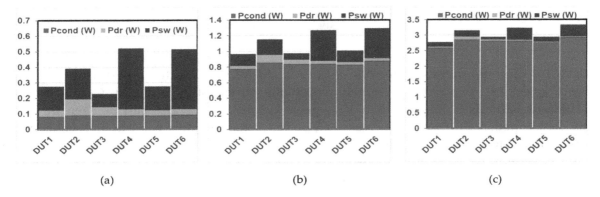

(a)	(b)	(c)

Figure 10. Measured power losses owing to driving (P_{DR}), switching (P_{SW}), and conduction (P_{ON}) at (**a**) 10%, (**b**) 50%, and (**c**) 100% load for each primary-side SJ-MOSFET.

In Figure 10a, at a low load, whereas low P_{ON} losses remain almost equal for all the DUTs, differences in P_{DR} losses have a small impact and P_{SW} losses are dominant. For heavy loads (Figure 10c), P_{ON} is by far the main factor of losses in the SJ-MOSFETs, yet disparity among the P_{SW} losses is discernible. At a medium load (Figure 10b), divergence in P_{SW} among transistors makes the difference (P_{ON} losses are the highest, but fairly the same value, but differences at P_{SW} have a great impact in the losses contribution). Even performing ZVS, P_{SW} losses are relevant and differences in the total power losses between DUTs are the result of $P_{SW} + P_{DR}$ at light loads (Figure 10a) and $P_{SW} + P_{ON}$ at heavy loads (Figure 10c). These P_{SW} losses under ZVS conditions are consistent with the existence of the E_i previously reported.

4.3. Validation of the Simulation Model Including C_{OSS} Hysteresis

The developed spice model of all the SJ-MOSFETs under test includes the definition of the C_{OSS} with two look-up-tables with pairs of values voltage-capacitance, one for the charge and one for the discharge (obtained using the procedure presented in Section 2). On the basis of the circuit proposed in Figure 6 and using the proposed spice models of the SJ-MOSFETs, some simulations of the LLC resonant converter are carried out using LTSpice. In these simulations, emphasis is on the primary side of the converter and the accurate definition of the C_{OSS} value. Experimental and simulated VSW waveforms are compared in Figure 11 and good agreement is obtained.

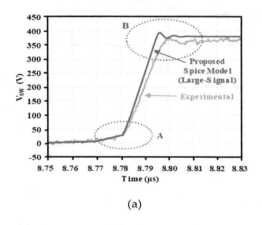
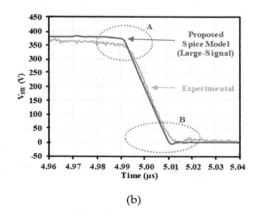

(a)   (b)

Figure 11. V_{SW} waveform extracted by experimental measurement (green) and simulation with the proposed large-signal spice model (red) during (**a**) the increase of V_{SW} and (**b**) the decrease of V_{SW}.

It should be noted that the equivalent capacitance seen from the port defined by V_{SW} is the parallel combination of the output capacitance of DUT_{LOW} and DUT_{HIGH} (two nonlinear capacitors), defined as

$$C_{eq} = \frac{C_{OSS_{DUT_{LOW}}}(V_{SW}) \cdot C_{OSS_{DUT_{HIGH}}}(V_{IN} - V_{SW})}{C_{OSS_{DUT_{LOW}}}(V_{SW}) + C_{OSS_{DUT_{HIGH}}}(V_{IN} - V_{SW})} \quad (3)$$

ff

Taking into account that the value of C_{OSS} of each SJ-MOSFET is di erent during its charge and discharge, C_{eq} is not symmetric (as presented in previous works not including the C_{OSS} hysteresis [15]) and a different value is obtained when V_{SW} goes up and down. As can be seen in Figure 12, the equivalent impedance during the increase of V_{SW} (C_{eq1}) has the same value at high voltage than the equivalence impedance during the decrease of V_{SW} (C_{eq2}) at low voltage. Consequently, in Figure 11, similar V_{SW} evolution can be seen in the corners marked as A and B during the increase and the decrease of V_{SW}. The proposed spice model captures the corner asymmetry (see corners A and B have different curvature) when V_{SW} goes up and down during DUT_{LOW} turn-off and turn-on transitions, thus being consistent with the existence of a C_{OSS} hysteresis and matching the experimental measurements.

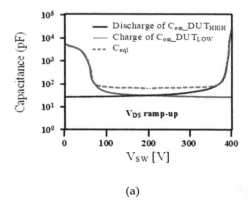
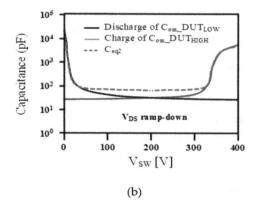

(a)   (b)

Figure 12. Derivation of C_{eq1} and C_{eq2}, which are asymmetric with respect to the charge and discharge of the C_{OSS} of DUT_{HIGH} and DUT_{LOW}. (**a**) Increase of V_{SW} and (**b**) decrease of V_{SW}.

4.4. Validation of the Proposed FoM including E_i

In Figure 9, there is not a clear trend regarding the efficiency that DUTs show for different load demands. Some of them might be suitable for low power, but, in contrast, their performance is worse at full load. That is the case of DUT5. FoMs based on the information provided by the datasheet do not always explain these differences in operation. For example, the worse performance at full load of DUT5 can be explained by its high on-resistance, but the performance for a light load of DUT5 is better than the performance of DUT6, while their switching characteristics are almost the same (even a bit better than those of DUT6).

Consequently, new FoMs are needed to know in detail where the power losses come from, as a great percentage of the converter total losses is attributable to the primary-side SJ-MOSFETs [13] for the whole load range. The proposed FoM should allow a proper selection of the SJ-MOSFETs in an LLC resonant converter. In the last column of Table 1, the value of the proposed FoM for all the DUTs is included, while in Figure 13, these values are compared with respect to DUT1.

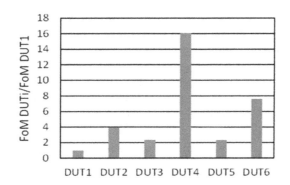

Figure 13. Comparison of the proposed FoM of the DUTs.

The best performance of the LLC is obtained with DUT1, which also has the lowest value of the proposed FoM. DUT2, DUT3, and DUT5 have low values of the proposed FoM and the performance of the LLC with them is also good, especially at medium and light loads. DUT4 has the highest value of the proposed FoM, and the LLC with this DUT has the lowest performance.

As can be clearly seen, the proposed FoM can predict the better performance of the LLC with DUT5 than with DUT6 (especially at light load), which cannot be explained using the characteristics from datasheets.

5. Conclusions

The existence of E_i in power devices, which produces significant switching energy losses in high switching frequency power converters, even under ZVS, has been shown in this paper. Moreover, E_i are related to a C_{OSS} hysteresis.

A simple methodology using a commercial system is suggested to manufacturers to provide E_i (which is not included in datasheets) and information about the C_{OSS} hysteresis. The relevance of the C_{OSS} hysteresis information is validated by developing simulation models that accurately match the experimental charge and discharge waveforms of the C_{OSS}. These new models will allow the designer to better predict the behavior of the power devices and their corresponding power losses, in order to to be able to design more efficient applications.

Efficiency experimental results on an LLC resonant converter are used to validate the suitability of the proposed FoM including E_i, to properly select the transistors used in soft-switching power converters operating at high frequencies.

The impact of these kinds of losses is important in high switching frequency power converters and should be properly modelled to be able to predict the performance of different commercial power transistors in case the designer needs to compare several of those for a certain application; consequently, new data and models are needed.

Author Contributions: Conceptualization, J.R.; methodology, M.R.R.; software, G.G.; validation, M.R.R.; formal analysis, A.R.; resources, P.V.; writing–original draft preparation, M.R.R. and A.R.; writing–review and editing, A.R., D.G.L., and J.R.; supervision, A.R., J.R., and D.G.L.; All authors have read and agreed to the published version of the manuscript.

Nomenclature

V_{IN}	input voltage of the LLC converter.
V_{OUT}:	output voltage of the LLC converter
DUT_{HIGH}/DUT_{LOW}	primary side MOSFETs under test
L_R	discrete series resonant inductance of the resonant tank
L_M	magnetizing inductance of the resonant tank
C_R	discrete resonant capacitor of the resonant tank
V_{GS}	gate-to-source voltage of the DUT_{LOW}
V_{SW}	drain-to-source voltage of the DUT_{LOW}. Switching voltage
I_{RES}	current through the resonant tank
I_{SHUNT}	drain-to-source current through DUT_{LOW}
I_{CH}	simulated current through the channel of the DUT_{LOW}
P_{INS}	instantaneous power dissipated in the DUT_{LOW}. Product of I_{SHUNT} and V_{SW}.
C_{OSS}	non-linear output capacitance of the MOSFETs
G, S, and D	gate, source and drain terminals of the MOSFETs
P	p-doped zone
N	n-doped zone
E_i	intrinsic energy losses after charging and discharging C_{OSS} up to a certain drain-to-source voltage (V_{DS}) in off-state
E_{CUM}	cumulative energy in the MOSFET when applying a voltage pulse on it during off-state.
P_{SW}, P_{DR}, and P_{ON}	switching, driving, and conduction power losses of the MOSFETs

References

1.   Chen, Y.; Wang, H.; Liu, Y.F. Improved hybrid rectifier for 1-MHz LLC-based universal AC-DC adapter. In Proceedings of the IEEE Applied Power Electronics Conference and Exposition (APEC), Tampa, FL, USA, 26–30 March 2017; pp. 23–30.

2.   Deng, J.; Li, S.; Hu, S.; Mi, C.C.; Ma, R. Design methodology of LLC resonant converters for electric vehicle battery chargers. *IEEE Trans. Veh. Technol.* **2014**, *63*, 1581–1592. [CrossRef]

3.   Hu, H.; Fang, X.; Chen, F.; Shen, Z.J.; Batarseh, I. A modified high-efficiency LLC converter with two transformers for wide input-voltage range applications. *IEEE Trans. Power Electron.* **2013**, *28*, 1946–1960. [CrossRef]

4.   Fedison, J.B.; Fornage, M.; Harrison, M.J.; Zimmanck, D.R. COSS related energy loss in power MOSFETs used in zero-voltage-switched applications. In Proceedings of the IEEE Applied Power Electronics Conference and Exposition (APEC), Fort Worth, TX, USA, 16–20 March 2014; pp. 150–156.

5.   Li, X.; Bhalla, A. Comparison of intrinsic energy losses in unipolar power switches. In Proceedings of the IEEE Wide Bandgap Power Devices and Applications (WiPDA), Fayetteville, AR, USA, 7–9 November 2016; pp. 228–232.

6.   Tong, Z.; Zulauf, G.; Rivas-Davila, J. A Study on off-state losses in silicon-carbide schottky diodes. In Proceedings of the IEEE 19th Workshop on Control and Modeling for Power Electronics (COMPEL), Padua, Italy, 25–28 June 2018.

7.   Rothmund, D.; Bortis, D.; Kolar, J.W. Accurate Transient Calorimetric Measurement of Soft-Switching Losses of 10-kV SiC mosfets and Diodes. *IEEE Trans. Power Electron.* **2018**, *33*, 5240–5250. [CrossRef]

8.   Zulauf, G.; Tong, Z.; Rivas-Davila, J. Considerations for active power device selection in high- and

very-high-frequency power converters. In Proceedings of the IEEE 19th Workshop on Control and Modeling for Power Electronics (COMPEL), Padua, Italy, 25–28 June 2018.

9. Fedison, J.B.; Harrison, M.J. Coss hysteresis in advanced superjunction mosfets. In Proceedings of the IEEE Applied Power Electronics Conference and Exposition (APEC), Long Beach, CA, USA, 20–24 March 2016; pp. 247–252.

10. Roig, J.; Bauwens, F. Origin of anomalous coss hysteresis in resonant converters with superjunction FETs. *IEEE Trans. Electron Devices* **2015**, *62*, 3092–3094. [CrossRef]

11. Zulauf, G.D.; Roig-Guitart, J.; Plummer, J.D.; Rivas-Davila, J.M. Coss measurements for superjunction MOSFETs: Limitations and opportunities. *IEEE Trans. Electron Devices* **2019**, *66*, 578–584. [CrossRef]

12. Roig, J.; Gomez, G.; Bauwens, F.; Vlachakis, B.; Rogina, M.R.; Rodriguez, A.; Lamar, D.G. High-accuracy modelling of ZVS energy loss in advanced power transistors. In Proceedings of the IEEE Applied Power Electronics Conference and Exposition (APEC), San Antonio, TX, USA, 4–8 March 2018; pp. 263–269.

13. Maria, R.; Rogina, A.; Rodriguez, D.; Lamar, G.; Roig, J.; Vanmeerbeek, P.; Bauwens, F. Novel selection criteria of primary side transistors for LLC resonant converters. In Proceedings of the IEEE Control and Modelling for Power Electronic (COMPEL), Padova, Italy, 25–28 June 2018.

14. Elferich, R. General ZVS half bridge model regarding nonlinear capacitances and application to LLC design. In Proceedings of the IEEE Energy Conversion Congress and Exposition (ECCE), Edinburgh, UK, 15–20 September 2012; pp. 4404–4410.

15. Costinett, D.; Maksimovic, D.; Zane, R. Circuit-oriented treatment of nonlinear capacitances in switched-mode power supplies. *IEEE Trans. Power Electron.* **2015**, *30*, 985–995. [CrossRef]

16. Christen, D.; Biela, J. Analytical switching loss modeling based on datasheet parameters for mosfets in a half-bridge. *IEEE Trans. Power Electron.* **2019**, *34*, 3700–3710. [CrossRef]

17. Miftakhutdinov, R. New ZVS analysis of PWM converters applied to super-junction, GaN and SiC power FETs. In Proceedings of the IEEE Applied Power Electronics Conference and Exposition (APEC), Charlotte, NC, USA, 15–19 March 2015; pp. 336–341.

18. Kasper, M.; Burkart, R.M.; Deboy, G.; Kolar, J.W. ZVS of power MOSFETs revisited. *IEEE Trans. Power Electron.* **2016**, *31*, 8063–8067. [CrossRef]

19. Available online: https://focus-microwaves.com/pulsed-iv/ (accessed on 13 April 2019).

20. Available online: http://www.80PLUS.org (accessed on 13 April 2019).

21. Steiner, A.; di Domenico, F.; Catly, J.; Stückler, F. *600 W Half Bridge LLC Evaluation Board with 600 V CoolMOS™ C7*; Infineon Technology: Neubiberg, Germany, 2015.

5

Disturbance Rejection Control Method for Isolated Three-Port Converter with Virtual Damping

Jiang You [1], Mengyan Liao [1], Hailong Chen [2,*], Negareh Ghasemi [3] and Mahinda Vilathgamuwa [4]

[1]   College of Automation, Harbin Engineering University, Harbin 150001, China; youjiang@hrbeu.edu.cn (J.Y.); liaomengyan417@163.com (M.L.)

[2]   College of Shipbuilding Engineering, Harbin Engineering University, Harbin 150001, China

[3]   School of Information Technology and Electrical Engineering, The University of Queensland, Brisbane 4072, Australia; n.ghasemi@uq.edu.au

[4]   School of Electrical Engineering and Computer Science, Queensland University of Technology, Brisbane 4000, Australia; mahinda.vilathgamuwa@qut.edu.au

*   Correspondence: chenhailong@hrbeu.edu.cn

Abstract: The high-power density and capability of three-port converters (TPCs) in generating demanded power synchronously using flexible control strategy make them potential candidates for renewable energy applications to enhance efficiency and power density. The control performance of isolated TPCs can be degraded due to the coupling and interaction of power transmission among different ports, variations of model parameters caused by the changes of the operation point and resonant peak of LC circuit. To address these issues, a linear active disturbance rejection control (LADRC) system is developed in this paper for controlling the utilized TPC. A virtual damping based method is proposed to increase damping ratio of current control subsystem of TPC which is beneficial in further improving dynamic control performance. The simulation and experimental results show that compared to the traditional frequency control strategy, the control performance of isolated TPC can be improved by using the proposed method.

Keywords: three-port converter; linear active disturbance rejection control; virtual damping; linear extended state observer

---

1. Introduction

The demand for three-port converters (TPCs) in renewable energy generation systems is increasing due to the compact structure of these converters and their ability to handle demanded power synchronously [1–5]. The TPCs not only facilitate multifunctional and multidirectional regulation for electrical power transmission but also provide flexibility in power control and power density enhancement in power conversion systems [6–10].

In an isolated three-port converter, the three windings of an isolated transformer share the same magnetic core, therefore there are unavoidable couplings of power transmission among the three ports of TPC. Decoupling control methods with proper decoupling factors are usually employed in three-port converters to achieve two single-input single-output (SISO) subsystems [11–14]. A classical frequency control theory is usually utilized to design controllers for each port respectively. Since the small signal models employed to design the controllers are produced by linearization of the nonlinear model of TPC at a steady-state operating point, the decoupling and dynamic performances of TPC control system can be degraded significantly by the variation of the operating point. Particularly, since the small signal models of TPC depend on a specific operating point, in a transient state process, a heavy change of the operating point parameters may affect decoupling of different ports and dynamic performance of the control system. Generally, the three-port converter is a multiple-input multiple-output (MIMO) system, several phase-shifting angles and equivalent duty cycles can be used as control signals, and several voltages and currents of different ports can be assigned as the output signals. A linear quadratic regulator (LQR) based method is applied in ref. [15] to develop a multivariable controller for a three-port converter. Though the LQR method seems capable of achieving performance balance of different ports, it has relatively high sensitivity to the accuracy of system parameters. The parameters of the control models will vary with the change in operating point as these small signal based models used in the control system design are derived at a specific steady state operating point. Also, the design of the parameters of the time domain based LQR method is relatively complex compared to the frequency domain design method.

The LADRC method was first proposed by Zhiqiang Gao, and it has advantages of tolerating changes in model parameters and possesses an inherent decoupling ability that is useful for control system design [16]. In the LADRC method, the influences of model parameter deviations and external interferences can all be regarded as a generalized disturbance [17]. Therefore, the linear extended state observer (LESO) [18,19] can be used to estimate the state variables and generalized disturbance, and the observed signals are used to synthesize control signal in the control system. Compared to conventional PI controller, the LADRC method is shown to enhance the dynamic performance of the control system in [20].

In order to improve the dynamic control performance of an isolated three-port converter in this study, the LADRC method is employed to decrease negative impact of reactions among different ports, and obtain high control performance under load change conditions.

The rest of the paper is organized as follows. In Section 2, the topology, modulation method, power delivery relationship, and control-oriented small signal models are presented. The design of a LADRC based control system for a three-port converter by utilizing its current and voltage control small signal models, and the proposed virtual damping method to suppress the resonant peak in the current control subsystem are given in Section 3. Also, in this section, the principle and the design procedure of decoupling control are briefly illustrated for comparison purposes. The simulation and experiment results are presented in Section 4. Finally, the conclusion is provided in Section 5.

2. Topology and Modeling of TPC

The circuit topology of an isolated TPC is presented in Figure 1a. In this figure, a DC power supply (e.g., it can be a fuel cell or a photovoltaic cell) is set in Port 1, and the power supply, v_{d1} is connected in series with an inductor L_{d1}, and r_e represents the equivalent series resistance (ESR) of L_{d1}. There is 180° phase shift between leg A and leg B, and the duty cycles of all switches in Port 1 are set to be 50% and the drive signals of the switches on the same leg are complementary. The Port 2 and Port 3 are connected with load and energy storage (ES) respectively and their switching patterns are as same as the switching mode of Port 1. A simplified equivalent Δ-connected circuit of the TPC by transferring the related parameters of Port 2 and Port 3 to Port 1 is given in Figure 1b. If the voltage between the middle points of leg A and leg B, v_1 is defined as a reference, the phase shifts of v_2 and v_3 relative to v_1 are denoted as φ_{12} and φ_{13} respectively, and they are shown in Figure 1c. Moreover, L_1, L_2, and L_3 are equivalent series inductances (including winding leakage inductance and additional inductance) of the three transformer windings. The expressions of L_{12}, L_{13}, and L_{23} of the Δ-connected circuit shown in Figure 1b are defined by (1).

$$\begin{cases} L_{12} = L_1 + L_2' + L_1 L_2'/L_3' \\ L_{23} = L_2' + L_3' + L_2' L_3'/L_1 \\ L_{13} = L_3' + L_1 + L_1 L_3'/L_2' \end{cases} \tag{1}$$

L_2' and L_3' are expressed by (2).

$$L_2' = \frac{N_1^2 L_2}{N_2^2}, L_3' = \frac{N_1^2 L_3}{N_3^2} \tag{2}$$

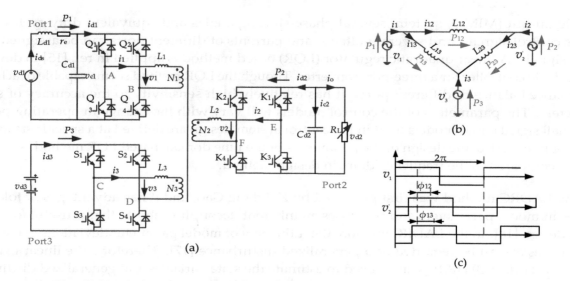

Figure 1. The isolated three-port converter: **(a)** topology; **(b)** equivalent Δ-connected circuit; **(c)** modulation scheme.

In each switching cycle, the total power transmitted between any two ports is approximated to its fundamental component. Therefore, the Fourier expansion based fundamental component analysis method is employed for theoretical analysis in this study. By utilizing the equivalent Δ-connection circuit in Figure 1b, the power equations of each port can be written as in (3).

$$\begin{cases} P_1 = P_{12} + P_{13} \\ P_2 = -P_{12} + P_{23} \\ P_3 = -P_{13} - P_{23} \end{cases}, \; (P_1 + P_2 + P_3 = 0) \tag{3}$$

where

$$\begin{cases} P_{12} = \frac{8N_1}{\pi^2 N_2 \omega_s L_{12}} V_{d1} V_{d2} \sin \varphi_{12} \\ P_{13} = \frac{8N_1}{\pi^2 N_3 \omega_s L_{13}} V_{d1} V_{d3} \sin \varphi_{13} \\ P_{23} = \frac{8N_1^2}{\pi^2 N_2 N_3 \omega_s L_{23}} V_{d2} V_{d3} \sin(\varphi_{13} - \varphi_{12}) \end{cases} \tag{4}$$

In (4), V_{d1}, V_{d2}, and V_{d3} are the rated amplitudes of v_1, v_2, and v_3, N_1, N_2, and N_3 are the transformer winding turns of respective ports, and ω_s is the switching angular frequency. Since the summation of P_1, P_2 and P_3 is kept at zero as shown in (3), that means the power of one port can be determined using the powers of the other two ports. From this point of view, the energy storage port is usually taken as an energy buffer that can be charged or discharged determined by the power delivery and load conditions of Port 1 and Port 2 respectively. According to (3) and (4), the power of Port 1 and Port 2 can be formulated as (5) and (6) respectively.

$$P_1 = \frac{8N_1 V_{d1} V_{d2}}{\pi^2 N_2 \omega_s L_{12}} \sin \varphi_{12} + P_1 = \frac{8N_1 V_{d1} V_{d3}}{\pi^2 N_3 \omega_s L_{13}} \sin \varphi_{13} \tag{5}$$

$$P_2 = -\frac{8N_1 V_{d1} V_{d2} \sin \varphi_{12}}{\pi^2 N_2 \omega_s L_{12}} + \frac{8N_1 V_{d2} V_{d3} \sin(\varphi_{13} - \varphi_{12})}{\pi^2 N_2 N_3 \omega_s L_{23}} \tag{6}$$

Therefore, the corresponding average currents in each switching cycle can be derived as (7) and (8) respectively.

$$\bar{i}_{d1} = \frac{P_1}{V_{d1}} = P_1 = \frac{8N_1 V_{d2}}{\pi^2 N_2 \omega_s L_{12}} \sin \varphi_{12} + \frac{8N_1 V_{d3}}{\pi^2 N_3 \omega_s L_{13}} \sin \varphi_{13} \tag{7}$$

$$\bar{i}_{d2} = \frac{P_2}{V_{d2}} = -\frac{8N_1 V_{d1} \sin \varphi_{12}}{\pi^2 N_2 \omega_s L_{12}} + \frac{8N_1^2 V_{d3} \sin(\varphi_{13} - \varphi_{12})}{\pi^2 N_2 N_3 \omega_s L_{23}} \tag{8}$$

By applying partial differential operation in (7) and (8) respectively, (9) can be obtained for a steady-state operating point A $(\varphi_{120}, \varphi_{130})$.

$$\begin{cases} G_{11} = \left.\dfrac{\partial \tilde{i}_{d2}}{\partial \varphi_{12}}\right|_A = -\dfrac{8N_1 V_{d1} \cos \varphi_{120}}{\pi^2 N_2 N_3 \omega_s L_{12}} - \dfrac{8N_1^2 V_{d3} \cos(\varphi_{130} - \varphi_{120})}{\pi^2 N_2 N_3 \omega_s L_{23}} \\[2mm] G_{12} = \left.\dfrac{\partial \tilde{i}_{d2}}{\partial \varphi_{13}}\right|_A = \dfrac{8N_1^2 V_{d3}}{\pi^2 N_2 N_3 \omega_s L_{23}} \cos(\varphi_{130} - \varphi_{120}) \\[2mm] G_{21} = \left.\dfrac{\partial \tilde{i}_{d1}}{\partial \varphi_{12}}\right|_A = \dfrac{8N_1 V_{d2}}{\pi^2 N_2 \omega_s L_{12}} \cos \varphi_{120} \\[2mm] G_{22} = \left.\dfrac{\partial \tilde{i}_{d1}}{\partial \varphi_{13}}\right|_A = \dfrac{8N_1 V_{d3}}{\pi^2 N_3 \omega_s L_{13}} \cos \varphi_{130} \end{cases} \tag{9}$$

Consequently, (9) can be simplified as (10).

$$\begin{bmatrix} \tilde{i}_{d2} \\ \tilde{i}_{d1} \end{bmatrix} = \begin{bmatrix} G_{11} & G_{12} \\ G_{21} & G_{22} \end{bmatrix} \begin{bmatrix} \tilde{\varphi}_{12} \\ \tilde{\varphi}_{13} \end{bmatrix} = G_A \begin{bmatrix} \tilde{\varphi}_{12} \\ \tilde{\varphi}_{13} \end{bmatrix} \tag{10}$$

As shown in (9), besides the circuit parameters, the value of any matrix element in G_A, G_{xy} (x = 1, 2 and y = 1, 2) is determined by the steady-state parameters (φ_{120} and φ_{130}). And it can be also seen from (10) that there are couplings between $\tilde{i}_{d1}$ and $\tilde{i}_{d2}$ caused by G_{21} and G_{12}.

The small signal linearization model of the three-port converter can be derived as in (11) by applying KCL and KVL laws in Figure 1.

$$\begin{cases} C_{d2} \dfrac{d\tilde{v}_{d2}}{dt} = -\dfrac{\tilde{v}_{d2}}{R_L} - G_{11}\tilde{\varphi}_{12} - G_{12}\tilde{\varphi}_{13} \\[2mm] L_{d1} \dfrac{d\tilde{i}_{ds}}{dt} = \tilde{v}_{d1} - \tilde{v}_{c1} - r_e \tilde{i}_{ds} \\[2mm] C_{d1} \dfrac{d\tilde{v}_{c1}}{dt} = \tilde{i}_{ds} - G_{21}\tilde{\varphi}_{12} - G_{22}\tilde{\varphi}_{13} \end{cases} \tag{11}$$

3. Control Strategy for TPC

3.1. Decoupling Control for TPC

Assuming the matrix, G_A given in (10) can be simplified to a diagonal matrix given in (12) by introducing a decoupling matrix H defined in (13).

$$\begin{bmatrix} G_{11} & G_{12} \\ G_{21} & G_{22} \end{bmatrix} \begin{bmatrix} H_{11} & H_{12} \\ H_{21} & H_{22} \end{bmatrix} = \begin{bmatrix} G_{11} & 0 \\ 0 & G_{22} \end{bmatrix} \tag{12}$$

$$H = \begin{bmatrix} H_{11} & H_{12} \\ H_{21} & H_{22} \end{bmatrix} = \begin{bmatrix} \dfrac{G_{11}G_{22}}{G_{11}G_{22}-G_{12}G_{21}} & \dfrac{-G_{12}G_{22}}{G_{11}G_{22}-G_{12}G_{21}} \\[3mm] \dfrac{-G_{11}G_{21}}{G_{11}G_{22}-G_{12}G_{21}} & \dfrac{G_{11}G_{22}}{G_{11}G_{22}-G_{12}G_{21}} \end{bmatrix} \tag{13}$$

The decoupling control block diagram is presented in Figure 2. In this figure, G_v and G_c represent the voltage controller and current controller respectively, which can be synthesized by using Bode plot based design method in frequency domain. In Figure 2, the open loop transfer functions of the voltage and current control subsystems are $G_{vo} = G_{11}G_1$ and $G_{co} = G_{22}G_2$ respectively, the transfer functions of G_1 and G_2 are given in (14). The resonant angular frequency of G_2 is $\omega_n = 1/\sqrt{L_{d1}C_{d1}}$.

$$\begin{cases} G_1(s) = \dfrac{R_L}{R_L C_{d2} s + 1} \\[2mm] G_2(s) = \dfrac{1}{L_{d1}C_{d1}s^2 + r_e C_{d1} s + 1} \end{cases} \tag{14}$$

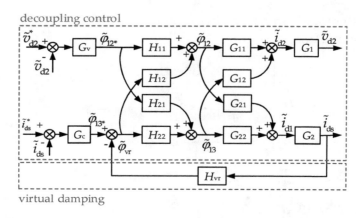

Figure 2. Decoupling control scheme of a three-port converter.

3.2. Virtual Damping Method

For Port 1 in Figure 1, L_{d1} and C_{d1} are utilized to constitute an LC filter to limit the amplitude of the double-switching frequency component of i_{ds} and reduce the negative impact of high-frequency ripple current on the power source (e.g., a fuel cell). However, this might cause performance degradation or an instability issue in current control subsystem due to the high resonant peak and a very weak damping ratio introduced by a pure LC circuit ($r_e = 0$) or with a very small value of r_e shown in (14). Though the resonant phenomenon can be addressed by decreasing the current control bandwidth, the dynamic performance cannot be guaranteed.

The block diagram of the current control subsystem, $G_{co} = G_{22}G_2$ is presented in Figure 3 according to (11). The transfer function, H_{vr} in Figure 2 is a compensation function that is proposed to implement virtual damping in this paper. The expression of H_{vr} is shown in Figure 3. In this figure, r_v is the desired virtual resistor. If the output of H_{vr}, which is $\widetilde{\varphi}_{vr}$, is moved from the point A to the point B, then, H_{vr} in Figure 3 is changed to $r_v / (s/\omega_p + 1)$, and it makes H_{vr} a rational fraction. ω_p is used to attenuate the high frequency noise, and if the value of ω_p is high enough, then, $r_v / (s/\omega_p + 1) \approx r_v$ becomes a resistor connected in series with r_e, and the damping ratio of G_2 becomes $(r_s + r_v)/2 \times \sqrt{C_{d1}/L_{d1}}$ with the introduction of r_v as a virtual resistor. In practical applications, the sampled i_{ds} is passed through H_{vr}, and then added to the output of the current controller to obtain the final phase shift between Port 1 and Port 3, and the value of ω_p can be selected between ω_s and $\omega_s/2$.

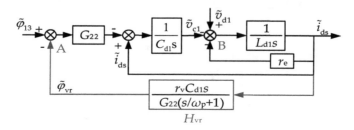

Figure 3. Schematic diagram of virtual damping implementation.

3.3. LADRC for TPC

The small signal model shown in (11) can be transformed into two subsystems to implement the LADRC based control method. The two differential equations for current control subsystem and voltage control subsystem are given by (15) and (16) respectively.

$$\ddot{\widetilde{i}}_{ds} = -\frac{1}{C_{d1}L_{d1}}\widetilde{i}_{ds} - \frac{r_e}{L_{d1}}\dot{\widetilde{i}}_{ds} + \frac{G_{22}}{C_{d1}L_{d1}}\widetilde{\varphi}_{12} + \frac{1}{L_{d1}}\dot{\widetilde{v}}_{d1} + (\frac{G_{21}}{C_{d1}L_{d1}} - b_c)\widetilde{\varphi}_{13} + b_c\widetilde{\varphi}_{13} = f_c + b_c\widetilde{\varphi}_{13} \quad (15)$$

$$\dot{\widetilde{v}}_{d2} = -\frac{1}{R_L C_{d2}}\widetilde{v}_{d2} - \frac{G_{12}}{C_{d2}}\widetilde{\varphi}_{13} + (-\frac{G_{11}}{C_{d2}} - b_v)\widetilde{\varphi}_{12} + b_v\widetilde{\varphi}_{12} = f_v + b_v\widetilde{\varphi}_{12} \quad (16)$$

In (15) and (16), $\tilde{v}_{d2}$ and $\tilde{i}_{ds}$ are taken as the output variables of the two subsystems, $\tilde{\varphi}_{13}$ and $\tilde{\varphi}_{12}$ are the control signals of the current control subsystem and the voltage control subsystem respectively.

$\tilde{\varphi}_{13}$ is considered as an external disturbance of voltage control subsystem, while $\tilde{\varphi}_{12}$ is considered as an external disturbance of the current control subsystem. Furthermore, f_c and f_v are used to represent the generalized disturbances that are associated with both inner and outer variable elements of the two subsystems (e.g., coupling, load, and operating point related parameter changes, etc.). In practical situations, the generalized disturbances, f_c and f_v are usually unknown and cannot be directly measured. Therefore, LESO is adopted to evaluate the generalized disturbances and relevant state variables in the LADRC method.

As for the current control subsystem, $x_c = \begin{bmatrix} \tilde{i}_{ds} & \dot{\tilde{i}}_{ds} & f_c \end{bmatrix}^T$ is selected as the state vector, the augmented state space model is formulated by (17)

$$\begin{cases} \dot{x}_c = A_c x_c + B_c \tilde{\varphi}_{13} + E_c \dot{f}_c \\ \tilde{i}_{ds} = C_c x_c \end{cases} \tag{17}$$

where

$$\begin{cases} A_c = \begin{bmatrix} 0 & 1 & 0 \\ 0 & 0 & 1 \\ 0 & 0 & 0 \end{bmatrix} \\ B_c^T = \begin{bmatrix} 0 & b_c & 0 \end{bmatrix} \\ E_c^T = \begin{bmatrix} 0 & 0 & 1 \end{bmatrix} \\ C_c = \begin{bmatrix} 1 & 0 & 0 \end{bmatrix} \end{cases} \tag{18}$$

The LESO of current control subsystem is constructed by (19).

$$\begin{cases} \dot{z}_c = [A_c - L_c C_c]z_c + \begin{bmatrix} B_c & L_c \end{bmatrix}w_c \\ y_c = z_c \end{cases} \tag{19}$$

where $y_c = z_c = \begin{bmatrix} z_{c1} & z_{c2} & z_{c3} \end{bmatrix}^T$ is the observed vector of x_c. $w_c = \begin{bmatrix} \tilde{\varphi}_{13} & \tilde{i}_{ds} \end{bmatrix}^T$ and L_c given in (20) is the observer gain that can be designed using the pole placement method [17].

$$L_c = \begin{bmatrix} 3\omega_{oc} & 3\omega_{oc}^2 & \omega_{oc}^3 \end{bmatrix}^T \tag{20}$$

where ω_{oc} is the equivalent bandwidth of the observer.

It should be noted that the disturbance caused by the resonance of $L_{d1}C_{d1}$ circuit is included in the generalized disturbance, f_c, therefore, the value of ω_{oc} should be at least twice as large as ω_n, that means $\omega_{oc}/\omega_n \geq 2$ should be satisfied to make the LESO obtain accurate value of f_c, otherwise, the control performance might be significantly degraded.

Similarly, the LESO used for voltage control subsystem is presented in (21).

$$\begin{cases} \dot{z}_v = [A_v - L_v C_v]z_v + \begin{bmatrix} B_v & L_v \end{bmatrix}w_v \\ y_v = z_v \end{cases} \tag{21}$$

In (21), $w_v = \begin{bmatrix} \tilde{\varphi}_{12} & \tilde{v}_{d2} \end{bmatrix}$. $y_v = z_v = \begin{bmatrix} z_{v1} & z_{v2} \end{bmatrix}^T$ is the output vector of (21) that corresponds to $x_v = \begin{bmatrix} \tilde{v}_{d2} & f_v \end{bmatrix}^T$. And the coefficient matrix in (21) are given in (22).

$$\begin{cases} A_v = \begin{bmatrix} 0 & 1 \\ 0 & 0 \end{bmatrix} \\ B_v^T = \begin{bmatrix} b_v & 0 \end{bmatrix} \\ E_v^T = \begin{bmatrix} 0 & 1 \end{bmatrix} \\ C_v = \begin{bmatrix} 1 & 0 \end{bmatrix} \end{cases} \tag{22}$$

The corresponding gain vector of the voltage LESO, L_v is shown in (23).

$$L_v = \begin{bmatrix} 2\omega_{ov} & \omega_{ov}^2 \end{bmatrix}^T \tag{23}$$

Assuming f_c and f_v can be observed accurately ($z_{c3} = f_c$, $z_{v2} = f_v$), and if $\tilde{\varphi}_{13}$ and $\tilde{\varphi}_{12}$ in (15) and (16) can be expressed as (24).

$$\begin{cases} \tilde{\varphi}_{13} = \frac{(u_c - z_{c3})}{b_c} = \frac{(u_c - f_c)}{b_c} \\ \tilde{\varphi}_{12} = \frac{(u_v - z_{v2})}{b_v} = \frac{(u_v - f_v)}{b_v} \end{cases} \tag{24}$$

The current and voltage control subsystems will be simplified to two simple cascaded integrators systems shown in (25).

$$\begin{cases} \ddot{\tilde{i}}_{ds} = u_c \\ \dot{\tilde{v}}_{d2} = u_v \end{cases} \tag{25}$$

The current and voltage control signals, u_c and u_v, for this cascaded integrator system can be proposed as (26).

$$\begin{cases} u_c = k_{pc}(u_{rc} - z_{c1}) - k_{dc}z_{c2} \\ u_v = k_{pv}(u_{rv} - z_{v1}) \end{cases} \tag{26}$$

where k_{pc}, k_{pv}, and k_{dc} are controller parameters, and u_{rc} and u_{rv} are current and voltage reference signals, respectively. In (26), it can be seen that u_c and u_v represent equivalent PD (proportional-derivative) and P (proportional) controllers respectively. The closed-loop transfer functions of the current and voltage control subsystem can be formulated as (27) and (28) which are obtained by substituting the two equations in (26) into the two equations of (25) respectively.

$$G_{cL} = \frac{k_{pc}}{s^2 + k_{dc}s + k_{pc}}, \ (k_{pc} = \omega_c^2, k_{dc} = 2\xi\omega_c) \tag{27}$$

$$G_{vL} = \frac{k_{pv}}{s + k_{pv}}, \ (k_{pv} = \omega_v) \tag{28}$$

In (27) and (28), ω_c and ω_v represent equivalent control bandwidths of the two closed-loop control subsystems with LADRC method, and ξ is the equivalent damping of the current control subsystem, which should be designed to guarantee smooth current change in transient state process (there is no intense oscillations in dynamic process). It can be seen from (27) and (28) that steady state errors are eliminated in the current and voltage closed-loop control systems (when $s = 0$, unity gain is obtained in (27) and (28) respectively) by utilizing (26) as control law. Furthermore, the closed-loop control performances of the two subsystems are completely determined by the designed controller parameters (k_{pc}, k_{pd} and k_{pv}) regardless of the changes of model parameters. This is a prominent characteristic of the LADRC method. (ω_c, ω_{oc}) and (ω_v, ω_{ov}) are the adjustable LADRC parameters in current and

voltage control subsystems, respectively. Since LADRC method is observer based, the bandwidth of the observer should be kept sufficiently larger than the bandwidth of the control system to realize effective compensation. Therefore, the two ratios, $\alpha_c = \omega_{oc}/\omega_c$ and $\alpha_v = \omega_{ov}/\omega_v$ should be larger than two at least in practical applications to get accurate observed values [21], otherwise, the control performance might not be guaranteed.

4. Simulation and Experimental Results

4.1. Simulation Results

In order to verify the theoretical analysis and design results of the proposed method, a simulation model of the isolated TPC is developed by using MATLAB/Simulink, and the main parameters of the simulation model are listed in Table 1.

Table 1. Parameters of The Simulation Model.

Symbol	Name	Value
v_{d1}	DC input voltage	24 V
v_{d2}	Output voltage	36 V
v_{d3}	Battery voltage	24 V
L_{d1}	Input filter inductor	100 μH
r_e	Input filter inductor ESR	0.1 Ω
C_{d1}	Input filter capacitor	1200 μF
C_{d2}	Output filter capacitor	1000 μF
R_L	Load resistor	45 Ω/22 Ω
f_s	Switching frequency	20 kHz

The steady-state operating point A (0.620, 0.379) is selected which corresponds to $R_L = 45$ Ω, $v_{d2} = 36$ V and $i_{ds} = 1.3$ A, and an extra 0.2 Ω virtual resistor is introduced. The open loop transfer functions of the two subsystems are obtained by substituting the parameters listed in Table 1 into G_{co} and G_{vo}, respectively. The controllers G_c and G_v, given by (29), are designed for the decoupled current and voltage subsystems respectively by using the frequency domain design method.

$$\begin{cases} G_c = \frac{500}{1\times10^{-4}s^2+s} \\ G_v = \frac{0.2813s+6.25}{1\times10^{-8}s^3+2\times10^{-4}s^2+s} \end{cases} \tag{29}$$

The Bode plots of the two subsystems with and without correction are shown in Figures 4a and 4b, respectively.

As can be seen in Figure 4a, the crossover frequency of the corrected current control subsystem with $r_v = 0.2$ (the curve B) is about 71 Hz, the phase margin is about 81°, and the gain margin is about 7 dB. The crossover frequency of the corrected voltage control subsystem is about 75 Hz, the phase margin is about 85°, and the gain margin is about 33 dB. As a comparison, if $r_v = 0.2$ is cancelled, the corrected current control subsystem (the curve C) will be unstable, since the resonant peak (the corresponding angular frequency is $\omega_n = 2887$ rad/s) of the curve C will intersect with 0 dB axis under this condition.

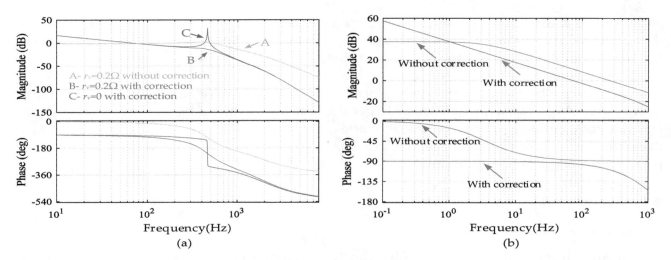

Figure 4. The bode plots of subsystems by using traditional frequency control method: (**a**) i_{ds} subsystem; (**b**) v_{d2} subsystem.

For the control systems with LADRC method, the equivalent control bandwidths of the current and voltage subsystems are the same, $\omega_c = \omega_v = 450$ rad/s (about 72 Hz) which are close to the designed crossover frequencies using the traditional frequency domain method. The observer bandwidths of the current and voltage subsystems are $\omega_{oc} = 4147$ rad/s (about 660 Hz), and $\omega_{ov} = 2000$ rad/s (about 318 Hz) respectively. The simulation results are shown in Figure 5.

Figure 5a shows simulation results of v_{d2} and i_{ds} with traditional frequency control under a step load change condition. As can be seen in Figure 5a, there is an obvious voltage drop (from 36 V to 35.2 V) at 0.5 s when the load is suddenly changed from 29 W to 52 W that resulted in transient fluctuations in i_{ds} (changed from 1.3 A to 1.25 A). When the load is suddenly reduced at 0.7s from 52 W to 29 W, current i_{ds} transiently increases from 1.3 A to 1.37 A, while v_{d2} increases to about 36.8 V, and then decreases gradually to its rated value after 100 ms.

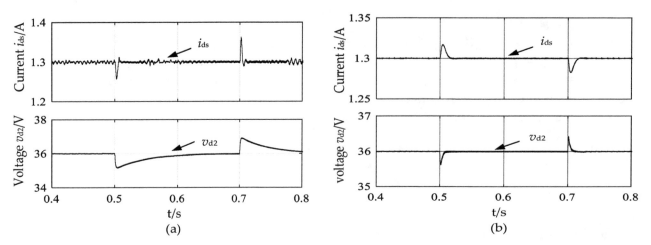

Figure 5. The simulation results of the system: (**a**) with traditional frequency control method; (**b**) with the LADRC method.

The simulation results of the system with the LADRC method and $r_v = 0.2\ \Omega$ are shown in Figure 5b. In this figure, it can be seen that for the same load change, v_{d2} drops from 36 V to 35.6 V at 0.5 s when the load increased, and it increases from 36 V to 36.4 V at 0.7 s when the load decreases. However, current is changed slightly with respect to the load change. For instance, the current drops from 1.3 A to 1.28 A when the load reduced and it increases from 1.3 A to 1.32 A when the load is increased. By comparing the result shown in Figure 5, it can be seen that the amplitude of i_{ds}

fluctuation in the system controlled with LADRC is lower than that of the system controlled with the traditional frequency control method. Also, the transient recovery time of v_{d2} is much shorter than that of the system with the traditional frequency control. These results imply that the information of the load change observed by LESO is effectively utilized in the control system.

4.2. Experimental Validation and Analysis

In order to further verify the effectiveness of the proposed method, an experimental platform was developed as shown in Figure 6. The circuit parameters and load change conditions of the experimental system are similar to the simulation model. The experimental results are shown in Figures 7–11.

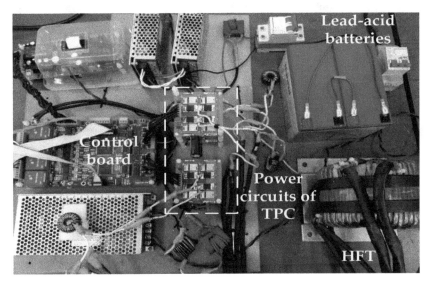

Figure 6. Hardware experiment circuit of the three-port converter.

Figure 7 shows the experimental results, i_{ds}, v_{d2}, Δv_{d2} (fluctuation of v_{d2}) and i_o achieved for the developed converter controlled with the traditional frequency control method. The results of i_{ds} and Δv_{d2} (v_{d2} was controlled to 36 V) for a sudden load increase are shown in Figure 7a. As shown in this figure, there is about 0.24 A drop in i_{ds} and 0.8 V drop in v_{d2} when the load is suddenly changed from 29 W to 52 W. Figure 7b shows the current and voltage changes with respect to sudden load drop where the current and voltage are increased by about 0.24 A and 0.9 V respectively.

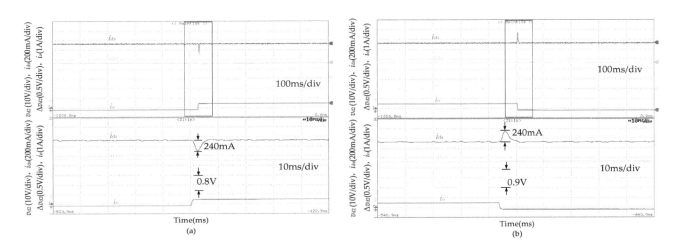

Figure 7. The experiment results of i_{ds}, v_{d2}, Δv_{d2} and i_o with the traditional frequency control method and $r_v = 0.2\ \Omega$: (**a**) sudden load increase; (**b**) sudden load decrease.

The effect of the proposed virtual resistor method on i_{ds} control is conducted by removing and re-adding the virtual resistor with the same current controller used in Figure 7. The experimental results are shown in Figure 8a, it can be seen that there are serious oscillations in i_{ds} (the current control subsystem is unstable in this case as indicated by the curve C in Figure 4 and voltage ripples (Δv_{d2}) of v_{d2} are also increased with the same oscillation frequency of i_{ds}. While the virtual resistor scheme is re-performed, the oscillation of i_{ds} can be suppressed soon, and the amplitude of voltage ripple in v_{d2} becomes lower.

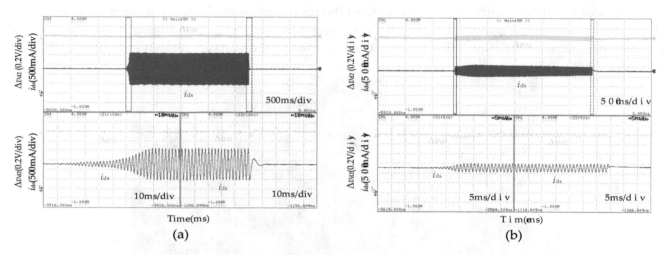

Figure 8. The experimental results with and without $r_v = 0.2\ \Omega$ virtual resistor ($\omega_n = 2887$ rad/s): (**a**) traditional frequency control; (**b**) LADRC.

The experimental results of i_{ds}, v_{d2}, Δv_{d2} and i_o with the LADRC method and $r_v = 0.2\ \Omega$ are shown in Figure 9. As shown in Figure 9a, when the same load increment (23 W) is experienced small changes are observed in i_{ds} (changes from 1.3 A to 1.35 A) and v_{d2} (changes from 36 V to 35.6 V) because of the coupling between current and voltage subsystems. As shown in Figure 9b, for a sudden load reduction, i_{ds} is changed from 1.3 A to 1.25 A, and v_{d2} is increased for about 0.5 V. Also, as illustrated in Figure 9, the fluctuations of i_{ds} and v_{d2} in the transient process with the LADRC method are lower than those with the traditional frequency control method as shown in Figure 7, and the voltage transient recovery time with the LADRC method is much shorter than that with the traditional frequency control. These results indicate that the control system with the LADRC has better decoupling performance and adaptability to the operating point changes compared to the control system with the traditional frequency control.

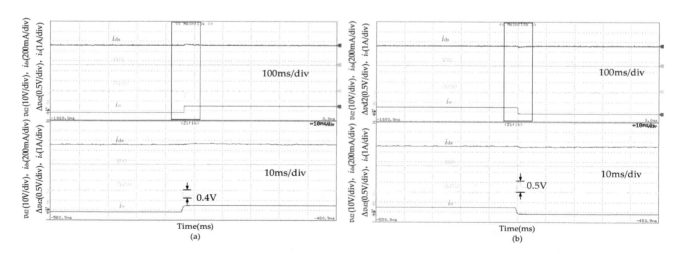

Figure 9. The experimental results of i_{ds}, v_{d2}, Δv_{d2} and i_o with LADRC method, $\omega_n = 2887$ rad/s and $r_v = 0.2\ \Omega$: (**a**) sudden load increase; (**b**) sudden load decrease.

The experimental result with and without 0.2 Ω virtual resistor with LADRC method and ω_n = 2887 rad/s is presented in Figure 8b. As shown in this figure, there are obvious current oscillations in i_{ds} when r_v = 0.2 Ω is removed, since the observer bandwidth (ω_{oc} = 4147 rad/s) is not sufficiently higher than the resonant angular frequency of $L_{d1}C_{d1}$ circuit ($\omega_{oc}/\omega_n \approx 1.43 < 2$). And it is similar to the case shown in Figure 8a, the current oscillations can be attenuated effectively if the virtual resistor method is reused.

For comparison study, the resonant angular frequency of $G_2(s)$ in (14) is reduced to ω_n = 1521 rad/s (by setting L_{d1} = 160 μH and C_{d1} = 2700 μF), then there is no oscillations in i_{ds} as shown in Figure 10, that means i_{ds} can be controlled well since the ratio of ω_{oc}/ω_n is about 2.73 which is larger than two in this case, it means that the impact caused by the resonance of $L_{d1}C_{d1}$ circuit can be much accurately observed by the LESO.

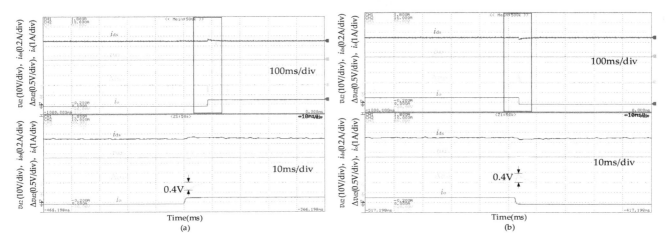

Figure 10. The experimental results of i_{ds}, v_{d2}, Δv_{d2} and i_o with LADRC method, ω_n = 1521 rad/s and r_v = 0 Ω: (**a**) sudden load increase; (**b**) sudden load decrease.

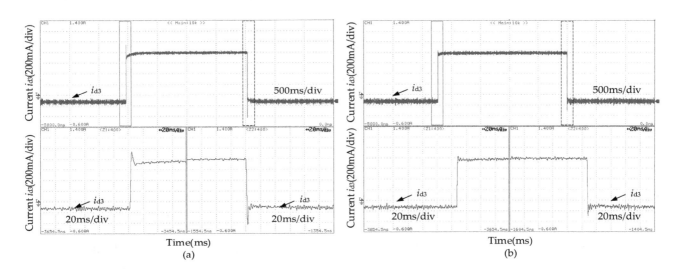

Figure 11. The experimental results of battery current i_{d3}: (**a**) traditional frequency control; (**b**) LADRC.

The battery current i_{d3} for the traditional frequency control and LADRC are shown in Figures 11a and 11b, respectively. In Figure 11, it can be seen that not only the overshoot of i_{d3} with the LADRC is relatively lower, but also the transient recovery time of i_{d3} is shorter than that obtained using the traditional frequency control. This indicates the LADRC method can provide better dynamic balance in the power delivery process. The results shown in Figure 11 have some internal relations with the experimental results presented in Figures 7 and 9. For instance, compared to Figure 11b with LADRC method applied, the battery overshoot current, i_{d3} shown in Figure 11a is larger when traditional frequency control method is applied. Larger overshoot current causes significant drop in i_{ds} shown in Figure 7a, whereas smaller overshoot of i_{d3} shown in Figure 11b with LADRC resulted in much smaller overshoot in i_{ds} shown in Figure 9a.

5. Conclusions

The model parameter variation caused by the operating point changes and couplings between different ports during power delivery in an isolated three-port converter has a negative impact on the converter control performance. In this study, the LADRC method is employed to control the three-port converter. In the LADRC method, the possible model parameter uncertainties, load changes and the negative impact of LC circuit resonance are all expressed as a generalized disturbance that is considered as a state variable and observed by the LESO which is utilized to synthesize the control signal. In this method, the bandwidth of the LESO is sufficiently higher than the equivalent control bandwidth and the resonant frequency of LC circuit that guarantee the system dynamics and generalized disturbance can be accurately observed. Therefore, the desired closed-loop control performance that is independent of parameters changes and external disturbances can be obtained. And a virtual resistor based method is proposed to increase damping ratio of the current control subsystem of TPC which is beneficial to further improve current control performance using LADRC method. The simulation and experimental results revealed that the proposed method is robust against model parameter changes and external disturbances. Therefore, the dynamic control performance of the three-port converter can be improved significantly.

Author Contributions: Conceptualization, J.Y., M.L. and H.C.; Validation, M.L., J.Y. and H.C.; Writing-Original Draft Preparation, J.Y. and H.C.; Writing-Review & Editing, N.G., M.V. and M.L.

References

1. Tao, H.; Kotsopoulos, A.; Duarte, J.L.; Hendrix, M.A.M. Design of a soft-switched three-port converter with DSP control for power flow management in hybrid fuel cell systems. In Proceedings of the 2005 European Conference on Power Electronics and Applications, Dresden, Germany, 11–14 September 2005.
2. Tao, H.; Duarte, J.L.; Hendrix, M.A.M. A Distributed Fuel Cell Based Generation and Compensation System to Improve Power Quality. In Proceedings of the 2006 CES/IEEE 5th International Power Electronics and Motion Control Conference, Shanghai, China, 14–16 August 2006.
3. Ling, Z.; Wang, H.; Yan, K.; Sun, Z. A new three-port bidirectional DC/DC converter for hybrid energy storage. In Proceedings of the 2015 IEEE 2nd International Future Energy Electronics Conference (IFEEC), Taipei, Taiwan, 1–4 November 2015.
4. Tao, H.; Duarte, J.L.; Hendrix, M.A.M. Line-Interactive UPS Using a Fuel Cell as the Primary Source. *IEEE Trans. Ind. Electron.* **2008**, *55*, 3012–3021. [CrossRef]
5. Wang, L.; Wang, Z.; Li, H. Optimized energy storage system design for a fuel cell vehicle using a novel phase shift and duty cycle control. In Proceedings of the 2009 IEEE Energy Conversion Congress and Exposition, San Jose, CA, USA, 20–24 September 2009.
6. Kim, S.Y.; Song, H.S.; Nam, K. Idling Port Isolation Control of Three-Port Bidirectional Converter for EVs. *IEEE Trans. Power Electron.* **2012**, *27*, 2495–2506. [CrossRef]
7. Jiang, Y.; Liu, F.; Ruan, X.; Wang, L. Optimal idling control strategy for three-port full-bridge converter. In Proceedings of the 2014 International Power Electronics Conference (IPEC-Hiroshima 2014—ECCE ASIA), Hiroshima, Japan, 18–21 May 2014.
8. Ajami, A.; Asadi Shayan, P. Soft switching method for multiport DC/DC converters applicable in grid connected clean energy sources. *IET Power Electron.* **2015**, *8*, 1246–1254. [CrossRef]
9. Tao, H.; Duarte, J.L.; Hendrix, M.A.M. Three-Port Triple-Half-Bridge Bidirectional Converter with Zero-Voltage Switching. *IEEE Trans. Power Electron.* **2008**, *23*, 782–792. [CrossRef]
10. Tao, H.; Kotsopoulos, A.; Duarte, J.L.; Hendrix, M.A.M. A Soft-Switched Three-Port Bidirectional Converter for Fuel Cell and Supercapacitor Applications. In Proceedings of the 2005 IEEE 36th Power Electronics Specialists Conference, Recife, Brazil, 12 June 2005.

11. Zhao, C.; Kolar, J.W. A novel three-phase three-port UPS employing a single high-frequency isolation transformer. In Proceedings of the 2004 IEEE 35th Annual Power Electronics Specialists Conference, Aachen, Germany, 20–25 June 2004.

12. Zhao, C.; Round, S.D.; Kolar, J.W. An Isolated Three-Port Bidirectional DC-DC Converter with Decoupled Power Flow Management. *IEEE Trans. Power Electron.* **2008**, *23*, 2443–2453. [CrossRef]

13. Tao, H.; Kotsopoulos, A.; Duarte, J.L.; Hendrix, M.A.M. Transformer-Coupled Multiport ZVS Bidirectional DC–DC Converter with Wide Input Range. *IEEE Trans. Power Electron.* **2008**, *23*, 771–781. [CrossRef]

14. Wang, L.; Wang, Z.; Li, H. Asymmetrical Duty Cycle Control and Decoupled Power Flow Design of a Three-port Bidirectional DC-DC Converter for Fuel Cell Vehicle Application. *IEEE Trans. Power Electron.* **2012**, *27*, 891–904. [CrossRef]

15. Falcones, S.; Ayyanar, R. LQR control of a quad-active-bridge converter for renewable integration. In Proceedings of the 2016 IEEE Ecuador Technical Chapters Meeting (ETCM), Guayaquil, Ecuador, 12–14 October 2016.

16. Gao, Z. Active disturbance rejection control: From an enduring idea to an emerging technology. In Proceedings of the 2015 10th International Workshop on Robot Motion and Control (RoMoCo), Poznan, Poland, 24 August 2015.

17. Gao, Z. Scaling and bandwidth-parameterization based controller tuning. In Proceedings of the 2003 American Control Conference, Denver, CO, USA, 4–6 June 2003.

18. Wang, W.; Gao, Z. A comparison study of advanced state observer design techniques. In Proceedings of the 2003 American Control Conference, Denver, CO, USA, 4–6 June 2003.

19. Miklosovic, R.; Radke, A.; Gao, Z. Discrete implementation and generalization of the extended state observer. In Proceedings of the 2006 American Control Conference, Minneapolis, MN, USA, 14–16 June 2006.

20. Sun, B.; Gao, Z. A DSP-based active disturbance rejection control design for a 1-kW H-bridge DC-DC power converter. *IEEE Trans. Ind. Electron.* **2005**, *52*, 1271–1277. [CrossRef]

21. Zhu, B. *Introduction of Active Disturbance Rejection Control*, 1st ed.; Beihang University Press: Beijing, China, 2017; pp. 42–45.

6

Analysis, Modeling and Control of Half-Bridge Current-Source Converter for Energy Management of Supercapacitor Modules in Traction Applications

Jorge Garcia [1,*], Pablo Garcia [1], Fabio Giulii Capponi [2] and Giulio De Donato [2]

[1]   LEMUR Group, Department of Electrical Engineering, University of Oviedo, 33204 Gijon, Spain; garciafpablo@uniovi.es

[2]   Department of Astronautical, Electrical and Energy Engineering, University of Roma "La Sapienza", 00184 Roma, Italy; fabio.giuliicapponi@uniroma1.it (F.G.C.); giulio.dedonato@uniroma1.it (G.D.D.)

*   Correspondence: garciajorge@uniovi.es

Abstract: In this work, an in-depth investigation was performed on the properties of the half-bridge current-source (HBCS) bidirectional direct current (DC)-to-DC converter, used to interface two DC-link voltage sources with a high-voltage-rating mismatch. The intended implementation is particularly suitable for the interfacing of a supercapacitor (SC) module and a battery stack in a hybrid storage system (HSS) for automotive applications. It is demonstrated that the use of a synchronous rectification (SR) modulation scheme benefits both the power-stage performance (in terms of efficiency and reliability) and the control-stage performance (in terms of simplicity and versatility). Furthermore, an average model of the converter, valid for every operating condition, is derived and utilized as a tool for the design of the control system. This model includes the effects of parasitic elements (mainly the leakage inductance of the transformer) and of the converter snubbers. A 3 kW prototype of the converter was used for experimental validation of the converter modeling, design, and performance. Finally, a discussion on the control strategy of the converter operation is included.

Keywords: hybrid storage systems; power electronic converters; half-bridge current-source converters; supercapacitors

---

1. Introduction

Among electrical energy storage passive devices, Supercapacitors (SC) outstand as one of the preferred solutions when very high power densities and long life cycles are required. These features also are used in Hybrid Storage Systems (HSS), where SC can be combined with other distinct storage devices, such as electrochemical batteries or fuel cells [1–19]. In these hybrid systems, SC usually operate as power sources, as they provide the required peak power commanded by the load. In turn, the combined device (e.g., the electrochemical battery) work as an energy source, providing the long-term energy required for the given operating conditions, such as islanding operation or back-up energy support after a fault ride-through sequence, etc.

The integration of SC into HSS has been covered extensively in the current state-of-the-art, by analyzing the modeling and operation of power electronic topologies, by proposing control algorithms and methods, and by investigating hierarchical energy management systems [20–27]. Although being the simplest and most inexpensive scheme, the direct connection between the storage systems prevents to fully exploit the high charging and discharging instant power ratings of SCs. This scheme also prevents a tight control on the independent power flows shared among the energy storage devices [28,29]. Therefore, practically every implemented solution found in the technical literature includes a bidirectional direct current (DC)-to-DC converter. These solutions can be categorized into three different classes: a series connection of the storage devices [6,7], a cascaded connection of the storage subsystems [8,9], and finally a parallel connection of the storage elements [10–12]. From a detailed analysis of the technical literature, the parallel configuration, depicted in Figure 1, outstands as the most practical solution that allows a full control of the power flow sharing scheme required for the application, as well as a complete SC voltage-range utilization, despite that in this case the bidirectional converter needs to be rated for the peak value of the required power [13,14].

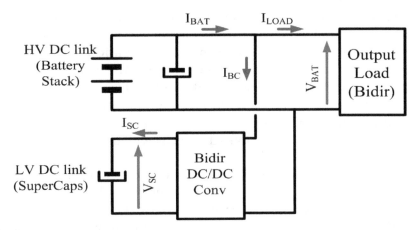

Figure 1. Parallel configuration for interfacing supercapacitors (SCs) with primary energy source (battery stack).

The bidirectional boost converter is formed by the typical connection present in a single phase leg of an inverter. It has two transistors, a series inductor connected between the mid-point of the transistors and the output, plus two capacitors intended to filter the input and the output of the converter. This scheme is generally adopted as the simplest and best-known solution. With this configuration, the peak power ratings required by most applications can be ensured by an assembly of few SC connected in series. This structure is also beneficial in terms of balancing the voltage at each SC module in the assembly. As a consequence, the SC side voltage ratings of the converter are usually much smaller than the voltage levels found at the DC-link. In addition, upon transient variations, the SC side voltage must allow variations from 80 to 20% of the SC ratings, in order to use efficiently the stored energy at the device. On the other side, the voltage at the DC-link may remain almost constant, as to optimize the full system design and operation. Hence, the range of the voltage gain between the input and the output of the bidirectional converter can typically surpass ratios of 1 to 10 and above, depending on the application.

Therefore, the boost converter is no longer a suitable solution, and so alternative topologies that allow for a high voltage gain need to be analyzed. Among these options, the most widely used include the use of a High-Frequency (HF) transformer [30–38]. Studies on the simplest isolated configurations, such as the bidirectional versions of the Flyback converter, the Forward Current-Source converter, the Push-Pull Current-Source converter, and the Half-Bridge Current-Source (HBCS) converter have been carried out [33–35]. These comparative analyses were carried out on the basis of the number of components, the switch stresses at the converters, and other figures of merit such as the utilization factor of the magnetic elements. This comparison shows how the bidirectional HBCS DC-to-DC converter is most suitable in low to medium power applications, for structures that directly interface the storage devices. Thus, the HBCS topology is of interest in HSS with storage devices arranged as shown in Figure 1 with the battery directly forming the DC bus at the converter. This is particularly interesting in traction applications, for which the weight and size of the full system are major concerns, and variable DC-link systems are considered as a suitable option [28,29].

When a given power profile is demanded by the load, the SC converter must react very quickly, in order to prevent the battery from providing the transient power. Given the current-source behavior of the low-voltage side of the HBCS converter, and provided that a tight, fast current control is implemented by generating the adequate SC current reference, the power delivered by this converter can successfully protect the battery from delivering fast, high power peaks, thus enhancing the performance and increasing the reliability of the full system. Other works have investigated the potential use of power converters for interfacing both the battery and the SC module, leading to more complex topologies and control methods [20,24,27,39].

The HBCS topology was introduced in an application to HSS in [11], but the power ratings of the laboratory demonstrator was only 100 W. Later, a 3 kW setup was reported in [35], being able to operate in open-loop, steady-state, both charging and discharging operation modes. A measured efficiency beyond 90% was then reported for the full range of operation. The authors in [35] analyzed the static gain of the HBCS converter, using conventioal switching patterns. Furthermore, they considered neither the modeling nor the control stage for closed-loop operation of the converter. On the other hand, the authors of [40,41] showed preliminary research conducted to unify the transfer function of the converter acting in a bidirectional operation scheme, as well as the dynamic behavior.

The present work investigates more in depth the performance of the HBCS converter in vehicle applications. Firstly, a unified control strategy of the converter, on the basis of a synchronous rectification switching pattern, is provided. By using this pattern, the converter can be controlled with a simple, unique control law valid for every operating condition. This allows the implementation of a full bidirectional SC current control, thus enabling a fast, tight power flow control in the storage system. Once this switching pattern is established and validated, the main contribution of this work is the extension of previously existing circuital models of the converter by including non-ideal effects of the parasitic elements in the transformer of the converter, as well as the dissipative effect of the losses in the system. The performance of the proposed model was validated through simulations and experiments carried out on a 3 kW laboratory prototype. The proposed model was also compared against the simplified ideal model. Finally, this work discusses the control strategy that can be implemented in the converter. A basic HBCS current loop was analyzed, designed, and implemented, with the aim to demonstrate its feasibility; finally, a discussion on the possibilities of the high-level control strategy is also included at the end of the paper.

This paper is organized as follows. Section 2 presents the conventional modulation strategy implemented in the state-of-the-art for the HBCS converter. After that, Section 3 evaluates the Synchronous Rectification (SR) scheme in the converter operation, considering both the charging and discharging modes of operation. It also demonstrates experimentally the benefits of SR operation for this topology. Section 4 further discusses the implications of SR in terms of the simplification of the control of the converter; most importantly, this section provides an averaged model of the HBCS converter, suitable to aid in the design of the control stage. Also, this model is validated through simulations. Subsequently, in Section 5 the current control algorithm is discussed and tested through simulations. This section also deals with the overall control scheme of the system. Section 6 validates the full model and the control scheme outlined in the previous sections by showing experimental results obtained with a laboratory prototype of a HBCS converter. Finally, Section 7 discusses the conclusions of the research and proposes some future developments.

2. Special Characteristics of the HBCS Converter

Figure 2 shows the HBCS converter. Because this converter is a bidirectional topology, no predefined input or output are established. For the coming discussion, the SC side is also referred to as the Low-Voltage (LV) side, signifying that this side generally has the lower of the rated voltages involved in the analysis. On the other hand, the alternate energy source (i.e., battery) and the DC-link are referred to as the High-Voltage (HV) side.

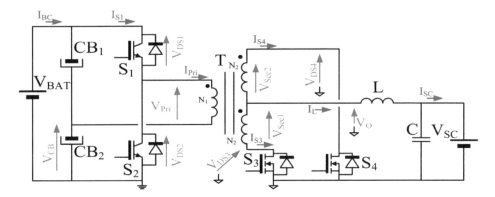

Figure 2. Layout of Half-Bridge Current-Source (HBCS) bidirectional converter. The references for the current and voltage considered for the analysis are shown.

The modulation scheme for the HBCS converter, as presented in [35], is obtained by shifting the command pulse waveforms of each switch in the primary- or secondary-side by 180°. Initially, the control parameters implemented at both charging and discharging operation modes were different. In charging mode, where the energy flows to the SC from the DC link side, switches S_3 and S_4 remain always turned off. This forces the current to flow only through the diodes of S_3 and S_4. Figure 3 shows the main current and voltage instant waveforms of the converter for charging (a) and discharging (b) modes. The switching pattern is defined by establishing the gate-to-emitter voltages of switches S_1 to S_4, denoted as V_{GE1}–V_{GE4}, respectively. The rest of the waveforms are consistent with the references in Figure 2. The modulation of switches S_1 and S_2 (Figure 3a) is carried out by shifting by 180° the control pulses, considering a given duty ratio, D:

$$D = \frac{T_{S1\ ON}}{T_S},\tag{1}$$

where $T_{S1\ ON}$ is the interval when switch S_1 remains turned on and T_S is the switching period. In order to avoid a simultaneous conduction of switches S_1 and S_2 at the battery side, and given that the switching pattern requires a fixed 180° phase shift between the firing signals of V_{GE1} and V_{GE2}, D is restricted to $0 < D < 0.5$. The voltage gain between the LV and HV sides, G_{CHRGE}, is

$$G_{CHRGE} = \frac{V_{SC}}{V_{BAT}} = D \cdot \frac{N_2}{N_1},\tag{2}$$

where N_1 and N_2 are the primary- and secondary-side number of turns of the HF transformer, respectively; V_{SC} is the voltage across the SC; and V_{BAT} is the DC-link voltage, that is, the battery voltage according to its state of charge. The magnetizing inductance of the transformer is neglected, because it is assumed that it is very high.

Figure 4 shows the current paths for the different switching modes of the converter, during the SC charging operation scheme. When switch S_1 is turned on and S_2 is turned off (Figure 4a), a positive current I_P flows flows through the primary side, considering the references in Figure 2. Current flows through the portion of the secondary winding connected to switch S_3, the output filter, and the SC; thus, the anti-parallel diode of switch S_3 carries the secondary-side current. Figure 4b, in turn, shows the switching mode when both transistors S_1 and S_2 are turned off. The inductor current I_L splits equally between the two parts of the secondary winding, flowing through the anti-parallel diodes of S_3 and S_4. Finally, Figure 4c depicts the current paths when S_2 is turned on and S_1 is turned off. This mode is similar to that shown in Figure 4a but considers the 180° phase shift between S_1 and S_2.

During the discharging mode, the energy flows back to the DC-link from the the SC. In this case, the switches under control are S_3 and S_4, while S_1 and S_2 are kept continuously turned off. Again, the modulation of S_3 and S_4 is implemented through a 180° shifting scheme, while the turn-on intervals are kept equal. Figure 3b shows the key waveforms obtained for this operation mode. The independent parameter for this mode, D', is defined as:

$$D' = \frac{T_{S3\ ON}}{T_S},$$

(3)

where $T_{S3\ ON}$ is the interval when S_3 is turned on. The constraint for this parameter is now $0.5 < D' < 1$, thus forcing an overlap in the conduction of S_3 and S_4. This overlapping is required because of the current-source behavior of the LV side of the converter. For this topology, unless at least one switch in the pair S_3–S_4 is turned on, an abrupt interruption of the current flowing through the inductor would lead to a dangerous overvoltage across the magnetic element, that might damage the converter. For the discharging operation mode, the voltage gain can be defined as:

$$G_{DISCHRGE} = \frac{V_{BAT}}{V_{SC}} = \frac{1}{1 - D'}\frac{N_1}{N_2}.$$

(4)

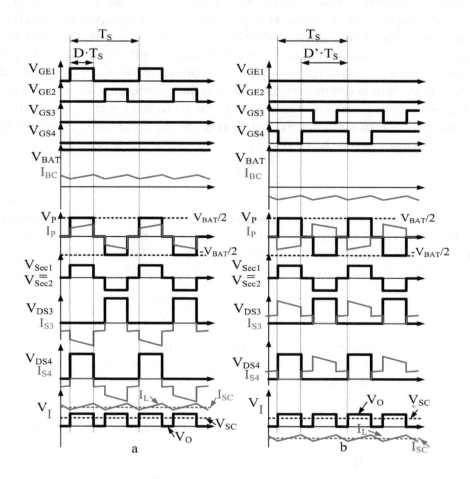

Figure 3. Main voltage (black) and current (red) waveforms of the half-bridge current-source (HBSC) for charging (**a**) and discharging (**b**) operation modes.

The switching modes for discharging the SC are represented in Figure 5. The current path when S_4 is turned off and S_3 is turned on is shown in Figure 5a. On the other hand, Figure 5b shows the current flowing through the switches when both S_3 and S_4 are turned on. Finally, it can be seen in Figure 5c how the current flows when switch S_4 is turned on and S_3 remains off. A close look at Figure 5a–c shows that the current paths are the same as those shown in Figure 4a–c, respectively, but with the directions of the currents reversed.

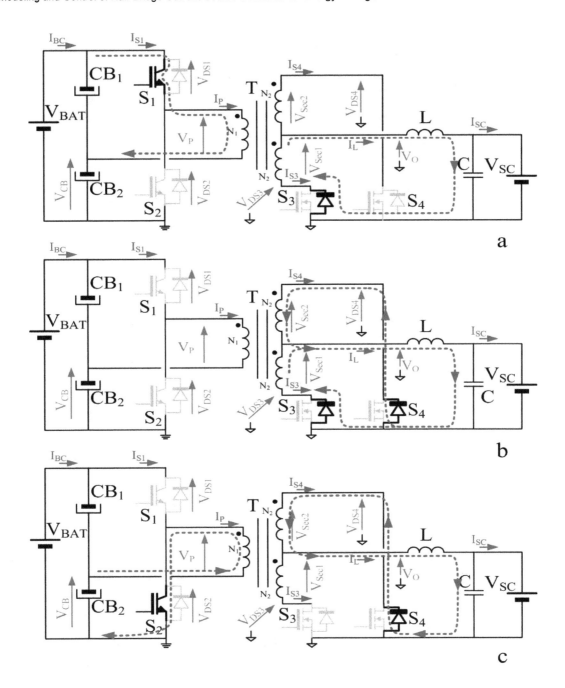

Figure 4. Switching modes of the half-bridge current-source (HBCS) converter operating during supercapacitor (SC) charging operation mode. (**a**) Switch S_1 on and switch S_2 off. (**b**) Both switches S_1 and S_2 off. (**c**) Switch S_1 off and switch S_2 on.

By using the control laws stated in Equations (2) and (4), both the charging and discharging control strategies can be easily designed. However, this modulation pattern requires the calculation of two independent control parameters, defined as the duty cycles of S_1 and S_3, depending on the operation mode under consideration, given by charging and discharging modes, respectively. This approach, though, yields to some issues at the boundary between these operation modes, given that the switches that must be controlled change abruptly. Therefore, if both control schemes are implemented independently, a smooth transition between modes of operation is prevented. However, this issue is easily avoided provided that SR is implemented. This SR pattern is introduced in the next section for the HBCS converter. The benefits for the control of the converter are discussed in the forthcoming sections.

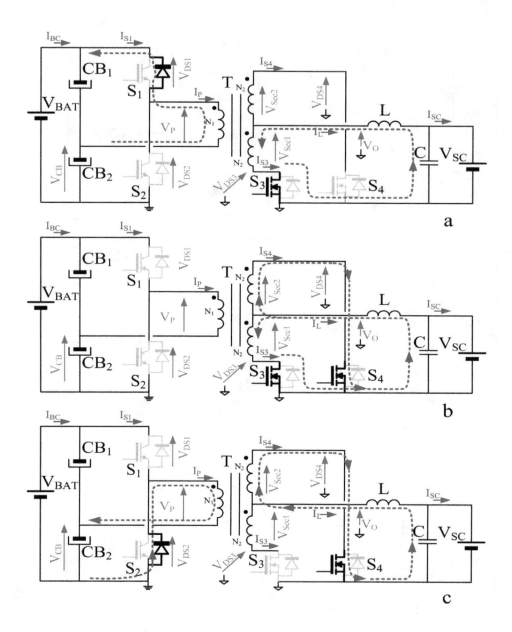

Figure 5. Switching modes of the half-bridge current-source (HBCS) converter operating during supercapacitor (SC) discharging operation mode. (**a**) Switch S_3 on and switch S_4 off. (**b**) Both switches S_3 and S_4 turned on. (**c**) Switch S_3 off and switch S_4 on.

3. Synchronous Rectification in the HBCS Converter

SR is a common strategy that establishes a given switching pattern for the switches in power electronic converters. Upon several conditions, it can decrease conduction losses, thus boosting the efficiency of the converter. This technique has already been described in the technical literature related with hybrid systems applications [36–38]. This technique takes advantage of the low on-state voltage of a MOSFET transistor, when compared to the forward voltage of its body diode. Given the equivalent series circuit of a diode, formed by a threshold voltage, V_{TH}, and a dynamic resistor, R_{DYN}, then the conduction losses, for the diode turned on, are given by:

$$P_{C-DIODE} \propto V_{TH} \cdot I_D + R_{DYN} \cdot I_D^2. \tag{5}$$

Moreover, the conduction losses in a MOSFET can be expressed as:

$$P_{C-MOSFET} \propto R_{DS\,ON} \cdot I_D^2, \tag{6}$$

where $R_{DS\,ON}$ is the drain-to-source on-resistance of the channel. As a result of an analysis of the equivalent circuit of the switch, the current tends to flow through the path with a lower voltage drop. In the HBCS converter depicted at Figure 2, S_3 and S_4 have already been implemented as MOSFET switches, with inherent anti-parallel body diodes. As a consequence, SR can be directly applied.

In order to validate the performance of the SR switching pattern, a reference design is considered for a 3 kW HBCS converter. The main characteristics of the setup are detailed in Table 1.

Table 1. Characteristics of half-bridge current-source (HBCS) design.

Parameter	Max.	Min.
SC voltage	45 V	25 V
HV DC-link voltage	350 V	
Switching frequency	20 kHz	
SC discharge-mode duty ratio	0.75	0.55
SC charge-mode duty ratio	0.45	0.25
Output power (SC disch. mode)	3 kW	1.6 kW
Output power (SC chrg mode)	3 kW	1.6 kW
SC current	65 A	−65 A
Transformer turns ratio	3.5:1:1	

Figure 6 shows a graphical representation of the expression of the drain-to-source voltage, V_{DS}, as well as the power dissipated through conduction losses, P_{COND}, both as a function of the current flowing through the drain of the transistor, I_D. These power and voltage values are given for both the transistor and diode elementary devices. For this study, the actual values given in the datasheets of the real components selected, in this case the $SKM121AR$ power MOSFET modules from SEMIKRON, have been used. The dotted black line shows the voltage at the MOSFET as a function of the current I_D. This line shows the typical load line of a resistive component. As a consequence, the power dissipated as a function of the I_D in the MOSFET is given by a quadratic relationship expressed in Equation (6), drawn as a filled black line in Figure 6. The voltage drop across the diode, in turn, is shown as an horizontal dotted gray line. The power dissipated in the diode, as a function of the current I_D, therefore follows a linear behavior, represented by a filled gray line. As can be seen, both power characteristics, those of the MOSFET and the diode, have an intersection point labeled as I_{LIM}. For I_D current values below I_{LIM}, the MOSFET losses are smaller than the diode losses, and therefore the synchronous rectification implies higher efficiencies compared to the standard switching pattern. For the real values of the devices under consideration, I_{LIM} is equal to

$$I_{LIM} = 70\ \text{A}. \tag{7}$$

From Table 1, the maximum value of the SC current, which is also the maximum currents allowed to flow through S_3 and S_4, has a value of 65 A. Therefore, and considering Figure 6, the conduction losses at the secondary side of the converter operating in charging mode will be decreased, provided that SR is implemented as the switching pattern. As a consequence, the whole efficiency of the converter will increase. It must be notice that in this case the switches at the DC link side have been implemented by means of IGBT switches. Therefore, and given that for this switch technology the current flowing from the emitter to the collector can only flow through the anti-parallel diode (and not through the transistor itself, as in the previous case), the use of a SR scheme during the SC discharge mode will not mean a decrease in the converter losses. In any case, a critical point for this analysis is to state that SR can still be used in this mode of operation without affecting the performance of the converter. This issue will be discussed in Section 4.

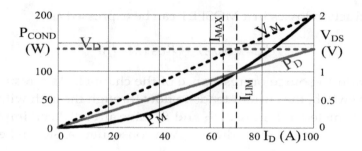

Figure 6. Drain-to-source voltage (V_{DS}, dashed lines) and conduction losses (P_{COND}, filled lines) for both diode conduction (gray) and MOSFET conduction (black), for the half-bridge current-source (HBCS) prototype.

A first set of experiments was carried to validate the feasibility of the SR switching pattern implementation in the HBCS converter. For this validation, and in order to register the steady-state operation of the converter, the setup was configured to supply a resistive load from a voltage source. The values of measured efficiencies in charging operation mode, for the standard and the SR schemes, are given in Figure 7. Three different resistive loads (1.0, 0.67 and 0.5 Ω, respectively), were used in the measurements. The voltage at the power source was kept constant at 300 V_{DC}. Due some practical constraints at the experimental setup, the voltage levels at the LV side were limited to three values, $V_{SC} = 25$ V, $V_{SC} = 30$ V, and $V_{SC} = 34$ V. From the plots at Figure 7, it can be deduced an increase in the converter efficiency in all cases.

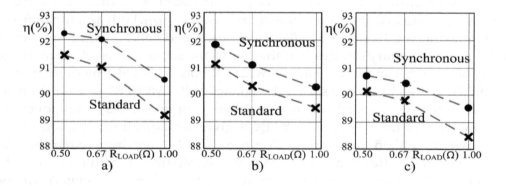

Figure 7. Graphical relationship of experimental efficiency measurements between standard and synchronous rectification switching schemes, as a function of load resistance. (a) $V_{SC} = 25$ V. (b) $V_{SC} = 30$ V. (c) $V_{SC} = 34$ V.

4. Full Model of the Converter under Synchronous Rectification

As a result of the use of SR, the HV and LV side switches, S_1–S_2 and S_3–S_4, present complementary control signals. In particular, the gating signals of S_3 and S_4, are the complementary signals of S_1 and S_2, respectively, as it was sketched in the waveforms shown in Figure 3a,b. This condition defines an obvious relationship upon SR scheme between the values of the parameter D of switches S_1–S_2 and D' of switches S_3–S_4. This can be expressed as

$$D = 1 - D' \quad . \tag{8}$$

As a result, a single, independent control parameter can be established for every operation mode in the converter, given that the remaining parameters are calculated automatically. This parameter is the duty ratio of switch S_1, noted by D. Equation (8), together with Equations (2) and (4), determines a unique equation to define the voltage gain of the topology, G_{HBCS}. This equation is, as mentioned, true for every mode of operation and charging condition:

$$G_{HBCS} = \frac{V_{SC}}{V_{BAT}} = D \cdot \frac{N_2}{N_1}. \tag{9}$$

Thus, with SR, the control law turns out to be unique, regardless of the mode of operation.

Once the switching strategy is defined, an accurate model of the converter needs to be established in order to properly design the control system. The simplest averaged large-signal model for the converter is shown in Figure 8. This basic model has already been explored in the definition and implementation of a control strategy for the topology [40]. The basic model has been implemented considering ideal switching and reactive devices. For the average circuit, the transistors and the HF transformer have been modeled by the two dependent sources within the dashed frames at Figure 8. The output value of these sources depend on the duty ratio D and on the turns ratio of the transformer, N_2/N_1. Downstream from these, the second order L–C filter is present.

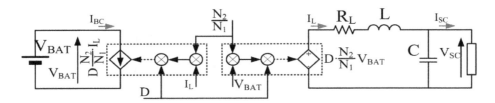

Figure 8. Large-signal averaged model based on ideal components.

In order to check this model, the performance of the circuit depicted in Figure 8 was compared with a full switching model of the HBCS converter, implemented in PSIM. In these simulations, a resistive load replaced the SCs, with the aim of having initial steady-state operation conditions. The dynamic behavior is obtained by imposing a small step in the duty ratio of the converter after reaching the steady state. In the tests shown in Figure 9, the step goes from 0.34 to 0.36. This picture shows the simulated waveforms of the output voltage (V_{SC}), input (I_{BC}) and output (I_{SC}) currents, and the inductor current (I_L), compared with their corresponding averaged values obtained from the basic model. The performance of the basic model differs significantly from the switching converter behavior. Even thought the natural frequency is mainly the same, there are two effects of mismatch in the behavior of the model. Firstly, the switching system is more damped than the dynamic response given by the model. Moreover, the equilibrium values after the transient are smaller in the switching circuit.

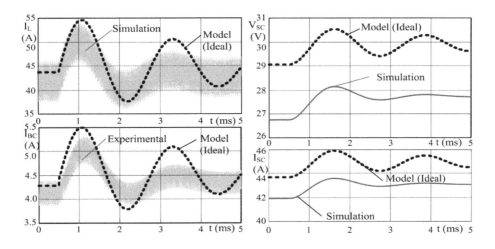

Figure 9. Simulation waveforms (gray) and averaged value of model based on ideal components (black), for filter current (I_L), load output current (I_{SC}), load voltage (V_{SC}), and input current (I_{BC}).

In order to solve this disagreement between the model and the switching circuit, an improved model has been developed. This model firstly accounts for the effects of the parasitic resistors of the transformer, the R_{DS} of the transistors at the HV side, and the snubbers. These effects are jointly modeled by a series resistance, R_{loss}, which adds a dissipative term in the model that contributes to increase the damping factor. However, the most significant effect is due to the leakage inductance of the transformer, depicted in Figure 10. Some waveforms are depicted in Figure 11 to describe the contribution of this leakage inductance in the full model. For simplicity in this analysis, the overall primary side leakage inductance is obtained as the sum of the measured primary- and reflected measured secondary-side leakage inductances.

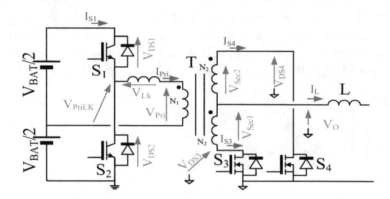

Figure 10. Half-bridge current-source (HBCS) converter equivalent circuit, including the overall leakage inductor at the primary side of the transformer.

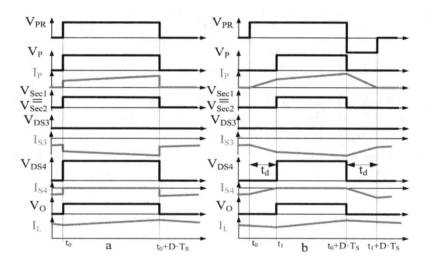

Figure 11. (a) Switching waveforms with ideal transformer. **(b)** Switching waveforms with overall leakage inductor at the primary side.

The waveform sequence starts when S_1 and S_2 are turned off, and thus there is no current flowing through the primary side of the transformer. In this situation, the output inductor current splits equally through the diodes at switches S_3 and S_4. At time t_0, S_1 is turned on, forcing the primary side voltage, V_P, to be half the voltage across the battery:

$$V_P = \frac{V_{BAT}}{2}. \tag{10}$$

The high inductance of the filter, L, causes I_L to keep flowing practically unchanged to the load. For the ideal transformer in Figure 11a, the current immediately flows through S_1, and thus the diodes of S_3 and S_4 are instantly turned-off. As a consequence, the secondary side voltages at the transformer instantly are equal to:

$$V_{Sec1} = V_{Sec2} = \frac{V_{BAT}}{2} \frac{N_2}{N_1}. \tag{11}$$

Nonetheless, for the real transformer case depicted in Figure 10, the leakage inductance prevents the secondary side current to change instantly (Figure 11b). Upon this situation current keeps flowing through the secondary sides of the transformer, S_3 and S_4. The secondary side voltages remain null immediately after turning on S_1, and thus the voltage at the leakage inductor equals the primary-side voltage. This issue affects the operation of the system. For instance, in charging mode, both HV-side switches, S_1 and S_2, are in the off-state just before S_1 is turned on. It can be seen in Figures 2 and 3a that each diode at the secondary side carries half I_L, while the voltages at the primary side, V_P, and at the common node of the secondary windings, V_O, are null:

$$I_{S3} = I_{S4} = -\frac{I_L(t)}{2} = -\frac{I_L}{2}, \tag{12}$$
$$V_0 = 0, \tag{13}$$
$$V_P = 0. \tag{14}$$

Equation (12) describes as well that the inductor current is considered as constant within a switching interval. When S_1 turns on at t_0, and given that current keeps on flowing through the LV side diodes, then the voltage V_P does not vary. Instead, because V_{DS1} equals zero, the voltage at the primary side of the real transformer (i.e., considering the parasitic inductance), V_{PR}, equals half the voltage at the battery:

$$V_{PR} = \frac{V_{BAT}}{2}. \tag{15}$$

Therefore, the values of the voltage across the leakage inductor, V_{Lk}, and the current at the primary side of the transformer, I_{Lk}, can be calculated as:

$$V_{Lk} = \frac{V_{BAT}}{2}, \tag{16}$$
$$I_{Lk} = I_{PR} = \frac{1}{L_{Lk}} \frac{V_{BAT}}{2} \cdot t. \tag{17}$$

As mentioned previously, the overall leakage inductance L_{Lk} takes into account the measured leakage inductor at the primary ($L_{Lk_{Pri}}$) and secondary ($L_{Lk_{Sec}}$) sides of the real transformer, referred to the primary side:

$$L_{Lk} = L_{Lk_{Pri}} + \frac{L_{Lk_{Sec}}}{2} \left(\frac{N_1}{N_2} \right)^2. \tag{18}$$

Upon these conditions, the inductance L_{Lk} is linearly charged. The currents through S_3 and S_4 are given by:

$$I_{S3} = -\frac{I_L}{2} - \frac{I_{PR}}{2} \frac{N_1}{N_2}, \tag{19}$$
$$I_{S4} = -\frac{I_L}{2} + \frac{I_{PR}}{2} \frac{N_1}{N_2}. \tag{20}$$

At instant t_1, I_{S4} is null, turning S_4 off, and thus from this instant the current I_L equals the current through the body diode of S_3:

$$I_{S3} = -I_L, \tag{21}$$

$$I_{S4} = 0. \tag{22}$$

As a consequence, the dead time t_d, defined as the interval between instants t_0 and t_1,

$$t_d = t_1 - t_0 \tag{23}$$

is the time it takes for the current I_{S3} to reach $-I_L$ after S_1 is turned on. This condition can be expressed as:

$$\frac{-I_L}{2} - \frac{1}{2}\frac{N_1}{N_2}\frac{V_{BAT}}{2}\frac{1}{L_{Lk}}(t_1 - t_0) = -I_L(t). \tag{24}$$

The explicit expression for t_d can thus be obtained:

$$t_d = 2 \cdot \frac{N_2}{N_1}\frac{I_L(t)}{V_{BAT}} \cdot L_{Lk}. \tag{25}$$

It is important to note that t_d represents a completely different behavior in the full model when compared to the basic one. For the basic model, when switch S_1 is turned on, V_O equals to half the battery voltage, obviously referred to the secondary side. Instead, for the full model, V_O is zero during t_d just after turning S_1 on, and then reaches the same $V_{BAT}/2$ for the rest of the interval when S_1 remains turned on. This situation affects the final voltage gain at the topology, given that the effective duty ratio of the converter, D_{eff}, is actually smaller for the full model than for the basic one. This D_{eff} can be defined now as:

$$D_{eff} = \frac{D_1 \cdot T_S - t_d}{T_S}. \tag{26}$$

As it can be deduced from Equation (25), the expression of the dead time is a non-linear function of a manifold of system parameters. Hence, the resulting model that takes this t_d into account is a non-linear model. The large-signal circuital model that considers t_d (and also R_{loss}) is shown in Figure 12. Additionally, the Equivalent Series Resistor, ESR_C, of the filter capacitor C has been included.

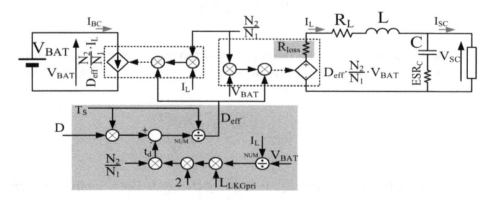

Figure 12. Full averaged large-signal model based on real components, parasitic elements, and snubbers.

The full average model has been simulated and compared with the switching circuit, in order to evaluate its performance. These simulations, shown in Figure 13, have been carried out in the same conditions that were established for the previous simulations of the basic model shown in Figure 9.

It can be seen how the new complete model tracks much more accurately the simulated waveforms at the switching model, again for a duty step from 0.34 to 0.36. Therefore it can be concluded that the dynamic behavior of the HBCS converter with the proposed SR scheme is truly represented by the obtained full model.

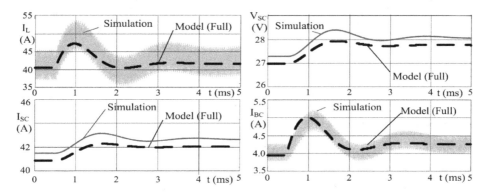

Figure 13. Simulation waveforms (gray) and averaged value of model based on the parameters of real components (black), for filter current (I_L), load output current (I_{SC}), load voltage (V_{SC}) and input current (I_{BC}).

An important remark on this behavior ir that this response corresponds to a second order system, as it is expected from a resistive load at the output of the L–C filter at the LV side. In the proposed application, the load is an assembly of SC. If this assembly is considered as an ideal pure capacitive load, given that the capacitance value of these devices is very large, then the output voltage would remain constant during the transient, and thus the dynamic behavior of the output voltage can be disregarded. For a more realistic approach, the SC assembly will be modeled by an equivalent circuit formed by a series resistor and the SC capacitor. Still, the SC will present a very large capacitance value, and therefore it behaves practically as a DC voltage source. Thus, the AC small signal model of the SC is given by the series resistor only, and the behavior of the full system is again corresponding to a second order system.

At this point, a small-signal model can be derived. Figure 14 shows the model that emerges after linearizing and perturbing the large signal full model, which has the following parameters:

$$R_1 = \frac{t_d}{T_S} \tag{27}$$

$$R_2 = \frac{2 \cdot D \cdot L_{Lk_{Pri}}}{T_S} \left(\frac{N_2}{N_1}\right)^2 \tag{28}$$

$$K_1 = \frac{N_2}{N_1} \cdot I_L \left(1 - \frac{t_d}{T_S}\right) \tag{29}$$

$$K_2 = 2 \cdot \frac{t_d}{T_S} \cdot D \cdot \frac{N_2}{N_1} \tag{30}$$

$$K_3 = \frac{N_2}{N_1} \cdot V_{BAT} - \frac{2 \cdot I_L^2 \cdot L_{Lk_{Pri}}}{T_S} \left(\frac{N_2}{N_1}\right)^2 \tag{31}$$

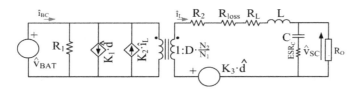

Figure 14. Full Small Signal Model of the system.

This last model finally allows the design and tuning of the control system, since the most significant dynamics of the system are taken into account. The second order behavior of the system, represented in the previous analysis, can be also derived from this model. In fact, the output voltage to duty ratio transfer function, can be calculated as:

$$\left.\frac{\hat{v}_{SC}}{\hat{d}}\right|_{\hat{v}_{BAT}=0} = \frac{K_3 \cdot R_0}{R_0 + ESR_C} \cdot \frac{1 + s \cdot C \cdot ESR_C}{\frac{R_{EQ} + R_0}{R_0 + ESR_C} + s \cdot \left[\frac{L}{R_0 + ESR_C} + C \cdot \left(R_{EQ} + \frac{R_0 \cdot ESR_C}{R_0 + ESR_C}\right)\right] + s^2 \cdot L \cdot C} \tag{32}$$

where

$$R_{EQ} = R_2 + R_{loss} + R_L \tag{33}$$

5. Proposed Control Scheme for the HBCS Converter

An adequate design of the control stage of the converter is needed in order to ensure the required power flows in the hybrid storage system under consideration [20–29]. In order to obtain the desired operating performance of the system, a High Level Control System provides the instant power references that each storage device must supply to the DC link. From this power reference, the HV current value of the HBCS converter, I_{BC}^* can be calculated. This scheme is depicted in Figure 15 [40].

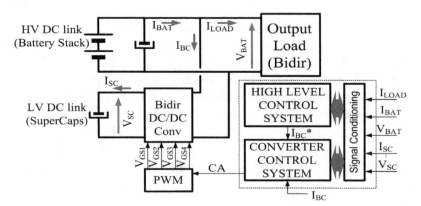

Figure 15. Control Strategy for the HBCS converter.

Considering that the final control parameter is the inductor current, I_L, the converter control stage covers two different aspects: Firstly, it is required that this stage generates a reference for such current, I_L^*, starting from the HV side current, I_{BC}^*. In addition, it implements a feedback control loop for such inductor current.

5.1. Feedback Control Loop for the SC current

This inductor current control loop is shown in Figure 16a. It must be noticed how the average value of I_L equals the average value of I_{SC}. In order to design and tune the regulator, the inductor voltage to inductor current transfer function $G(s)$ can be defined as:

$$G(s) = \frac{1}{R_L + s \cdot L} \tag{34}$$

For the current loop, a standard PI regulator has been designed. This regulator is tuned using the zero-pole cancellation method, and setting the desired bandwidth of the current loop. The regulator consists of a pure integrator and a zero that cancels the pole given by the inductance L and the ESR of the filter inductor, R_L. This ensures that the final bandwidth of the controlled system can be designed to a target value. The characteristic frequency of this zero might change upon eventual variations of the equivalent series resistor due temperature excursions, ageing, or even manufacturing tolerances.

However, assuming an adequate design of this inductor filter, then the ESR of this magnetic element will be relatively small; therefore, the variations in this parameter will not significantly affect the final performance of the controlled system. Once the regulator is tuned, the final implementation needs to obtain the expression of the duty ratio, D, from the control action of the current loop, CA. Considering Equations (25) and (26) and Figure 12, the relationship between D and V_L can be calculated:

$$V_L = D_{eff} \cdot \frac{N_2}{N_1} \cdot V_{BAT} - V_{SC} = D \cdot \left(1 - \frac{t_d}{T_S}\right) \frac{N_2}{N_1} \cdot V_{BAT} - V_{SC}. \tag{35}$$

Finally,

$$D = \frac{N_1}{N_2} \cdot \frac{V_L + V_{SC}}{V_{bat} - \frac{N_2}{N_1} \cdot \frac{2 \cdot I_L \cdot L_{Lk_{Pri}}}{T_S}}. \tag{36}$$

Figure 16b shows the implemented control scheme. The shaded box implements Equation (36), obtaining the duty ratio D from the control action of the current loop, CA, which is V_L. Figure 17 shows simulations of the performance of this inner loop for a bandwidth of 500 Hz. This plot provides a sequence of current steps as a reference to the current loop, I_L^*. It can be seen how the actual current value tracks perfectly the reference. Moreover, the current can swiftly change its direction, changing automatically from charging to discharging operation modes, even for large relative current steps. This feature is a consequence of using the discussed SR modulation scheme in the topology.

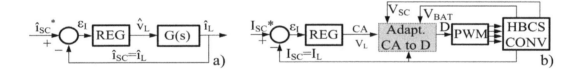

Figure 16. Inner control loop approach (**a**) for designing and tuning the regulator and (**b**) for implementing the control scheme.

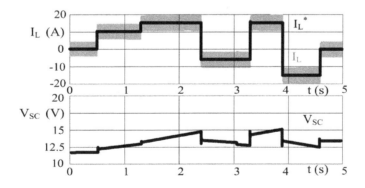

Figure 17. Simulation of inductor current I_L (gray, upper plot) and supercapacitor (SC) voltage V_{SC} (black, lower plot) during step changes in the commanded current (black, lower plot). Positive current corresponds to charging mode, negative to discharging mode.

5.2. Generation of the SC Current Reference

The complete control scheme is shown in Figure 18. An external estimator stage is designed to generate the feedback loop current reference, I_L^*, from the original I_{BC}^* reference. The relationship between the HV-side current and the inductor current can be obtained from the model in Figure 14 as

$$\hat{i}_{BC} = K_1 \cdot \hat{d} + \frac{1}{R_1} \cdot \hat{v}_{BAT} + \left(\frac{N_2}{N_1} \cdot D - K_2\right) \cdot \hat{i}_L, \tag{37}$$

$$\left.\frac{\hat{i}_{BC}}{\hat{i}_L}\right|_{\substack{\hat{d}=0 \\ \hat{v}_{BAT}=0}} = \frac{N_2}{N_1} \cdot D - K_2. \tag{38}$$

The value of the inductor current reference, I_L^*, is estimated from the HV-side current, I_{BC}^*, by means of the model, as is given in Equation (38). Other control strategies based on additional feedback loops, such as a cascaded control loop to generate the inductor current reference, would imply a more complex implementation of the controller, in terms of computation, number of sensors, filtering and signal conditioning stages, and so on. In addition, they would provide a decrease in the dynamics, given that the external loop dynamics requires a bandwidth significantly smaller than that of the inner loop, in order to ensure a good overall performance.

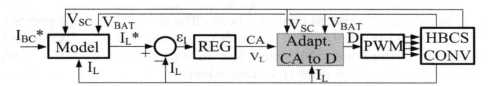

Figure 18. Complete control scheme for the half-bridge current-source (HBCS) converter.

6. Experimental Results

Figure 19 shows a 3 kW laboratory prototype of the HBCS, used in this work for the practical demonstration of the system performance. The main characteristics of the setup are detailed in Table 1. Preliminary results of the performance of this setup were reported in [35].

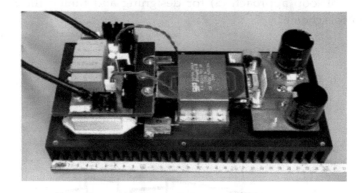

Figure 19. Laboratory prototype of the half-bridge current-source (HBCS) converter.

Figure 20 shows the experimental waveform of the converter, with the same operating conditions as reported in Figure 9. It can be seen how the experimental waveforms were very similar to the simulated waveforms for the switching converter. It also can be seen how the ideal model failed to track the real waveforms properly.

Similarly, Figure 21 compares the proposed complete model, including the effects of the parasitic leakage inductance of the transformer and the resistance to account for the losses, against the experimental waveforms, resembling the conditions stated for Figure 13. This validates the obtained model for the HBCS converter, as the complete model tracked the real converter waveforms very accurately.

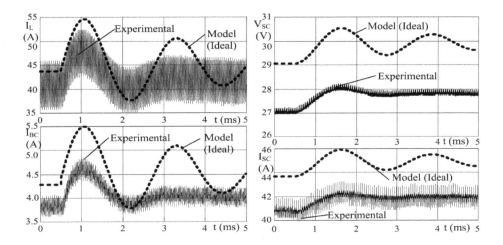

Figure 20. Experimental waveforms (gray) and averaged value of model based on ideal components (black, dashed), for filter current (I_L), load output current (I_{SC}), load voltage (V_{SC}), and input current (I_{BC}).

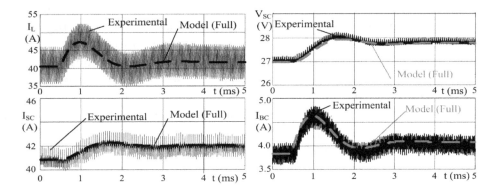

Figure 21. Experimental waveforms (gray) and averaged value of model based on real components (black), for filter current (I_L), load output current (I_{SC}), load voltage (V_{SC}), and input current (I_{BC}).

In order to validate experimentally the implementation of the control scheme, a series of current reference steps were implemented in the converter. The reference steps that were provided to the simulations, shown in Figure 17, were also supplied to the laboratory prototype. Figure 22 shows the performance of the experimental setup. It can be seen how the real converter performed as expected, thus validating the converter control scheme.

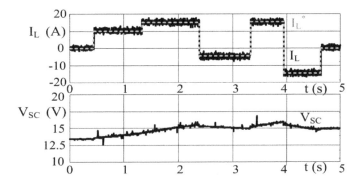

Figure 22. Inductor current I_L (black, upper plot) and supercapacitor (SC) voltage V_{SC} (black, lower plot) during step changes in the commanded current I_L^* (referenced by dashed gray lines). Positive current corresponds to charging mode, negative to discharging mode.

In order to check the dynamics of the inner control loop, a detailed experimental response of one of such current steps is shown in Figure 23.

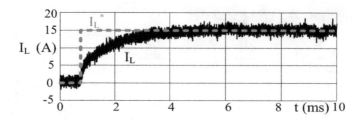

Figure 23. Inductor current I_L (black) and reference I_L^* (grey) during one step change.

7. Conclusions and Future Developments

This research has demonstrated the feasibility of the HBCS converter as a bidirectional power converter for HSS systems in traction applications, where the DC-link is directly connected to the battery bank. The main topics covered include the analysis and design of the power and control stages, the modeling including parasitic elements of the topology, and a validation procedure through a laboratory prototype.

As a first contribution, it has been demonstrated that the use of SR is a suitable switching scheme in the converter, given the demonstrated benefits in the efficiency and the enhancement in the control system implementation. By means of this switching pattern, a full, bidirectional power flow control can be simply implemented in the converter. Moreover, this feature is obtained at no cost, since the required hardware elements to implement SR pattern are present in the conventional HBCS topology.

Another key contribution is to include the parasitic elements in the topology aiming to obtain a highly representative circuital model of the converter. This model has been demonstrated by means of simulations and through experiments in a 3 kW laboratory setup. With the information derived from this dynamic model, the design of the control stage for the HSS can be easily implemented. This research also has provided a design example of a current control loop in order to govern the power flow in the HSS outlined.

A series of future developments arise from this research. One of these developments is the implementation of the HSS in a full powertrain for vehicle applications, including the integration of the HBCS converter control with the complete high-level control system. Another development covers the optimization of the converter power topology and control system, including the definition of the control strategy for the complete system. Additionally, the extension of the application of the HBCS converter to other types of loads (e.g., inductive loads) can be considered in future research.

Author Contributions: J.G. and F.G.C. conceived the research and designed and performed the experiments; all the authors analyzed the data and contributed to the discussion and conclusions.

Abbreviations

The following abbreviations are used in this manuscript:

CA	Control action
DC	Direct current
ESR	Equivalent series resistor
LV	Low-voltage
HBCS	Half-bridge current-source
HF	High-frequency
HSS	Hybrid storage system
HV	High-voltage
IGBT	Insulated-gate bipolar transistor
MOSFET	Metal-oxide-semiconductor field-effect transistor
SC	Supercapacitor
SR	Synchronous rectification

References

1.   Ostadi, A.; Kazerani, M.; Chen, S.K. Hybrid Energy Storage System (HESS) in vehicular applications: A review on interfacing battery and ultra-capacitor units. In Proceedings of the 2013 IEEE Transportation Electrification Conference and Expo (ITEC), Detroit, MI, USA, 16–19 June 2013; pp. 1–7.
2.   Amjadi, Z.; Williamson, S.S. Digital Control of a Bidirectional DC/DC Switched Capacitor Converter for Hybrid Electric Vehicle Energy Storage System Applications. *IEEE Trans. Smart Grid* **2014**, *5*, 158–166. [CrossRef]
3.   Kuperman, A.; Mellincovsky, M.; Lerman, C.; Aharon, I.; Reichbach, N.; Geula, G.; Nakash, R. Supercapacitor Sizing Based on Desired Power and Energy Performance. *IEEE Trans. Power Electron.* **2014**, *29*, 5399–5405. [CrossRef]
4.   Cohen, I.J.; Kelley, J.P.; Wetz, D.A.; Heinzel, J. Evaluation of a Hybrid Energy Storage Module for Pulsed Power Applications. *IEEE Trans. Plasma Sci.* **2014**, *42*, 2948–2955. [CrossRef]
5.   Farooque, M.; Maru, H.C. Fuel cells-the clean and efficient power generators. *Proc. IEEE* **2001**, *89*, 1819–1829. [CrossRef]
6.   Di Napoli, A.; Caricchi, F.; Crescimbini, F. Fuel cells-the clean and efficient power generators. In Proceedings of the 6th European Conference on Power Electronics, Seville, Spain, 19–21 September 1995.
7.   Mestre, P.; Astier, S. Application of supercapacitors and influence of the drive control strategies on the performances of on electric vehicle. In Proceedings of the 15th Electric Vehicle Symposium (EVS-15), Bruxelles, Belgium, 29 September–3 October 1998.
8.   Grbovic, P.J.; Delarue, P.; Moigne, P.L.; Bartholomeus, P. Modeling and Control of the Ultracapacitor-Based Regenerative Controlled Electric Drives. *IEEE Trans. Ind. Electron.* **2011**, *58*, 3471–3484. [CrossRef]
9.   Schmidt, M. EV Mini-Van Featuring Series Conjunction of Ultracapacitors and Batteries for Load Leveling of its Batteries. In Proceedings of the 15th Electric Vehicle Symposium (EVS-15), Bruxelles, Belgium, 29 September–3 October 1998.
10. Itani, K.; Bernardinis, A.D.; Khatir, Z.; Jammal, A.; Oueidat, M. Regenerative Braking Modeling, Control, and Simulation of a Hybrid Energy Storage System for an Electric Vehicle in Extreme Conditions. *IEEE Trans. Transp. Electr.* **2016**, *2*, 465–479. [CrossRef]
11. King, R.; Schwartz, J.; Cardinal, M.; Garrigan, K. Development and system test of high efficiency ultracapacitor/battery electronic interface. In Proceedings of the 15th Electric Vehicle Symposium (EVS-15), Bruxelles, Belgium, 29 September–3 October 1998.
12. Di Napoli, A.; Crescimbini, F.; Solero, L.; Caricchi, F.; Giulii Capponi, F. Multiple-input DC-DC power converter for power-flow management in hybrid vehicles. In Proceedings of the Conference Record of the 2002 IEEE Industry Applications Conference, 37th IAS Annual Meeting (Cat. No.02CH37344), Pittsburgh, PA, USA, 13–18 October 2002; Volume 3, pp. 1578–1585.
13. Di Napoli, A.; Giulii Capponi, F.; Solero, L. Power Converter Arrangements with Ultracapacitor Tank for Battery Load Leveling in EV Motor Drives. In Proceedings of the 8th European Conference on Power Electronics, Lausanne, Switzerland, 7–9 September 1999.
14. Herath, N.; Binduhewa, P.; Samaranayake, L.; Ekanayake, J.; Longo, S. Design of a dual energy storage power converter for a small electric vehicle. In Proceedings of the 2017 IEEE International Conference on Industrial and Information Systems (ICIIS), Peradeniya, Sri Lanka, 15–16 December 2017; pp. 1–6.
15. Kouchachvili, L.; Yaïci, W.; Entchev, E. Hybrid battery/supercapacitor energy storage system for the electric vehicles. *J. Power Sources* **2018**, *374*, 237–248. [CrossRef]
16. Song, Z.; Li, J.; Hou, J.; Hofmann, H.; Ouyang, M.; Du, J. The battery-supercapacitor hybrid energy storage system in electric vehicle applications: A case study. *Energy* **2018**, *154*, 433–441. [CrossRef]
17. Cabrane, Z.; Ouassaid, M.; Maaroufi, M. Analysis and evaluation of battery-supercapacitor hybrid energy storage system for photovoltaic installation. *Int. J. Hydrogen Energy* **2016**, *41*, 20897–20907. [CrossRef]
18. Hernández, J.; Sanchez-Sutil, F.; Vidal, P.; Rus-Casas, C. Primary frequency control and dynamic grid support for vehicle-to-grid in transmission systems. *Int. J. Electr. Power Energy Syst.* **2018**, *100*, 152–166. [CrossRef]
19. Thounthong, P.; Piegari, L.; Pierfederici, S.; Davat, B. Nonlinear intelligent DC grid stabilization for fuel cell vehicle applications with a supercapacitor storage device. *Int. J. Electr. Power Energy Syst.* **2015**, *64*, 723–733. [CrossRef]
20. Sun, L.; Feng, K.; Chapman, C.; Zhang, N. An Adaptive Power-Split Strategy for Battery-Supercapacitor Powertrain-Design, Simulation, and Experiment. *IEEE Trans. Power Electron.* **2017**, *32*, 9364–9375. [CrossRef]
21. Bougrine, M.; Benalia, A.; Delaleau, E.; Benbouzid, M. Minimum time current controller design for

two-interleaved bidirectional converter: Application to hybrid fuel cell/supercapacitor vehicles. *Int. J. Hydrogen Energy* **2018**, *43*, 11593–11605. [CrossRef]

22. He, H.W.; Xiong, R.; Chang, Y.H. Dynamic Modeling and Simulation on a Hybrid Power System for Electric Vehicle Applications. *Energies* **2010**, *3*, 1821–1830. [CrossRef]

23. Shah, N.; Czarkowski, D. Supercapacitors in Tandem with Batteries to Prolong the Range of UGV Systems. *Electronics* **2018**, *7*, 6. [CrossRef]

24. Zhang, Q.; Deng, W. An Adaptive Energy Management System for Electric Vehicles Based on Driving Cycle Identification and Wavelet Transform. *Energies* **2016**, *9*, 341. [CrossRef]

25. Passalacqua, M.; Lanzarotto, D.; Repetto, M.; Marchesoni, M. Advantages of Using Supercapacitors and Silicon Carbide on Hybrid Vehicle Series Architecture. *Energies* **2017**, *10*, 920. [CrossRef]

26. Wang, Y.; Hu, H.; Zhang, L.; Zhang, N.; Sun, X. Real-Time Vehicle Energy Management System Based on Optimized Distribution of Electrical Load Power. *Appl. Sci.* **2016**, *6*, 285. [CrossRef]

27. Wang, B.; Xu, J.; Wai, R.J.; Cao, B. Adaptive Sliding-Mode With Hysteresis Control Strategy for Simple Multimode Hybrid Energy Storage System in Electric Vehicles. *IEEE Trans. Ind. Electron.* **2017**, *64*, 1404–1414. [CrossRef]

28. Thounthong, P.; Rael, S. The benefits of hybridization. *IEEE Ind. Electron. Mag.* **2009**, *3*, 25–37. [CrossRef]

29. Lai, J.; Nelson, D.J. Energy Management Power Converters in Hybrid Electric and Fuel Cell Vehicles. *Proc. IEEE* **2007**, *95*, 766–777. [CrossRef]

30. Trovão, J.P.; Silva, M.A.; Dubois, M.R. Coupled energy management algorithm for MESS in urban EV. *IET Electr. Syst. Transp.* **2017**, *7*, 125–134. [CrossRef]

31. Jain, M.; Daniele, M.; Jain, P.K. A bidirectional DC-DC converter topology for low power application. *IEEE Trans. Power Electron.* **2000**, *15*, 595–606. [CrossRef]

32. Kazimierczuk, M.K.; Vuong, D.Q.; Nguyen, B.T.; Weimer, J.A. Topologies of bidirectional PWM dc-dc power converters. In Proceedings of the IEEE 1993 National Aerospace and Electronics Conference-NAECON 1993, Dayton, OH, USA, 24–28 May 1993; Volume 1, pp. 435–441.

33. Giulii Capponi, F.; Cacciato, F. Using Super Capacitors in Combination with Bi-Directional DC/DC Converters for Active Load Management in Residential Fuel Cell Applications. In Proceedings of the 1st European Symposium on Supercapacitors (ESSCAP'04), Belfort, France, 4–5 November 2004.

34. Cacciato, M.; Caricchi, F.; Giulii Capponi, F.; Santini, E. A critical evaluation and design of bi-directional DC/DC converters for super-capacitors interfacing in fuel cell applications. In Proceedings of the Conference Record of the 2004 IEEE Industry Applications Conference, 39th IAS Annual Meeting, Seattle, WA, USA, 3–7 October 2004; Volume 2, pp. 1127–1133.

35. Giulii Capponi, F.; Santoro, P.; Crescenzi, E. HBCS Converter: A Bidirectional DC/DC Converter for Optimal Power Flow Regulation in Supercapacitor Applications. In Proceedings of the 2007 IEEE Industry Applications Annual Meeting, New Orleans, LA, USA, 23–27 September 2007; pp. 2009–2015.

36. Zhang, Y.; Guo, Z.; Zhang, Y.; Zhan, T.; Jin, L. Active battery/ultracapacitor hybrid energy storage system based on soft-switching bidirectional converter. In Proceedings of the 2013 International Conference on Electrical Machines and Systems (ICEMS), Busan, South Korea, 26–29 October 2013; pp. 2177–2182.

37. Wai, R.J.; Duan, R.Y. High-Efficiency Bidirectional Converter for Power Sources with Great Voltage Diversity. *IEEE Trans. Power Electron.* **2007**, *22*, 1986–1996. [CrossRef]

38. Yamamoto, K.; Hiraki, E.; Tanaka, T.; Nakaoka, M.; Mishima, T. Bidirectional DC-DC converter with full-bridge/push-pull circuit for automobile electric power systems. In Proceedings of the 2006 37th IEEE Power Electronics Specialists Conference, Jeju, South Korea, 18–22 June 2006; pp. 1–5.

39. Dusmez, S.; Hasanzadeh, A.; Khaligh, A. Comparative Analysis of Bidirectional Three-Level DC DC Converter for Automotive Applications. *IEEE Trans. Ind. Electron.* **2015**, *62*, 3305–3315. [CrossRef]

40. Garcia, J.; Giulii Capponi, F.; Borocci, G.; Garcia, P. Control strategy for Bidirectional HBCS Converter for supercapacitor applications. In Proceedings of the 2014 IEEE 23rd International Symposium on Industrial Electronics (ISIE), Istanbul, Turkey, 1–4 June 2014; pp. 1794–1799.

41. Garcia, J.; Garcia, P.; Giulii Capponi, F.; Borocci, G.; De Donato, G. Analysis, modeling and control of half-bridge current-source converter for supercapacitor applications. In Proceedings of the 2014 IEEE Energy Conversion Congress and Exposition (ECCE), Pittsburgh, PA, USA, 14–18 September 2014; pp. 3786–3793.

7

High Efficiency Solar Power Generation with Improved Discontinuous Pulse Width Modulation (DPWM) Overmodulation Algorithms

Lan Li [1], Hao Wang [1], Xiangping Chen [2,3,*], Abid Ali Shah Bukhari [4], Wenping Cao [4,*], Lun Chai [1] and Bing Li [1]

[1] College of Electrical and Power Engineering, Taiyuan University of Technology, Shanxi 030024, China; lilan@tyut.edu.cn (L.L.); wanghaom93@163.com (H.W.); chailun_tyut@126.com (L.C.); biang1221@163.com (B.L.)

[2] Electrical Engineering School, Guizhou University, Guiyang 550025, China

[3] Faculty of Engineering, Cardiff University, Cardiff CF24 3AA, UK

[4] School of Engineering and Applied Science, Aston University, Birmingham B4 7ET, UK; bukhars2@aston.ac.uk

* Correspondence: ee.xpchen@gzu.edu.cn (X.C.); w.p.cao@aston.ac.uk (W.C.)

Abstract: The efficiency of a photovoltaic (PV) system strongly depends on the transformation process from solar energy to electricity, where maximum power point tracking (MPPT) is widely regarded as a promising technology to harvest solar energy in the first step. Furthermore, inverters are an essential part of solar power generation systems. Their performance dictates the power yield, system costs and reliable operation. This paper proposes a novel control technology combining discontinuous pulse width modulation (DPWM) and overmodulation technology to better utilize direct current (DC) electrical power and to reduce the switching losses in the electronic power devices in conversion. In order to optimize the performance of the PV inverter, the overmodulation region is refined from conventional two-level space vector pulse width modulation (SVPWM) control technology. Then, the turn-on and turn-off times of the switching devices in different modulation areas are deduced analytically. A new DPWM algorithm is proposed to achieve the full region control. An experimental platform based on a digital signal processing (DSP) controller is developed for validation purposes, after maximum power is achieved via a DC/DC converter under MPPT operation. Experimental results on a PV system show that the DPWM control algorithm lowers the harmonic distortion of the output voltage and current, as well as the switching losses. Moreover, better utilization of the DC-link voltage also improves the PV inverter performance. The developed algorithm may also be applied to other applications utilizing grid-tie power inverters.

Keywords: DPWM; MPPT; photovoltaic power system

---

1. Introduction

Currently, there is great concern about global warming due to the rapid depletion of fossil fuels [1]. Thus, the utilization of renewable energy has received increasing attention in industry and research communities. Solar energy is one of the most promising renewable energy sources in the world, and photovoltaic (PV) power generation systems are a growing area for research and development [2].

Conventionally, the efficiency of direct-coupled PV systems could be very low due to the high dependence on the irradiance and temperature conditions. This can be overcome by continuously tracking the maximum power point (MPP) of the system at varied conditions of irradiance and temperature [3–5]. This method is known as maximum power point tracking (MPPT). In order to realize MPPT, DC/DC converters are commonly used in the PV power system. Moreover, the PV modules are usually connected in series to raise the output direct current (DC) voltage and, in parallel, to increase the output power. However, this will lead to a multi-peak effect, which poses a challenge to maintain the equal terminal voltage across PV modules.

On the other hand, inverters are a key component in solar photovoltaic systems, and their performance determines the power yield, system costs and reliable operation [6–8]. Most PV inverters adopt a two-level control technique based on the space vector pulse width modulation (SVPWM). The multi-peak effect potentially increases the switching losses in the power switching devices [9–12]. In order to reduce this effect, a synchronous pulse width modulation (PWM) method is often utilized in cascaded inverters [13]. When the DC voltage is low, an overmodulation control mode is adopted for the two inverters. Therefore, a T-type three-level overmodulation strategy is developed [14–16]. By doing so, PV inverters can still achieve maximum power point tracking (MPPT) which can prolong the running time of PV inverters and improve the output power. However, these technologies increase the complexity and the cost of photovoltaic power generation systems, as well as the switching losses and total harmonic distortion (THD) of the PV systems [17–20].

In this paper, a new discontinuous pulse width modulation (DPWM) scheme is proposed to achieve optimal control of PV inverters along with MPPT in a boost DC/DC converter. It combines DPWM and overmodulation to reduce device conducting periods, based on two-level SVPWM control technology [21]. As a result, the device power losses are reduced effectively. Moreover, the overmodulation segment control method can also reduce the harmonic distortion of the output voltage and improve the DC-link voltage utilization. Thereby, the overall system efficiency will be increased.

The contents are organized as follows: An equivalent circuit of a PV power system under an MPPT scheme is presented in Section 2, followed by the introduction of a DPWM overmodulation algorithm in Section 3. Section 4 demonstrates both simulation and experimental results in detail. The key findings are summarized in Section 5.

2. Equivalent Circuit of a PV Power System

Figure 1 shows the structure of a typical photovoltaic system. The MPPT control of PV panels is achieved by a DC/DC converter, which can also maintain the voltage stability of the DC bus. The PV inverter (DC/AC converter) can achieve the active/reactive power control.

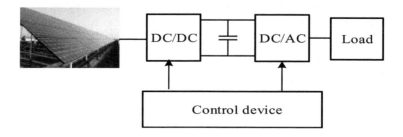

Figure 1. Structure of the photovoltaic (PV) power.

2.1. Perturbation and Observation (P&O) MPPT Algorithm

Conventionally, MPPT is embedded in a converter to determine the duty cycle that maximizes the PV power yield [22]. A perturbation and observation method (P&O) is used to find the maximum power point [23] by altering the array terminal voltage and then comparing the PV output power with its previous value. If the power increases while voltage increases, the PV array is operating in the correct direction; otherwise, the operational point should be adjusted to its the opposite direction [24,25]. The main advantage of P&O lies in its simplicity. This method shows its effectiveness, provided that solar irradiation does not change very quickly. As shown in Figure 2, there are four operational points,

A, B, C and D. From point A to point B, the PV power increases while the voltage of point B is higher than that of point A. Therefore, the next perturbation voltage keeps increasing, or vice versa. If the operation starts from point C to D, PV power decreases while the voltage of point C is higher than the voltage of point D. The next operation should be changed to the opposite direction. Therefore, the next perturbation reduces the voltage so as to redirect the trajectory towards the maximum power point. Accordingly, four scenarios are summarized in Table 1.

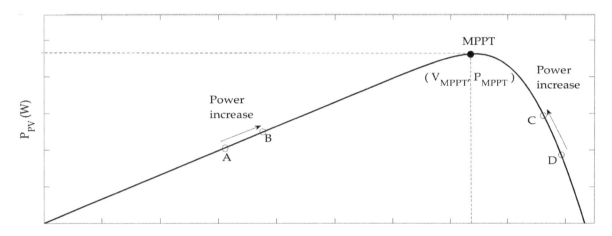

Figure 2. Maximum power point tracking (MPPT) from different trajectories.

Table 1. Trajectory analysis with perturbation and observation (P&O) MPPT.

No.	Scenario	Example Route	Action
1	$P_{current} > P_{previous}$ & $V_{current} > V_{previous}$	A → B	Increase voltage
2	$P_{current} > P_{previous}$ & $V_{current} < V_{previous}$	D → C	Decrease voltage
3	$P_{current} < P_{previous}$ & $V_{current} > V_{previous}$	C → D	Decrease voltage
4	$P_{current} < P_{previous}$ & $V_{current} < V_{previous}$	B → A	Increase voltage

A boost DC/DC converter is adopted to adjust the terminal voltage by regulating the duty ratio D. It will therefore incur the change of the equivalent power output which in turn realizes the maximum power output of a PV array. A flowchart of the P&O method is shown in Figure 3.

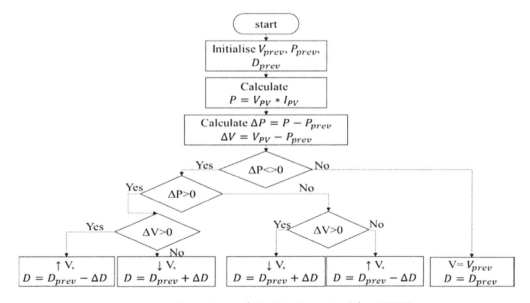

Figure 3. Flowchart of the P&O method for MPPT.

2.2. A Boost DC-DC Converter and the Equivalent Circuit

In this study, a boost converter is adopted to operate the voltage via changing D. A boost converter is expected to connect with the output of PV arrays to produce the equivalent output voltage equal to the voltage V_{mpp} at the maximum power point, with equivalent resistance equal to R_{mpp} with the current I_{mpp} over the output circuit, as shown in Figure 4. Selecting a proper DC-DC converter with reasonable circuit parameters is essential.

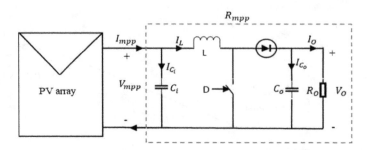

Figure 4. Equivalent circuit including a PV array and a boost DC-DC converter.

To design a boost converter for a PV system, key elements in the selection include the inductance of an inductor L', the capacitance of an input capacitor C_i, the input resistance R_{mpp}, capacitance of an input capacitor C_i, and the load resistance R_O. The purpose of this boost converter is to realize the equivalent circuit resistance to R_{mpp}. Therefore, the maximum output power under the current condition can be produced via this boost converter from a PV array.

The maximum power point resistance R_{mpp} is calculated based on the maximum power point voltage V_{mpp} and the maximum power point current I_{mpp} by using a simple calculation in Equation (1). R_O can be calculated by Equation (2). The duty ratio D can be derived from Equation (3). Assuming there is no power loss in circuit, an energy balance equation can be established (Equation (4)).

$$R_{mpp} = \frac{V_{mpp}}{I_{mpp}} \tag{1}$$

$$R_O = \frac{R_{mpp}}{(1-D)^2} \tag{2}$$

$$\text{Or} \qquad D = 1 - \sqrt{\frac{R_{mpp}}{R_O}} \tag{3}$$

$$p_{mpp} = \frac{V_{mpp}^2}{R_{mpp}} = \frac{V_O^2}{R_O} \tag{4}$$

$$\text{Or} \qquad \frac{V_{mpp}^2}{V_O^2} = \frac{R_{mpp}}{R_O} \tag{5}$$

Equation (5) presents the relationship between resistance and voltage. The output voltage is calculated from Equation (5) to Equation (6).

$$V_O = V_{mpp} \sqrt{\frac{R_O}{R_{mpp}}} \tag{6}$$

The inductance can be estimated by many methods [26,27]. A general calculation is given by Equation (7).

$$L' = \frac{V_{mpp} \cdot D}{I_{mpp} \cdot \gamma_{IL} \cdot f} \tag{7}$$

where f and γ_{IL} refer to the switching frequency and the inductor current ripple factor.

Input capacitor C_i can be calculated according to Equation (8).

$$C_i = \frac{D}{8 \cdot L \cdot \gamma_{Vmpp} \cdot f^2} \tag{8}$$

Output capacitor C_O can be calculated according to Equation (9) [28].

$$C_O = \frac{D}{R_O \cdot \gamma_{VO} \cdot f} \tag{9}$$

where the current ripple factor γ_{IL} and the voltage ripple factors γ_{Vmpp}, γ_{VO} are refined within 5%.

2.3. Division of the Overmodulation Area

Figure 5 presents the main circuit topology of the three-phase grid-tie inverter, where U_d is the DC-link voltage, i_a, i_b, and i_c are the inverter output currents, L is the filter inductance, and R is the filter inductance equivalent series resistance, respectively.

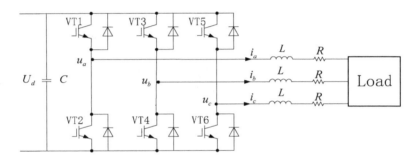

Figure 5. Topology of the three-phase grid-tie inverter.

In the inverter, the on-state of the upper arm switches and the off-state of the lower arm switches are defined as "1"; otherwise, they are "0". Therefore, the three bridge arms of the inverter have eight switch states, corresponding to eight basic voltage space vectors: $u_0(000)$, $u_1(100)$, $u_2(110)$, $u_3(010)$, $u_4(011)$, $u_5(001)$, $u_6(101)$ and $u_7(111)$. The SVPWM modulation voltage vector and the sector distribution of the PV inverter are demonstrated in Figure 6. The amplitude of $u_1 \sim u_6$ is $\frac{2}{3}U_d$, and the phase angles of $u_1 \sim u_6$ differ by 60°.

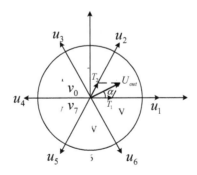

Figure 6. Space voltage vectors and sector distribution.

According to the volt-second balance principle:

$$U_{out} \cdot T_s = u_1 \cdot T_1 + u_2 \cdot T_2 \tag{10}$$

where U_{out} is the given output voltage vector, T_1 is the action time for u_1, T_2 is the action time for u_2, and T_s is the switching period.

The output space rotating vector with a constant rotating speed and a constant amplitude is achieved by the vector addition of adjacent vectors. The output range is in the inscribed circle of the regular hexagon constituted by the vectors $u_1 \sim u_6$, which are also known as the linear modulation areas. The maximum amplitude of the output phase voltage is given by:

$$u_{m\_line} = \frac{2 \cdot U_d}{3} \cdot \cos 30° = \frac{U_d}{\sqrt{3}} \tag{11}$$

If the PV inverter is controlled by the six-step wave mode outside of the linear modulation area, the amplitude of the phase voltage can be obtained [29].

$$u_{m\_max} = \frac{2 \cdot U_d}{\pi} \tag{12}$$

The region from outside the linear modulation area to the six-step maximum output voltage area is called the overmodulation area. The modulation coefficient m is defined as:

$$m = \frac{\pi \cdot U_{out}}{2 \cdot U_d} \tag{13}$$

There are three different regions as per the modulation coefficient. In the linear modulation area, $m < 0.907$; in the overmodulation area I, $0.907 < m < 0.952$; in the overmodulation area II, $0.952 < m < 1$. Figure 7 shows the trajectory of the synthesized voltage vector in the overmodulation areas. The simplified formulas for overmodulation areas I and II are given by [30–33].

$$\alpha_r = \frac{\pi}{6} - \arccos\left(\frac{U_d}{\sqrt{3} \cdot U_{out}}\right) \tag{14}$$

$$\begin{cases} \alpha_h = 6.40 \cdot m - 6.09 & (0.952 \leq m < 0.9800) \\ \alpha_h = 11.57 \cdot m - 11.34 & (0.9800 \leq m < 0.9975) \\ \alpha_h = 48.96 \cdot m - 48.43 & (0.9975 \leq m < 1) \end{cases} \tag{15}$$

Take sector I for example. In overmodulation area I, the rotational speed of the output voltage vector remains constant and the amplitude is limited by the hexagon. The vertex of the trajectory follows the thick solid line of ABCD. In overmodulation area II, the rotation speed of the output voltage vector changes and the amplitude is limited. When $\alpha < \alpha_h$, the output voltage vector is u_1. When $\alpha_h \leq \alpha < \frac{\pi}{3} - \alpha_h$, the trajectory of the output voltage vector is the BC solid line. When $\alpha \geq \frac{\pi}{3} - \alpha_h$, the output voltage vector is u_2. When $\alpha_h = \frac{\pi}{6}$, the output voltage vector traces at the vertex of the regular hexagon, and the modulation coefficient is the maximum ($m = 1$). The remaining sectors are the same.

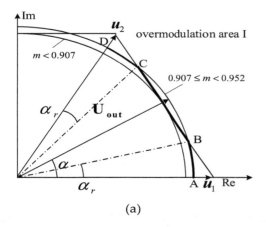

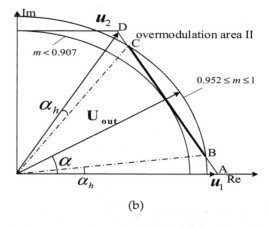

(a)   (b)

Figure 7. Synthesized voltage vector locus. (a) Overmodulation area I; (b) Overmodulation area II.

2.4. Full Modulation Region Voltage Vector

The key to controlling the voltage vector in the full modulation area is to determine the action time of the voltage vector according to the modulation coefficient m.

Take sector I for example again: U_{out} is synthesized by two basic voltage space vectors $u_1(100)$ and $u_2(110)$ and it is known that $u_1 = \frac{2}{3}U_d$, $u_2 = \frac{2}{3}U_d \cdot e^{j\frac{\pi}{3}}$. According to the sine theorem:

$$\frac{U_{out}}{\sin\frac{2\pi}{3}} = \frac{u_1 \cdot \frac{T_1}{T_s}}{\sin\left(\frac{\pi}{3} - \alpha\right)} = \frac{u_2 \cdot \frac{T_2}{T_s}}{\sin(\alpha)} \tag{16}$$

The action time T_1 and T_2 can be further calculated. When $T_1 + T_2 < T_s$, the zero vector $u_0(000)$ or $u_7(111)$ is used to fill the remaining time T_0.

(1) SVPWM linear modulation area ($m < 0.907$)

It can be obtained from Equation (16):

$$\begin{cases} T_1 = \sqrt{3} \cdot T_s \cdot \frac{U_{out}}{U_d} \cdot \sin\left(\frac{\pi}{3} - \alpha\right) \\ T_2 = \sqrt{3} \cdot T_s \cdot \frac{U_{out}}{U_d} \cdot \sin(\alpha) \\ T_0 = T_s - T_1 - T_2 \end{cases} \tag{17}$$

(2) Overmodulation area I ($0.907 \leq m < 0.952$)

(i) When $0 \leq \alpha \leq \alpha_\gamma$ or $\frac{\pi}{3} - \alpha_\gamma \leq \alpha \leq \frac{\pi}{3}$, T_1, T_2 and T_0 are calculated in the same way as Equation (17).

(ii) When $\alpha_\gamma \leq \alpha < \frac{\pi}{3} - \alpha_\gamma$,

$$\begin{cases} T_1 = T_s \cdot \frac{\sin\left(\frac{\pi}{3} - \alpha\right)}{\sin\left(\frac{\pi}{3} + \alpha\right)} \\ T_2 = T_s \cdot \frac{\sin(\alpha)}{\sin\left(\frac{\pi}{3} + \alpha\right)} \\ T_0 = 0 \end{cases} \tag{18}$$

(3) Overmodulation area II ($0.952 \leq m < 1$)

(i) When $0 \leq \alpha < \alpha_h$,

$$T_1 = T_s, \ T_2 = 0, \ T_0 = 0 \tag{19}$$

(ii) When $\alpha_h \leq \alpha < \frac{\pi}{3} - \alpha_h$,

$$\gamma = \frac{\pi}{6} \cdot \frac{(\alpha - \alpha_h)}{\left(\frac{\pi}{6} - \alpha_h\right)}$$

$$\begin{cases} T_1 = T_s \cdot \frac{\sin\left(\frac{\pi}{3} - \gamma\right)}{\sin\left(\frac{\pi}{3} + \gamma\right)} \\ T_2 = T_s \cdot \frac{\sin\gamma}{\sin\left(\frac{\pi}{3} + \gamma\right)} \\ T_0 = 0 \end{cases} \tag{20}$$

(iii) When $\frac{\pi}{3} - \alpha_h \leq \alpha < \frac{\pi}{3}$,

$$T_1 = 0, \ T_2 = T_s, \ T_0 = 0 \tag{21}$$

3. DPWM Overmodulation Algorithm

In traditional SVPWM modulation algorithms, A, B, and C from the given voltage vector U_{out} are calculated from the $\alpha - \beta$ coordinate plane.

$$\begin{cases} A = U_\beta \\ B = U_\alpha \cdot \sin 60° - U_\beta \cdot \cos 60° \\ C = -U_\alpha \cdot \sin 60° - U_\beta \cdot \cos 60° \end{cases} \tag{22}$$

where U_α and U_β are the α, β components of U_{out} in the $\alpha - \beta$ coordinate plane, respectively. S is defined as:

$$S = \text{sign}(A) + 2 \cdot \text{sign}(B) + 4 \cdot \text{sign}(C) \tag{23}$$

Thus, the relationship between sectors N and S can be obtained, as shown in Table 2.

Table 2. Relationship between S and N.

S	1	2	3	4	5	6
Sector N	II	VI	I	IV	III	V

As for the DPWM scheme, the switches are controlled only by the zero vector u_0 or u_7 in a triangular carrier cycle, as shown in Figure 8. u_7 is selected in the 60° range centred around the basic vectors u_1, u_3 and u_5, while u_0 is selected in other sectors. The DPWM sector number and its action conditions are tabulated in Table 3. There are twelve 30° sectors in the DPWM, as compared to six sectors in conventional SVPWM.

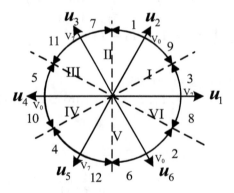

Figure 8. The sector distribution of the discontinuous width pulse modulation (DPWM).

Table 3. DPWM sector number and its action conditions.

Action condition	Sector N	S for the 30°	Sector
$\frac{U_\beta}{U_\alpha} < \tan\left(\frac{\pi}{6}\right)$	I	3 (True)	9 (False)
	IV	10 (True)	4 (False)
$U_\alpha > 0$	II	1 (True)	7 (False)
	V	6 (True)	12 (False)
$\frac{U_\beta}{U_\alpha} < -\tan\left(\frac{\pi}{6}\right)$	III	11 (True)	5 (False)
	VI	2 (True)	8 (False)

As for the DPWM modulation, the power devices switch four times in one sector because only u_0 or u_7 is used as the zero vector in one sector. Meanwhile, for the SVPWM modulation, the power devices switch six times in one sector because the vectors u_0 and u_7 are simultaneously applied (each action time is $T_0/2$).

Therefore, the switching frequency of the power devices in the DPWM modulation is reduced by one-third compared to the SVPWM modulation, and the switching sequence of the first sector is shown in Figure 9. According to the literature [34], the switching losses of the inverter power device include turn-on loss P_{on} and turn-off loss P_{off}:

$$P_{on} = \frac{1}{8} \cdot U_d \cdot t_{rN} \cdot \frac{I_{CM}^2}{I_{CN}} \cdot f_s \tag{24}$$

$$P_{off} = U_d \cdot I_{CM} \cdot t_{fN} \cdot f_s \cdot \left(\frac{1}{3\pi} + \frac{1}{24} \cdot \frac{I_{CM}}{I_{CN}} \right) \tag{25}$$

where t_{rN} and t_{fN} are the turn-on time and turn-off time, respectively; f_s is the switching frequency of power devices; I_{CN} is the forward current of IGBT; I_{CM} is the amplitude of the sinusoidal current.

Switching losses are proportional to f_s according to Equations (26) and (27), so the switching losses of the DPWM modulation can be reduced by one-third compared to the switching losses of the SVPWM modulation.

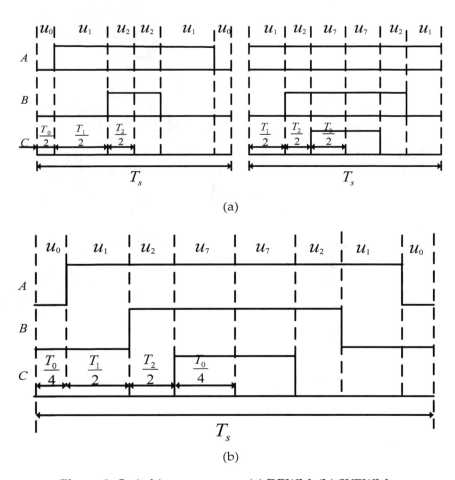

Figure 9. Switching sequences. (a) DPWM; (b) SVPWM.

4. Simulation and Experimental Validation of the Proposed Scheme

A PV inverter experimental test rig based on digital signal processing (DSP) is developed to verify the DPWM control algorithm as Figure 10a shown. DSP takes a high-performance 32-bit fixed-point TMS320F2812 as the controller (clock frequency 150 MHz, working voltage 3.3 V, Texas Instruments Incorporated, TX, USA); its chip contains 128 K 16-bit FLASH, 16-way 12-bit A/D conversion, two event managers and so on. Experimental tests are carried out by five photovoltaic panels (as shown in Figure 10b) in series under an ambient temperature of 25 °C and a daylight illuminance of 1000 W/m^2; the parameters of a single photovoltaic module are presented in Table 4. The output voltage range of the DC/DC converter (5 kW) is 100-400 V. The power module model of the DC/AC converter (2 kW) is PM30RSF060 (as shown in Figure 10c), and its maximum switching frequency is 20 kHz.

Before the proposed system is validated by the experimental tests, an MPPT algorithm with the associated equivalent circuit is simulated.

Table 4. Photovoltaic module parameters.

Open-Circuit Voltage (V)	Short-Circuit Current (A)	Max Voltage (V)	Max Current (A)	Max Power (W)
45.2	5.36	37.1	5.11	190

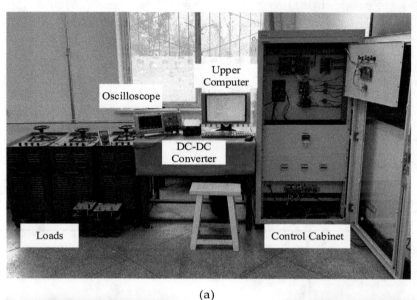

(a)

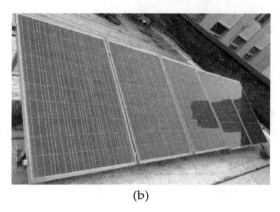

(b)

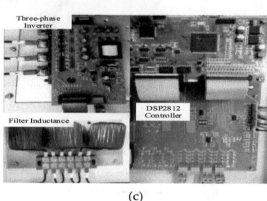

(c)

Figure 10. Experimental system. (**a**) Test rig; (**b**) Photovoltaic modules; (**c**) DC/AC converter.

4.1. Simulation Results of Different Varied Solar Irradiations

In a PV power system, the weather conditions, such as cloudy, rainy, dust, etc., will influence the power conversion. In most applications in this field, solar irradiation is considered as one of the most important factors in power generation, where weather variation will be mostly reflected in solar irradiation. We therefore test the PV power system performance as the solar irradiation varies. A series of waveforms in Figure 11 provides the auto-tracking process when the MPPT algorithm works with the DC/DC converter. Through the DC/DC converter, the output DC voltage of photovoltaic modules is kept at 140 V while the duty ratio D is 0.735 in the reference condition. In order to validate the performance of a DC/DC converter with the MPPT algorithm, varied solar irradiations are applied to the PV array. The performance is shown in Figure 11a–d.

As Figure 11a–d show, the current, the voltage and the power of the PV panel spontaneously follow the rapid variation of the solar irradiation. The proposed PV power system reaches the target of maximal utilisation of solar energy with limited oscillation, which in turn validates the excellent performance of the MPPT scheme applied in this investigation.

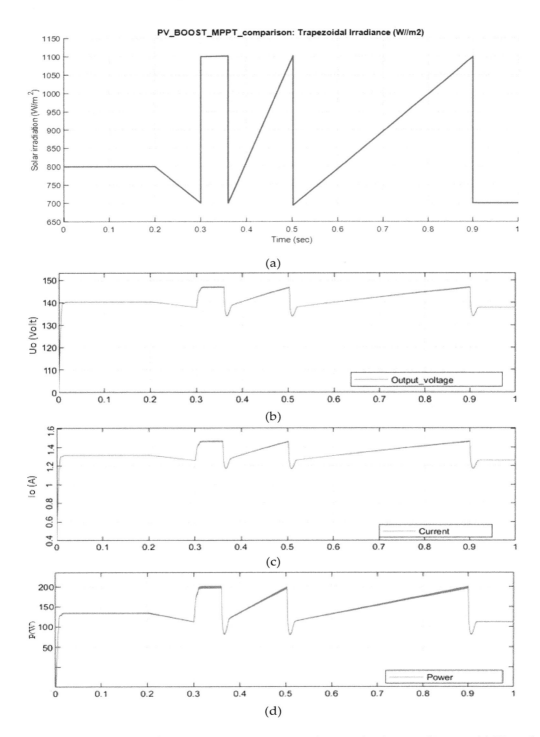

Figure 11. The performance of the DC/DC converter with varied solar irradiation. (**a**) Varied solar irradiation; (**b**) the output voltage at maximum power point (MPP); (**c**) the output current at MPP; (**d**) the output power at MPP.

4.2. DC/AC under the DPWM scheme

The DC/AC converter runs with a three-phase symmetrical resistive load ($R_L = 40\ \Omega$, $L_L = 10$ mH). The switching period is selected as $T_s = 0.0002$ s. The test waveforms are collected by the Tektronix oscilloscope TDS2024 (Test equipment Solutions Ltd, Berkshire, UK).

The control algorithm is designed to implement the proposed DPWM scheme. The flow chart of the DPWM algorithm can be obtained according to Equations (17)-(21), as shown in Figure 12.

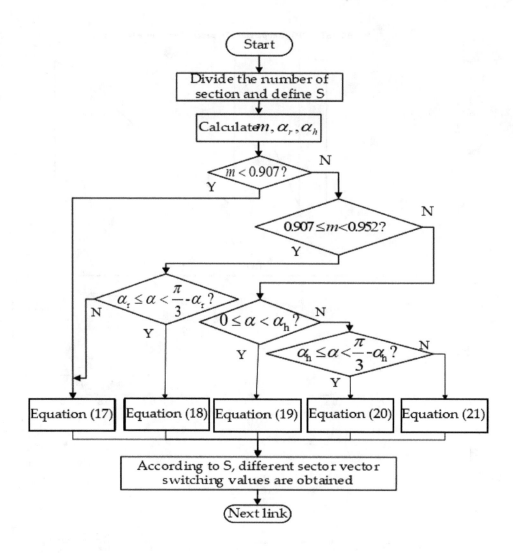

Figure 12. Flowchart of the proposed DPWM regions algorithm.

The transformation sequence of the DPWM overmodulation sector is shown in Figure 13.

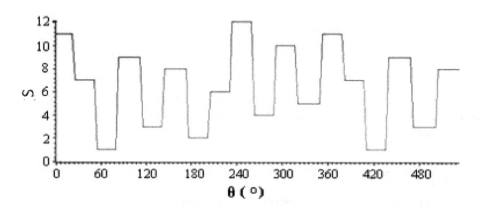

Figure 13. Sector transformation sequence.

The SVPWM modulation waveform in the linear modulation area and the DPWM modulation waveform with different modulation coefficients are shown in Figure 14.

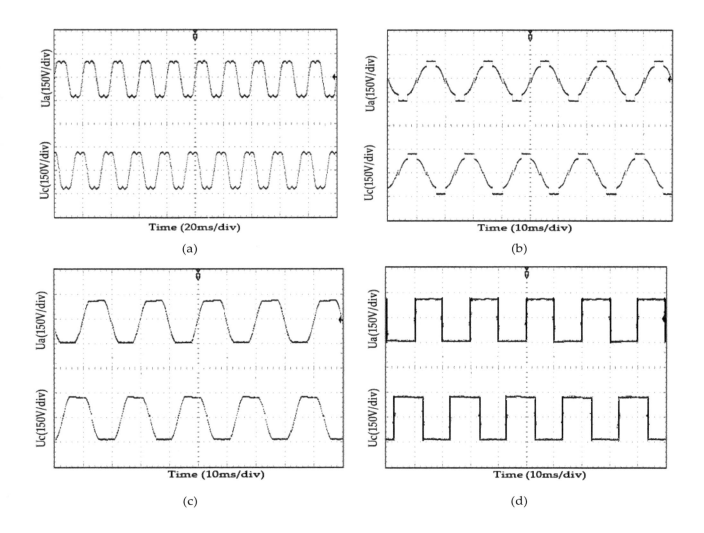

Figure 14. Modulation waveforms under different modulation schemes. (**a**) Space vector pulse width modulation (SVPWM); (**b**) DPWM (m = 0.778); (**c**) DPWM (m = 0.916); (**d**) DPWM (m = 1).

It can be seen from the figures that as the modulation coefficient increases, the peak of the DPWM modulation wave is gradually flattened and finally operates in the square wave operation state, thereby achieving linear control of the inverter output fundamental voltage over the entire modulation range.

When the modulation coefficient m = 0.92, the output pulse waveforms of TMS320F2812 under different strategies are shown in Figure 15. The waveforms of the output phase voltage and current are presented in Figure 16 and the spectral analysis results of the current are shown in Figure 17.

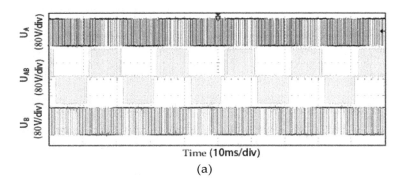

(a)

Figure 15. *Cont.*

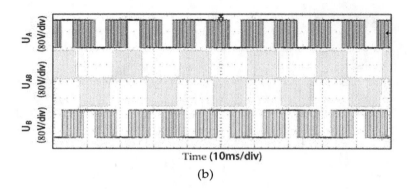

Figure 15. Pulse waveforms under different modulation schemes. (**a**) SVPWM; (**b**) DPWM.

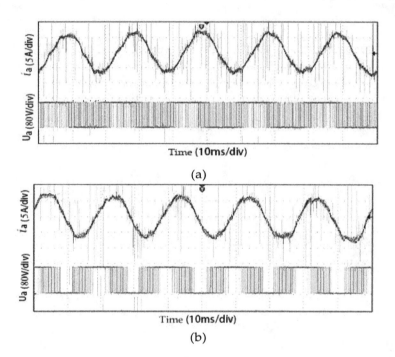

Figure 16. Output voltage and current waveforms. (**a**) SVPWM; (**b**) DPWM.

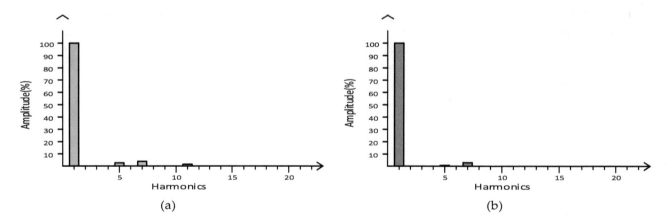

Figure 17. Spectral analysis results of the current. (**a**) SVPWM; (**b**) DPWM.

From Figures 15–17, it can be seen that the upper and lower arms are not operated in the one-third period under the proposed strategy. Compared with the switching control pulse of conventional SVPWM with the same carrier frequency, the switching time is reduced by one-third. The switching losses can be effectively reduced so as to increase the efficiency. The harmonic contents of the load current measured by the spectrum analyzer are presented in Figure 17 and Table 5.

Table 5. Harmonic contents in the load current.

Harmonics	SVPWM	DPWM
5th	3.55%	0.93%
7th	4.84%	3.68%
11th	1.61%	0
THD	6.58%	4.40%

Clearly, the amplitude of the 5th, 7th and 11th harmonics under the DPWM modulation strategy are lower than that of the SVPWM modulation strategy. The THD of the conventional SVPWM modulation strategy is 6.58%, while that of the DPWM overmodulation strategy is reduced to 4.4%. The current waveforms of the proposed DPWM algorithm are close to sinusoidal, and it appears to have a higher utilization ratio of the DC voltage than the SVPWM. In turn, this leads to reduced switching losses and improved THD.

5. Conclusions

This paper has proposed a high-efficiency PV power generation system by combining an MPPT algorithm and a new control technology evolving from DPWM and overmodulation. It can realize the modulation of a full area on the basis of traditional SVPWMs. A DC/DC converter with an inverter simulation model and an experimental test rig are developed to justify the proposed method. The main contributions of this work are:

(i)   A P&O MPPT algorithm is applied to a boost DC/DC converter so as to effectively harvest solar energy and transform to DC electricity;

(ii)   A novel control technology is proposed, combining discontinuous pulse width modulation (DPWM) and overmodulation technology to better utilize the DC-link voltage.

(iii)   It has been shown by measurements that through implementing this algorithm, the switching losses in the power electronic devices are reduced.

(iv)   The test results have confirmed that the DPWM overmodulation algorithm can effectively reduce harmonic distortion of the three-phase output voltage and current. It has also improved the conversion efficiency of photovoltaic systems.

(v)   The proposed technology is simple to implement in practical PV inverters as there are no alterations to existing hardware design. It may also be applied to other grid-tie inverters to improve their performance.

Author Contributions: In this article, L.L. and W.C. conceived and designed the system; H.W., L.C. and B.L. performed the experiments; A.A.S.B. and X.C. built the simulation models; X.C. and L.L. analyzed data and wrote the paper.

Nomenclature

α	Angle between the output voltage vector and the horizontal axis
α_γ	Angle between the intersection of the output voltage vector and the hexagon boundary, and the vertex of hexagon
α_h	Control angle to determine how long the output voltage vector stays at the vertex of hexagon
γ_{IL}	Current ripple factor of the inductor
γ_{VO}	Voltage ripple factor of the inductor

C_i,	Capacitance of the input capacitor in DC/DC converter
C_o,	Capacitance of the output capacitor in DC/DC converter
D	Duty ratio of DC/DC converter
f	Switching frequency of a DC/DC converter
i_a, i_b, i_c	Inverter output currents
I_{mpp}	Equivalent output current at maximum power point
L	Filter inductance
L_L	Symmetrical load inductance
L'	Inductance of a DC/DC converter
m	Modulation coefficient
N	Sector
p_{mpp}	Maximum power of a PV module
R	Filter inductance
R_L	Symmetrical load resistance
R_{mpp}	Equivalent resistance at maximum power point
R_O	Load resistance of the DC/DC converter
S	Sector number
T_1, T_2, T_0	Action time of adjacent fundamental voltage vectors and zero vector
T_s	Switching period
$u_0 \sim u_7$	Basic voltage space vectors
U_α, U_β	Two components of the output voltage vector in the $\alpha - \beta$ coordinates
U_d	DC-link voltage
u_m	Amplitude of the phase voltage
u_{m_max}	Maximum phase voltage in linear modulation area
U_{out}	Output voltage
V_{mpp}	Equivalent output voltage at maximum power point
V_O	Load voltage of the DC/DC converter
t_{rN}	Turn-on time
t_{fN}	Turn-off time
f_s	Switching frequency of power devices
I_{CN}	Forward current of IGBT
I_{CM}	Amplitude of the sinusoidal current

References

1. Bukhari, S.A.A.S.; Cao, W.P.; Soomro, T.A.; Guanhao, D. Future of microgrids with distributed generation and electric vehicles. In *Development and Integration of Microgrids*; Cao, W.P., Yang, J., Eds.; InTech: Rijeka, Croatia, 2017.

2. Aly, M.; Ahmed, E.M.; Shoyama, M. Modulation Method for Improving Reliability of Multilevel T-type Inverter in PV Systems. *IEEE J. Emerg. Sel. Top. Power Electron.* **2019**. [CrossRef]

3. Gosumbonggot, J.; Fujita, G. Global Maximum Power Point Tracking under Shading Condition and Hotspot Detection Algorithms for Photovoltaic Systems. *Energies* **2019**, *12*, 225. [CrossRef]

4. Baimel, D.; Tapuchi, S.; Levron, Y.; Belikov, J. Improved Fractional Open Circuit Voltage MPPT Methods for PV Systems. *Electronics* **2019**, *8*, 321. [CrossRef]

5. Afzal Awan, M.M.; Mahmood, T. A Novel Ten Check Maximum Power Point Tracking Algorithm for a Standalone Solar Photovoltaic System. *Electronics* **2018**, *7*, 327. [CrossRef]

6. Boobalan, S.; Dhanasekaran, R. Hybrid topology of asymmetric cascaded multilevel inverter with renewable energy sources. In Proceedings of the 2014 International Conference on Advanced Communication Control and Computing Technologies (ICACCCT), Ramanathapuram, India, 8–10 May 2014; pp. 1046–1051.

7. Ali, W.H.; Cofie, P.; Fuller, J.H.; Lokesh, S.; Kolawole, E.S. Performance and efficiency simulation study of a smart-grid connected photovoltaic system. *Energy Power Eng.* **2017**, *9*, 71. [CrossRef]

8. Djokic, S.; Langella, R.; Meyer, J.; Stiegler, R.; Testa, A.; Xu, X. On evaluation of power electronic devices efficiency for nonsinusoidal voltage supply and different operating power. *IEEE Trans. Instrum. Meas.* **2017**, *66*, 2216–2224. [CrossRef]

9.   Zhang, X.; Li, J.; Zhao, W.; Tao, L. Review of high efficiency photovoltaic inverter. *Chin. J. Power Sources* **2016**, *4*, 931–934.

10.   Wilkinson, S.; Gilligan, C. *PV Inverter Service Plans & Extended Warranties Report—2013*; HIS Technology: London, UK, 2013.

11.   Suresh, K.; Prasad, M.V. Performance and evaluation of new multilevel inverter topology. *Int. J. Adv. Eng. Technol.* **2012**, *2*, 485–494.

12.   Zhu, X. Three level and efficiency of photovoltaic Inverter. *Electr. Age* **2016**, *8*, 46–47.

13.   Griva, G.; Oleschuk, V. Synchronous operation of dual-inverter-based photovoltaic system with low DC-voltages. In Proceedings of the 2010 International Conference on Electrical Machines (ICEM), Rome, Italy, 6–8 September 2010; pp. 1–6.

14.   Park, Y.; Sul, S.K.; Hong, K.N. Overmodulation strategy for current control in photovoltaic inverter. *IEEE Trans. Ind. Appl.* **2016**, *1*, 322–331. [CrossRef]

15.   Chen, W.; Sun, H.; Gu, X.; Xia, C. Synchronized space-vector PWM for three-level VSI with lower harmonic distortion and switching frequency. *IEEE Trans. Power Electron.* **2016**, *31*, 6428–6441. [CrossRef]

16.   Beig, A.; Kanukollu, S.; Hosani, K.; Dekka, A. Space-vector-based synchronized three-level discontinuous PWM for medium-voltage high-power VSI. *IEEE Trans. Ind. Electron.* **2014**, *61*, 3891–3901. [CrossRef]

17.   Mathew, J.; Mathew, K.; Azeez, N.B.; Rajeevan, P.P.; Gopakumar, K. A hybrid multilevel inverter system based on dodecagonal space vectors for medium voltage IM drives. *IEEE Trans. Power Electron.* **2013**, *28*, 3723–3732. [CrossRef]

18.   Carrasco, G.; Silva, C.A. Space vector PWM method for five-phase two-level VSI with minimum harmonic injection in the overmodulation region. *IEEE Trans. Ind. Electron.* **2013**, *5*, 2042–2053. [CrossRef]

19.   Holmes, D.; Lipo, T. *Pulse Width Modulation for Power Converters: Principles and Practice*; Wiley IEEE Press: Piscataway, NJ, USA, 2003.

20.   Elmelegi, A.; Aly, M.; Ahmed, E.M. Developing Phase-Shift PWM-Based Distributed MPPT Technique for Photovoltaic Systems. In Proceedings of the 2019 International Conference on Innovative Trends in Computer Engineering (ITCE), Aswan, Egypt, 2–4 February 2019; pp. 492–497.

21.   Faranda, R.; Sonia, L. Energy comparison of MPPT techniques for PV Systems. *WSEAS Trans. Power Syst.* **2008**, *3*, 446–455.

22.   Espi, J.M.; Castello, J. A Novel Fast MPPT Strategy for High Efficiency PV Battery Chargers. *Energies* **2019**, *12*, 1152. [CrossRef]

23.   Tan, B.; Ke, X.; Tang, D.; Yin, S. Improved Perturb and Observation Method Based on Support Vector Regression. *Energies* **2019**, *12*, 1151. [CrossRef]

24.   Narendiran, S. Grid tie inverter and mppt—A review. In Proceedings of the IEEE 2013 International Conference on Power and Computing Technologies (ICCPCT), Nagercoil, India, 20–21 March 2013.

25.   Mohan, N.; Undeland, T.M.; Robbins, W.P. *Power Electronics: Converters, Applications, and Design*; John, Wiley & Sons: Hoboken, NJ, USA, 2003.

26.   Razman, A.; Tan, C.W. Design of boost converter based on maximum power point resistance for photovoltaic applications. *Sol. Energy* **2018**, *160*, 322–335.

27.   Femia, N.; Petrone, G.; Spagnuolo, G.; Vitelli, M. Optimization of Perturb and Observe Maximum Power Point Tracking Method. *IEEE Trans. Power Electron.* **2005**, *20*, 963–973. [CrossRef]

28.   Wu, X.; Liu, W.; Ruan, Y.; Zhang, L. SVPWM over-modulation algorithm and its application in two-level Inverter. *Electr. Mach. Control* **2015**, *1*, 76–81.

29.   Wu, D.; Xia, X.; Zhang, Z.; Li, C. A SVPWM overmodulation method based on three-phase bridge arm coordinates. *Trans. China Electrotech. Soc.* **2015**, *1*, 150–158.

30.   Zhou, X.W.; Liu, W.G.; Lang, B.H. Study on generalized discontinuous PWM algorithm and harmonic. *Small Spec. Electr. Mach.* **2007**, *11*, 5–7.

31.   Jia, L.; Li, H.M.; Liang, J. Research on switching losses of an inverter under discontinuous PWM strategies. *Mech. Electr. Eng. Mag.* **2006**, *1*, 51–53.

32.   Liang, W.H.; Du, X.T.; You, L.R. Realization of expend linear modulation area and overmodulation algorithm in SVPWM. *Power Electron.* **2013**, *5*, 10–12.

33.   Fan, S.L.; Zhao, J.A. Novel overmodulation method based on optimized SPWM. *Power Electron.* **2013**, *6*, 26–28.

34.   Mao, P.; Xie, S.J.; Xu, Z.G. Switching transients model and loss analysis of IGBT module. *Proc. Chin. Soc. Electr. Eng.* **2010**, *15*, 40–47.

Methods of Modulation for Current-Source Single-Phase Isolated Matrix Converter in a Grid-Connected Battery Application

Goh Teck Chiang * and Takahide Sugiyama

Toyota Central R&D Labs Inc., Nagakute City 480-1192, Japan; t-sugiyama@mosk.tytlabs.co.jp
* Correspondence: tcgoh@mosk.tytlabs.co.jp

Abstract: This paper discusses three methods of modulation for a single-phase isolated matrix converter. The matrix converter is combined with a transformer integration to perform power decoupling control in order to reduce the number of component and capacitor volumes. Due to the reason of (i) Alternating current (AC/AC) direct conversion and (ii) transformer integration, obtaining a clean sinusoidal grid current waveform in the modulation of matrix converter (MC) is important. Three methods of modulation are compared in terms of control complexity, quality waveform, and inductive-capacitive-inductive (LCL) filter sizing. The principal control of each method is described. Finally, a prototype was tested to verify the validity and the effectiveness of grid current control and power decoupling in the spoken circuit structure.

Keywords: AC/AC conversion; decoupling control; modulation

1. Introduction

The rapidly expanding growth of battery storage system (BSS) has urged high demands for a single-phase power converter. Figure 1a shows the applications such as home energy management system (HEMS), uninterruptible power supply (UPS), and small-scale datacenter, which uses a single-phase power converter as an interface between BSS and grid (AC 80–240 V 50/60 Hz). These applications require isolation and typically rate from 1–3 kW with a high voltage battery (100–300 V). As the price of battery is expected to reach a new low in the near future, a low cost and small size single-phase power converter is highly demanded.

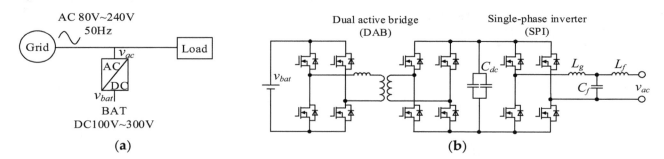

Figure 1. (a) Single-phase power converter for a grid-connected battery application. (b) Conventional circuit structure consists of a dual-active bridge (DAB) and a single-phase inverter (SPI).

Figure 1b shows the conventional circuit for a single-phase power converter. The circuit is composed of a dual-active bridge (DAB), a single-phase inverter (SPI), and an LCL filter [1,2]. The size reduction of a single-phase power converter is challenging because of using many passive components.

Several studies have focused on reducing the size of inductive component such as inductor and transformer by using a high frequency technique [3–5]. However, the major size of the converter is occupied by the capacitors C_{dc} that are used to absorb the single-phase power fluctuation. Due to the reason of current limitation in the electrolytic capacitor, capacitors are connected in parallel to form a big capacitor bank in order to absorb the single-phase power fluctuation.

The matrix converter (MC) shows a promising solution for size reduction because the capacitor can be removed [6,7]. Hence, MC can convert high frequency transformer voltage (i.e., 50 kHz) to low frequency voltage (i.e., 50 Hz), at the same time controlling the current flow bidirectional. However, for a single-phase application, a low frequency current that contains twice of the grid frequency occurrs in the battery side due to the direct AC/AC conversion. Depending on the type of battery, such as lithium battery, the single-phase fluctuation in the battery needs to be eliminated in order to protect the battery from overvoltage.

Single-phase active power decoupling techniques have been discussed and reported for compensating the single-phase power fluctuation [8,9]. A power decoupling circuit can be considered to add between the battery and full-bridge inverter (FBI) in order to compensate the single-phase low frequency current. Figure 2 shows a conventional circuit structure that consists of a power decoupling circuit, a FBI, and a MC. The power decoupling circuit consists of an inductor L_b, a capacitor C_b, a diode, and two switching devices. The single-phase power fluctuation is compensated by charging and discharging the capacitor C_b according to the grid phase angle. As a result, the single-phase power fluctuation can be eliminated with a smaller capacitor than the conventional capacitor bank.

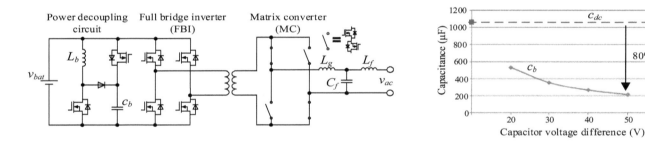

Figure 2. Circuit structure consists of a power decoupling circuit, a full-bridge inverter (FBI), and a matrix converter (MC).

The required capacitance in a single-phase converter can be defined by Equation (1).

$$c_b \geq \frac{p_{cap}}{2 \times \pi \times 2 \times f_g \times \Delta v_{cb} \times v_{cb}}; \Delta v_{cb} = v_{cbmax} - v_{cbmin} \tag{1}$$

where v_{cb} is the average battery voltage, Δv_{cb} is the capacitor voltage difference, and f_g is grid frequency. Figure 2 shows that a 1 kW calculation (capacitor power (p_{cap}) is half of the rated power), by using the power decoupling control to increase the capacitor voltage difference Δv_{cb} to 50 V, the required capacitance can be reduced by 80% comparing that to the conventional circuit at the same rated power.

However, the major drawback is that this circuit requires extra switching devices and passive component. Here, an integration technique that utilized the center-tapped of a transformer has been discussed, as shown in Figure 3 [10,11]. The center-tapped of the transformer is utilized by connecting the passive components (L_b and C_b) in order to perform the power decoupling control. Then, the FBI controls the high frequency transformer voltage and the capacitor voltage at the same time, therefore the switching devices in the power decoupling circuit can be reduced.

Without the transformer integration, the power decoupling circuit can be individually controlled and the method of modulation for the matrix converter is rather simple. Literature reviews [12–14] have demonstrated several valid modulations for the MC, where a good quality waveform can be obtained without the need for concern for the power decoupling. However, when the transformer integration is applied with MC, the modulation for MC has to be changed in order to synchronize with the FBI to obtain a proper voltage period. The failure of obtaining a clean sinusoidal waveform can distort the battery current inherently, because this capacitor cannot absorb the current fluctuation.

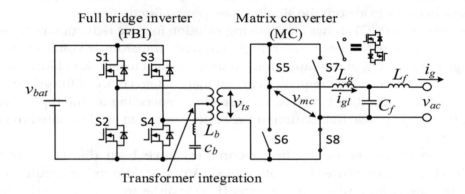

Figure 3. Circuit structure consists of a full-bridge inverter (FBI) with a transformer integration decoupling control and a matrix converter (MC).

This paper discusses three methods of modulation for the MC that is applied with the transformer integration. The first conventional method is a carrier comparison with a D-FF and the second conventional method is a delta-sigma conversion based on pulse density modulation (PDM) which have been addressed in [15–17]. This paper introduces a third method which is a carrier comparison with a zero-vector commutation, and further discusses the difference of each method. The details of each method is described individually. Then, the comparisons among these methods in term of (i) control complexity, (ii) waveform quality, and (iii) LCL sizing are discussed. Then, the validity of the modulation along with comparison results is shown. Finally, a 1 kW prototype was tested to show the validity of the power decoupling with a MC.

2. Control Scheme

2.1. System Control

The system control block diagram is shown in Figure 4. The control is divided into two parts: (i) Voltage and current closed-loop controls in FBI and (ii) grid voltage and current with a phase locked loop (PLL) in MC. The FBI is performed as a voltage source to control the high frequency transformer voltage v_{ts} and capacitor voltage v_{cb} at the same time. Then, the high frequency transformer voltage is fed into the MC, and therefore MC is performed as a current source converter to control the grid current.

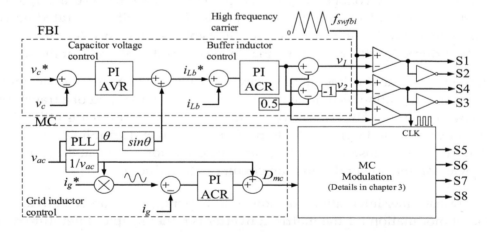

Figure 4. System control block diagram, full-bridge inverter (FBI) controls the voltage level and matrix converter (MC) controls the current for the single-phase power converter.

In FBI, a low time response of automatic voltage regulator (AVR) is applied to control the capacitor voltage proportionally to half of the battery voltage. Then, the grid phase angle which is calculated from the PLL in the grid control is added to control the phase angle of the inductor current. Then, a high time response current control (automatic current regulator ACR) is applied into the inductor current control. Note that the input of phase angle in the FBI control is also used to enable or disable the power decoupling control (where 0 is disabled).

Since the MC is a current source controller, a normalized grid voltage command is feed-forwarded into the controller. Then, a high time response current control is applied into the grid inductor current, where a sinusoidal duty command (D_{mc}) is used to generate the corresponding gate signals.

Figure 5 shows the principal control of the power decoupling. The relationship among the grid power p_g, battery power p_{bat}, and capacitor power p_{cb} is defined in Equation (2), where p_{avg} is the average power and μ_o is the grid phase angle. When subjected to the frequency of the grid power, the battery current contains twice the grid frequency.

$$p_g = p_{bat} - p_{cb}; \; p_{cb} = p_{avg} \cos(2\omega_o t) \qquad (2)$$

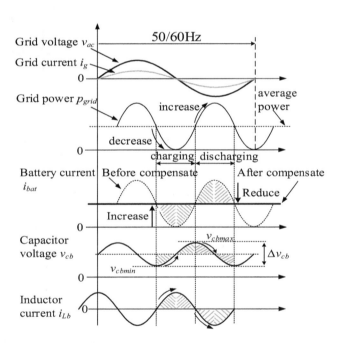

Figure 5. Principle control of power decoupling, charging and discharging states in the capacitor are used to compensate the single-phase current that occurs in the battery current.

In the power decoupling, the capacitor power is divided into a charging and discharging state. When the grid power is lower than the average power, this period is known as a charging state. During the charging state, the battery power is loaded into the capacitor by controlling the center-tapped inductor current. Then, the capacitor voltage difference Δv_{cb} increases from v_{cbmin} to v_{cbmax} during this period.

When the grid power is higher than the average power, this period is known as a discharging state. During the discharging state, the previously charged power in the capacitor is discharged by the center-tapped inductor current. Then, the capacitor voltage difference Δv_{cb} decreases from v_{cbmax} to v_{cbmin} during this period. Since no power delivery is needed from the battery, the battery current remains at its average value. By repeating these two cycles according to the grid phase angle, the capacitor voltage difference is controlled to compensate the single-phase current in the battery.

2.2. Switching Behaviors in FBI

The switching behavior and current relationships in FBI are described. Figure 6 shows the relationships between the two current components (DC and AC) in the FBI. The DC current component occurs when the DC/DC conversion is performed between the battery voltage and capacitor voltage. In this case, the FBI is equivalent to a buck converter with a 180 degree phase shift. Two DC currents, i_{Lb1} and i_{Lb2}, are manipulated with the duty to control the center-tap connected inductor current and capacitor voltage, which can be defined in Equations (3) and (4).

$$v_{cb} = D_{fbi} \times v_{bat};$$
(3)

$$i_{Lb} = i_{Lb1} + i_{Lb2};$$
(4)

where v_{cb} is the capacitor voltage, v_{bat} is the battery voltage, D is the duty of FBI, and i_{Lb} is the inductor current.

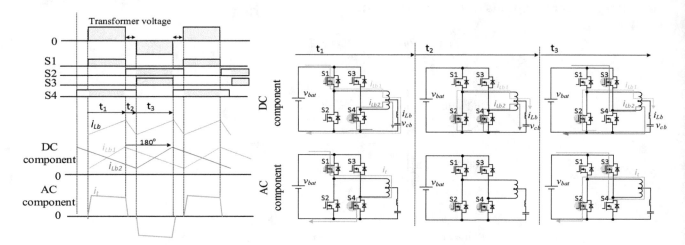

Figure 6. Relationship between the two current components in FBI.

On the other hand, the AC current component occurs when the battery power delivers the grid via the transformer, which is the transformer current. The AC current component is controlled with corresponding to the modulation of MC, which is equivalent to the grid inductor current. Therefore, the relationship between the transformer current and grid inductor current can be expressed as Equation (5).

$$i_{tp} \times N = i_{ts} = i_{gl};$$
(5)

where i_{tp} is the primary side current (FBI), i_{ts} is the secondary side current (MC), N is the transformer ratio, and i_{gl} is the grid inductor current.

According to the state of the capacitor (charging or discharging) and the amplitude of the battery current, the total of four switching behaviors can be summarized as shown in Figure 7. The zero-voltage periods of FBI (S1S3 or S2S4 are turned on) are utilized to discharge and charge the inductor current.

During the discharging state, the battery current needs to be reduced and therefore the charged energy in the capacitor C_b discharges to the battery side. When S2S4 are turned on, the current circulates via S2 and S4 to keep the charged energy. When S1S3 are turned on, the charged energy in the inductor is released to the battery via S1 and S3. The current cancellation between the battery current and inductor current reduces the high peak of the battery current.

During the charging state, the battery current needs to be increased and therefore the low peak of the battery current charges into the capacitor C_b. Here, when S1S3 are turned on, the battery current flows via S1 and S3 to charge inductor Lb. Then, when S2S4 are turned on, the charged energy in the inductor circulates via switching devices.

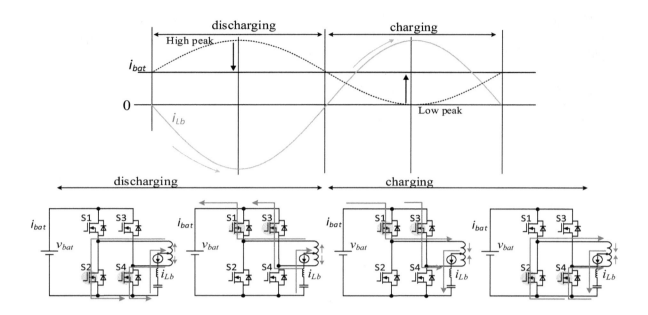

Figure 7. Switching behavior of full-bridge inverter (FBI) according to the state of capacitor and amplitude of the battery current.

2.3. Modulation in MC

In MC, a low frequency voltage pulse width is formed to control the grid inductor by accumulating from the high frequency transformer voltage v_{ts}. The high frequency transformer is controlled by FBI which is magnetized from the battery voltage. As shown in Figure 8, the switching sequence is divided into positive and negative voltage periods according to the polarity of the grid voltage. Then, each of these voltage periods is implemented with zero-vector periods in order to discharge the grid inductor current. As a result, the method of modulation is used to control the length of these voltage periods in order to control the 50 Hz grid inductor current sinusoidal.

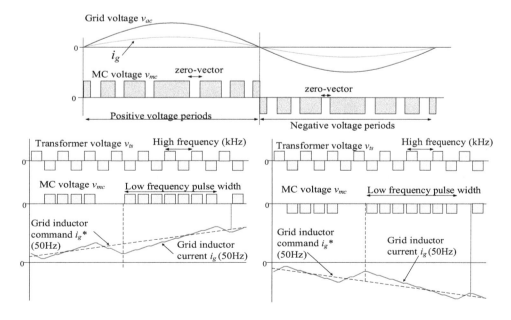

Figure 8. Modulation in matrix converter (MC), zero-vector periods are implemented in both positive and negative voltage periods.

Figure 9a shows the switching behavior in MC, which can differ to normal switching states and zero-vector periods. When the gate signals S5S8 are turned on, the grid inductor is induced by the positive transformer voltage to charge the grid inductor. Otherwise, when the gate signals S6S7 are turned on, the grid inductor is then induced by the negative transformer voltage. On the other hand, when the gate signals S5S7 (or S6S8) are turned on, known as the zero-vector periods, a circulating loop is created inside the switching devices to allow the grid inductor current to circulate and discharge the energy in the grid inductor.

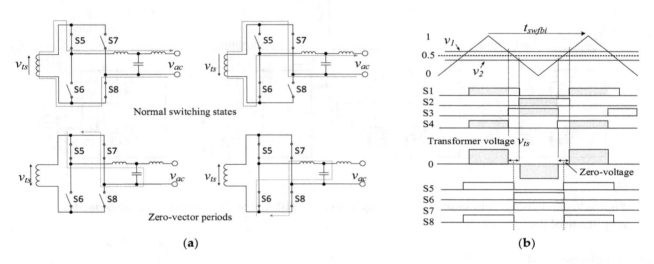

Figure 9. Switching behaviors in matrix converter (MC). **(a)** Normal switching states and zero-vector periods. **(b)** Zero-voltage switching (ZVS) relationship between full-bridge inverter (FBI) and matrix converter (MC).

Furthermore, Figure 9b illustrates the zero-voltage switching (ZVS) relationships between FBI and MC. Since the transformer voltage is magnetized from the battery voltage, a three-level high frequency transformer voltage can be produced. That is, when gate signals S1 S3 or S2 S4 in FBI are turned on, no voltage-product occurrs in the MC. These zero-voltage periods are utilized in the switching intervals of MC to reduce the switching loss. During the gate-off transition, the drain-source voltage of switching devices drops to zero before the gate signal is turned off. Then, during the gate-on transition, the gate signal is turned on before the voltage is applied to the switching devices. Therefore, both of the transitions can achieve ZVS.

However, the leakage inductance of the transformer needs to be taken into consideration during the switching intervals. The energy in the leakage inductance needs to be discharged while the transformer current changes the direction. Here, the approach is to use the grid inductor current to cancel out with the leakage inductance current during the switching intervals. Figure 10 explains and illustrates the phenomenon, where the positive transformer voltage is changed to the negative transformer voltage while the grid side produces a positive voltage.

As shown in Figure 10, the transformer voltage becomes zero before the switching intervals start. Then, following that the S6AB and S7AB are turned on in the next switching interval. During this state, the leakage inductance current is used to discharge the capacitance S5A and also charge the capacitance S6B. At the same time, the grid inductor current is flowing via S6AB in a reverse direction, as a result the leakage inductance current and grid inductor current cancel out each other. The same phenomenon applies to S7AB and S8AB, capacitance S7B is charged and capacitance S8A is discharged by the leakage inductance current. Then, the grid inductor current is flowing via S7BA in an opposite direction to achieve the current canceling.

As a result, forming an accurate voltage period, achieving ZVS, and current cancelling at the same time is important in the method of modulation.

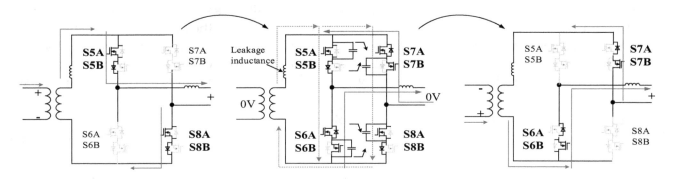

Figure 10. Current cancelling in matrix converter (MC) to discharge the leakage inductance current during switching intervals.

3. Methods of Modulation

3.1. Carrier Comparison with D-FF (D-FlipFlop)

The first method is to use a D-flipflop (D-FF) function, the control block diagram is shown in Figure 11. A carrier comparison with D_{mc} is used to generate two sets of switching signals SPQ and SNQ. When SPQ and SNQ are both turned on zero-vector periods are formed. These two switching signals are inputted to a D-FF, where the D-FF is synchronized with the CLK, and a XNOR logic is applied to produce gate signals for S5–S8. The CLK is used to synchronize the switching intervals of MC with the zero-voltage periods of FBI in order to achieve ZVS.

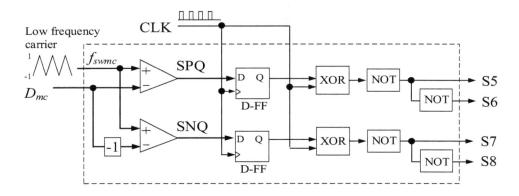

Figure 11. Carrier comparison with the D-flip flop block diagram.

However, D-FF creates a voltage error due to the occurrence of improper time length. Figure 12 shows the relationships among switching signals SPQ SNQ, gate signals S5–S8, and voltage pulse width. First, SPQ and SNQ form the required voltage pulse width accordingly based on the carrier comparison. After the SPQ and SNQ are aligned with D-FF, the voltage pulse width applied to the grid inductor either becomes longer or shorter than the original voltage pulse width. These improper pulse widths create voltage errors and the average grid inductor current is misadjusted. As a result the grid current fluctuates irregularly.

3.2. Delta-Sigma Conversion with Pulse Density Modulation (PDM)

In order to eliminate the voltage error, a delta-sigma conversion which is based on pulse density was discussed. Figure 12 shows the control block diagram and Figure 13 shows the relationship between duty D_{mc} and quantization error Q_r. The carrier comparison is not applied because the integral changes corresponding to the quantization error. One cycle of the quantization level is equivalent to one cycle of the CLK. The Q_r is obtained based on the differential value between the D_{mc} and D_{mc}. Note that the amplitude of D_{mc} does not change according to the grid current command ($i_g{}^*$) but the level of quantization error changes depending on the pulse density. As shown in Figure 14, the original middle point is $D_{mc} = 0.5$. Then, the level of Q_r changes depending on the pulse density that is used to form the grid current command, which is $Q_r > D_{mc}$ or $Q_r < D_{mc}$. The comparison between the D_{mc} and Q_r produces the corresponding voltage signals Sa and Sb in order to produce the desired voltage pulse width. EXOR logic is applied to Sa and Sb to synchronize with CLK in order to produce gate signals.

That is, when i_g needs to increase, a longer voltage pulse width is required and therefore Q_r gets higher than D_{mc}. On the other hand, when i_g needs to decrease, a shorter voltage pulse is required and Q_r gets lower than D_{mc}.

These phenomenon are illustrated in Figure 15, where (a) $Q_r < 0.5$ and (b) $Q_r > 0.5$. In Figure 15a, in order to decrease the grid current, most of the Q_r periods are lower than D_{mc}, then Sa produces a short voltage signal only when Q_r is higher than D_{mc}. On the other hand, in Figure 15b, in order to increase the grid current, a longer voltage pulse is required. Notice that the level of Q_r increases, and most of the Q_r periods are higher than D_{mc} to produce the desired voltage pulse width. As a result, the grid current can be controlled sinusoidal without the voltage error, and ZVS can be achieved by synchronizing to CLK.

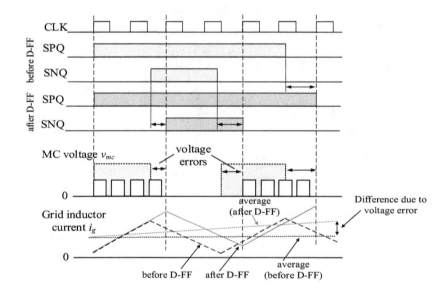

Figure 12. Relationship among CLK, SPQ, SNQ, MC voltage, and the grid inductor current are shown to demonstrate the voltage error in D-flip flop.

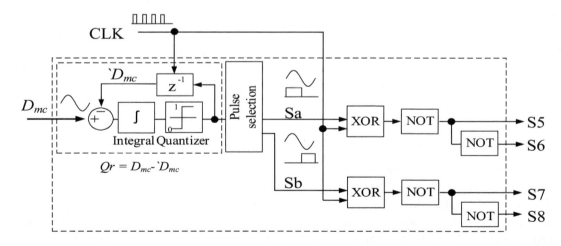

Figure 13. Delta-sigma conversion block diagram.

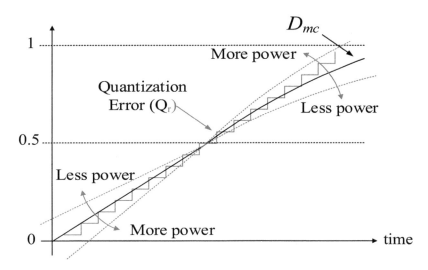

Figure 14. Principle control of delta-sigma, where quantization error changes according to the power level in order to obtain the desired pulse width ($Q_r > D_{mc}$ or $Q_r < D_{mc}$).

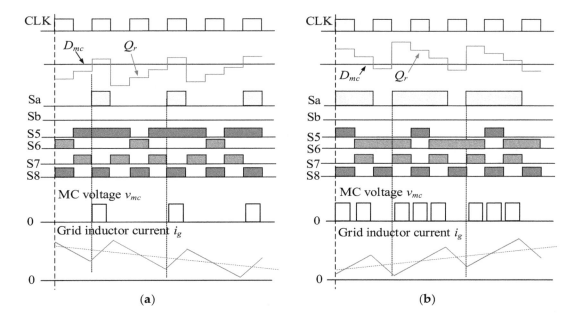

Figure 15. Relationship among CLK, D_{mc}, Q_r, MC voltage, and the grid inductor to demonstrate the control of delta-sigma. (**a**) $Q_r < 0.5$. (**b**) $Q_r > 0.5$.

However, without the carrier comparison the integral resets the value depending on the quantization level at a random frequency, as shown in Figure 16. As a result, the grid current ripple has an inconsistent frequency which causes a resonance problem during the low output power [18]. Furthermore, the resonance also occurs in the battery current due to the AC/AC direct conversion. Note that this resonance cannot be compensated in the single-phase power decoupling, therefore one approach is to decrease the cut-off frequency of the LCL; however, the size of LCL needs to increase as a drawback.

3.3. Carrier Comparison with Zero-Vector Commutation

The method of carrier comparison with zero-vector commutation is shown in Figure 17. This control is implemented with a constant frequency and a commutation to eliminate voltage error. A carrier comparison which is based on the pulse width modulation (PWM) is used and compared with D_{mc} to generate a constant frequency voltage pulse width, similar to D-FF. Then, a zero-vector determination (FS-SYN) is used to distinguish between the normal switching states and zero-vector periods. During the normal switching states, the CLK synchronizes the switching timing so that each of the switching intervals of MC can achieve ZVS.

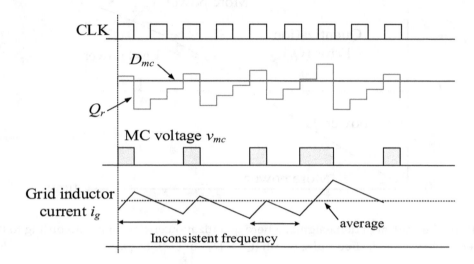

Figure 16. Inconsistent frequency in the grid inductor current due to the quantization error.

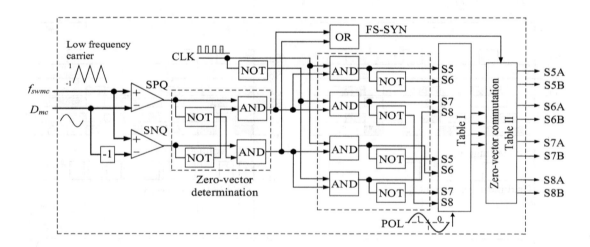

Figure 17. Carrier comparison with zero-vector commutation is introduced to overcome the voltage error and the inconsistent frequency problem.

During the zero-vector periods, since the transformer voltage is applied on the switching devices, hard-switching will cause the voltage at the switching device. In order to prevent the short-circuit state, the transformer current first needs to be blocked before switching. Furthermore, a current circulating path must first be created in order to achieve current cancelling.

The zero-vector commutation is applied only to the first and last switching intervals of the zero-vector periods. A total of six categories are divided in the zero-vector commutation which depends on the polarity of the transformer voltage, as shown in Figure 18. That is, if the zero-vector period occurs from a positive transformer voltage and ends on a positive voltage or ends on a negative voltage, it is known as PV-to-Z, PVZ-to-PV or PVZ-to-NV, respectively. On the other hand, if the zero-vector period occurs from a negative transformer voltage and ends on a positive voltage or ends on a negative voltage, it is known as NV-to-Z, NVZ-to-PV or NVZ-to-NV, respectively. Then, a two-step commutation is performed to circulate and cancel out the leakage inductance current. The switching algorithms of normal switching states are summarized in Table 1, and the switching algorithms of zero-vector commutation are summarized in Table 2.

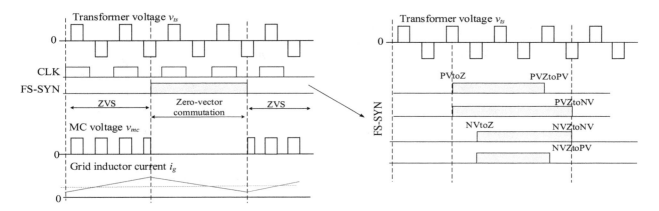

Figure 18. Categories of zero-vector commutation depending on the polarity of transformer voltage.

Table 1. Switching algorithms of normal switching states (ZVS).

Symbols	Polarity = 1			Polarity = 0		
SPQ	1	1	1	0	0	0
SNQ	0	0	1	1	1	1
Vts	+	−	+/−	+	−	+/−
S5	1	0	0	0	1	1
S6	0	1	1	1	0	0
S7	0	1	0	1	0	1
S8	1	0	1	0	1	0
Vmc	+	+	0	−	−	0

Table 2. Switching algorithms of zero-vector commutation.

Actions	PV-to-Z		PVZ-to-PV		PVZ-to-NV		NV-to-Z		NVZ-to-PV		NVZ-to-NV	
Sequence	1st	2nd	1st	2nd	1st	2nd	1st	2nd	1st	2nd	1st	2nd
S5A	1	1	1	1	1	0	0	0	1	1	0	0
S5B	1	1	1	1	0	0	0	0	0	1	0	0
S6A	0	0	0	0	1	1	1	1	0	0	1	1
S6B	0	0	0	0	0	1	1	1	1	0	1	1
S7A	0	1	0	0	1	1	0	0	0	0	1	1
S7B	1	1	1	0	1	1	1	0	0	0	0	1
S8A	1	0	0	1	0	0	1	1	1	1	1	0
S8B	0	0	1	1	0	0	0	1	1	1	0	0

Figure 19 shows the switching sequence of the zero-vector commutation for the case PV-to-Z. First, S7B is turned on and S8B is turned off, the transformer current continues to flow in the same direction via the S8A and S8B diode. Then, as S8A is turned off, the capacitance in S8A is charged by the leakage inductance current in order to build up the blocking voltage. Thus, the capacitance in S7A is discharged to reduce the blocking voltage. At the same time, the grid inductor current flows in the opposite direction via S7AB and therefore both currents are cancelled out with each other in S7AB. After S8 is completely turned off the grid inductor current starts circulating via S5AB and S7AB, the zero-vector period is created in the grid side.

Figure 20 shows the switching sequence of the zero-vector commutation for the case PVZ-to-NV. First, S6A is turned on and S5B is turned off, the circulating current continues to flow in the same direction via S5AB and S7AB. Then, as S6B is turned on, the capacitance in S5B is charged by the transformer current to build up the blocking voltage and capacitance in S6B is discharged. At the same time, the grid inductor current is cancelled out with the transformer current in S5AB. Since S5B is completely turned on, S5A is turned off at no loss and the transformer current starts to flow to the grid via S6AB and S7AB.

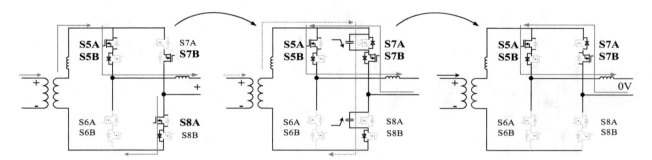

Figure 19. Switching sequence for PV-to-Z, current cancelling in S7AB.

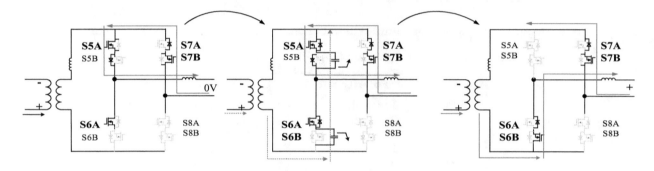

Figure 20. Switching sequence for PVZ-to-NZ, current cancelling in S5AB.

On one hand, for the case of PVZ-to-PV, current cancelling cannot be performed due to the polarity of transformer voltage. The differential current is creating a voltage surge but a short-circuit state is not created and therefore a breakdown of the device does not happen. Figure 21 explains the phenomenon of the switching sequence. First, S7A is turned off and S8B is turned on. Next, the capacitance in S8A needs to discharge in order to allow the current flows. Therefore, as S7B is turned off, the transformer current is charging the capacitance in S7A at the same time discharging the capacitance in S8A. Note that the only circulating path for the grid inductor current is via S5AB and S8AB and therefore the grid inductor current flows with the transformer current via S8AB in the same direction. As a result, the voltage surge occurs while turning on S8AB.

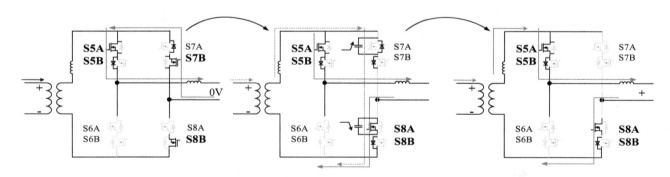

Figure 21. Switching sequence for PVZ-to-PV, current cancelling is not achieved.

Figures 22–24 illustrate the switching sequence for the case of negative transformer voltage. The principle control of current cancelling is similar, Figure 22 shows the switching sequence for NV-to-Z. While the transformer current is flowing via S6AB and S7AB, S8A is first turned on. Then, as S7B is turned off, the capacitance in S7B is charged and the capacitance S8B is discharged by the transformer current. At the same time, the grid inductor current flows via S8AB and current cancelling can be achieved in S8AB during this state.

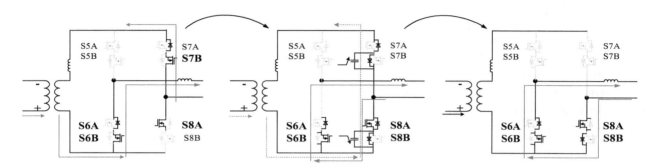

Figure 22. Switching sequence of NV-to-Z, current cancelling in S8AB.

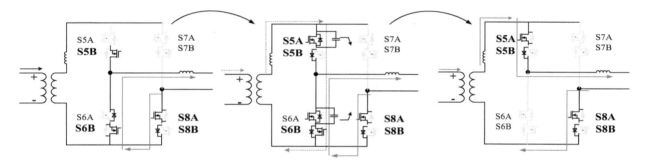

Figure 23. Switching sequence of NVZ-to-PV, current cancelling in S6AB.

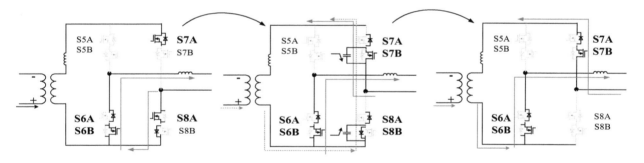

Figure 24. Switching sequence of NVZ-to-NZ, current cancelling is not achieved.

Figure 23 shows the switching sequence for NVZ-to-PV. While the grid inductor current is circulating via S6AB and S8AB, S5B is turned on and S6A is turned off. Then, as S5A is turned on, the capacitance in S5A is discharged and the capacitance in S6A is discharged by the transformer current. The grid inductor current flows in an opposite direction in S6AB to achieve current cancelling. Since the S6A is completely turned on, the transformer current starts to flow via S5AB and S8AB. On the other hand, similar to PVZ-to-PV, current cancelling cannot be achieved in NVZ-to-NV. As shown in Figure 24, as S8B is turned on, the capacitance in S8B is charged and the capacitance in S7B is discharged by the transformer current. Due to this reason, the grid inductor current flows in the same direction with the transformer current in S7AB, and the voltage surge occurs during this interval.

Note that this zero-vector commutation differs from the traditional commutation in MC [19,20]. The traditional commutation is applied to form the voltage pulse width, however the zero-vector commutation is applied during the zero-vector periods only (no voltage-product). Therefore, the voltage error that occurred in the traditional commutation is not a concern. The purpose of the zero-vector commutation is to cancel the current while charging and discharging the capacitance in the switching devices, which is simpler than the traditional commutation.

4. Simulation Results

The comparisons among the methods of modulation are demonstrated in the simulation results. The simulation parameters of each method are summarized in Table 3, which is similar to the experimental parameters. Moreover, the proportional-integral (PI) gain control for each method has been tuned to provide the best result. Figure 25 shows the relationships between transformer voltage, MC voltage, and grid inductor current based on the three modulations: (a) Carrier comparison with D-FF, (b) delta-sigma conversion with PDM, and (c) carrier comparison with zero-vector commutation, respectively.

Table 3. Simulation/experimental parameters.

Names	Symbol	Value
Battery voltage	v_{bat}	100–200 V
Grid voltage	v_{ac}	100 V 50 Hz
Transformer voltage ratio	$N_{fbi}:N_{mc}$	1:1.75
FBI switching frequency	fsw_fbi	100 kHz
Capacitor	C_b	400 μF
Inductor	L_b	10 μH
MC switching frequency	fsw_mc	10 kHz
Filter inductor	L_f	50 μH
Filter capacitor	C_f	22 μF
Grid inductor	L_g	425 μH

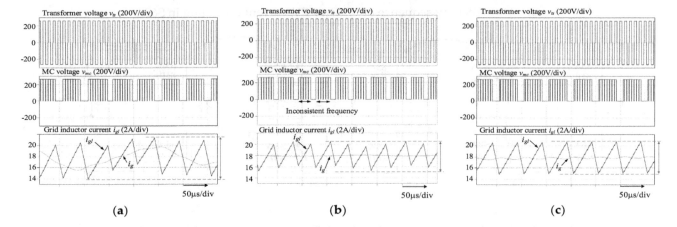

Figure 25. Simulation results that demonstrate the relationship between the matrix converter (MC) voltage pulse width and grid inductor current with different modulation methods. (**a**) D-flip flop, (**b**) delta-sigma conversion, (**c**) zero-vector commutation.

In the D-FF, it can be noticed that due to the misalignment of pulse width, the average of the grid current cannot be constantly controlled. As a result, the fluctuation of grid inductor current is the largest among the three methods. In the delta-sigma conversion, the voltage error can be resolved and therefore the fluctuation of grid inductor current is smaller than D-FF. However, it can be confirmed from the voltage pulse width that it has an inconsistent frequency due to the level of quantization error. With the zero-vector commutation, the voltage error can be eliminated and the fluctuation of grid inductor current is removed due to containing a consistent frequency in the current ripple. Therefore, the waveform quality can be improved as compared to the other two methods.

Figure 26 shows the simulation results at a low output power (300 W) that demonstrates the waveform of the battery current with all the three methods. Due to the direct AC/AC conversion, the distortion of grid current directly affects the waveform of the battery current. Notice that in Figure 26a, the battery current is heavily distorted in the D-FF because of the voltage error. In the case of delta-sigma conversion as shown in Figure 26b, the distortion in the battery current can be greatly reduced because the voltage error has resolved. However, a resonant frequency of the LCL filter occurrs at the battery current because of inconsistent frequency in the grid inductor current. As shown in Figure 26c, the zero-vector commutation can solve the two above problems. The distortion and resonance frequency in the battery current can both be removed due to a clean sinusoidal waveform that can be achieved in the grid inductor current.

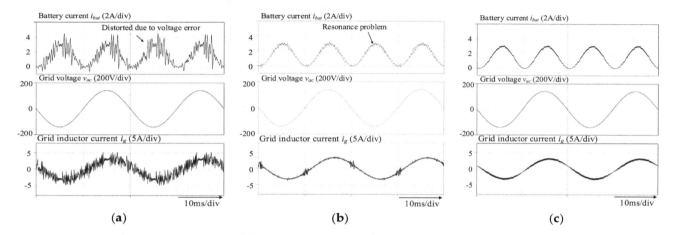

Figure 26. Low power (300 W) simulation result to demonstrate the differences of each method of modulation. (**a**) D-flip flop, (**b**) delta-sigma conversion, (**c**) zero-vector commutation.

Figure 27 shows another operating waveform at larger power (3.3 kW) to demonstrate the waveform of battery current with all the three modulation methods. D-FF is shown to have the worst distortion in battery current among the three methods. On the other hand, delta-sigma can achieve a clean sinusoidal waveform in the grid current nearly to the zero-vector commutation. This is because the peak-peak current is limited by the cut-off frequency of LCL filter, as the amplitude of the grid current becomes larger, the ripple current that is caused by the resonant frequency has lesser effect compared to the low power.

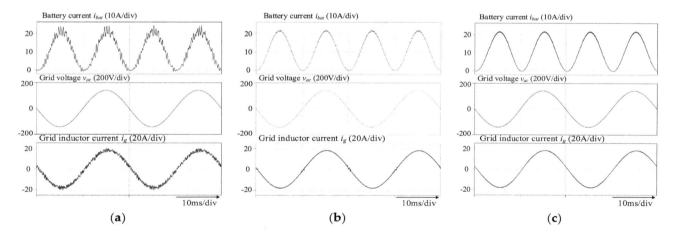

Figure 27. High power (3.3 kW) simulation result to demonstrate the difference of each method of modulation. (**a**) D-flip flop, (**b**) delta-sigma conversion, (**c**) zero-vector commutation.

Figure 28 shows the comparison of grid current THD among these modulations at low and high output power, respectively. The LCL cut-off frequency is regulated from 3 to 7.5 kHz by adjusting inductor L_f and the C_f capacitor while keeping the same impedance percentage. The results in Figure 28a (low power) shows that D-FF has the highest THD, and only the zero-vector commutation can reach the THD below 5% at a cut-off frequency of 5 kHz. In Figure 28b (high power), both the delta-sigma conversion and zero-vector commutation can reach the THD below 5% within the cut-off frequency from 3 to 7.5 kHz.

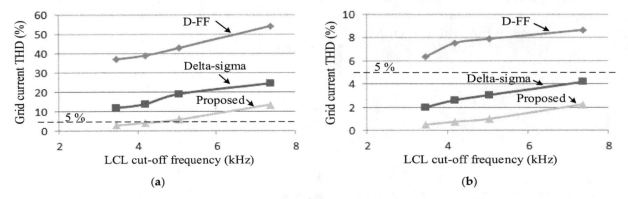

(a)   (b)

Figure 28. Comparison of grid current total harmonic distortion (THD) with the different inductive-capacitive-inductive (LCL) cut-off frequency between low power and high power. Zero-vector commutation achieves the lowest total harmonic distortion (THD) regardless of power level. (**a**) Low power (300 W). (**b**) High power (3k W).

As a result, in order for the delta-sigma control to achieve 5% THD at low output power, the cut-off frequency needs to be reduced with the penalty of a larger size in the LCL filter. Therefore, the zero-vector commutation can achieve the smallest size in the LCL filter and also achieve THD below 5% for both low and high output power, among the three methods.

5. Experimental Results

Figure 29 shows the layout of the 1 kW prototype (203 × 113 × 10 mm). Switching devices are placed on both sides (left: MC; right: FBI), then a planer transformer is placed in the middle and the capacitor C_b (400 μF) and an inductor L_b (10 μH) that is connected to the center point of transformer is placed on the top side of the prototype.

Figure 29. Layout of a 1 kW prototype.

Figure 30 shows the effectiveness of ZVS and zero-vector commutation. Figure 30a shows the result before applying zero-vector commutation. During the first (PV-to-Z) and last (PVZ-to-PV) switching intervals of zero-vector periods because hard-switching happens at a short-circuit state, therefore over-voltage occurrs at the transformer voltage. On the other hand, during the ZVS periods it can be confirmed that voltage spikes did not occur at the transformer voltage because switching devices are aligned to the zero-voltage of the transformer to achieve ZVS.

Figure 30b shows the result after applying the zero-vector commutation. As shown in the result, the voltage spike at the transformer voltage can be greatly reduced in PV-to-Z, comparing that to Figure 30a. The zero-vector commutation enables the switching state to go into the zero-vector periods to allow the current to circulate inside a loop and achieve current cancelling. As a result, the voltage spike of the switching device during the zero-vector periods can be resolved. On one hand, current cancelling cannot be achieved in PVZ-to-PV and therefore the voltage surge occurs on the transformer voltage. Since the short-circuit state can be prevented, the voltage surge is smaller than that compared to Figure 30a.

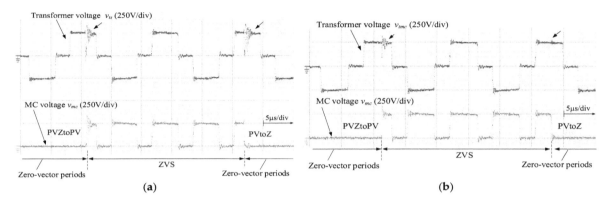

Figure 30. Experimental result to demonstrate the effectiveness of the zero-voltage switching (ZVS) and zero-commutation. (**a**) Without the zero-vector commutation. (**b**) With the zero-vector commutation.

The comparison of experimental results at low output power (300 W) between the delta-sigma and zero-commutation is shown in Figure 31a,b. In this result, the power decoupling was disabled in order to validate the effectiveness of the method of modulation. The capacitor voltage v_{cb} is controlled at 90 V constantly with a battery voltage of 180 V. In Figure 31a, we noticed that the grid inductor current i_{gl} has a huge current ripple due to the inconsistent frequency. As a result, the battery current fluctuates at a resonance frequency of 1.25 kHz. Figure 30b shows the results obtained by the zero-vector commutation, the distortion in the battery current can be nearly eliminated because a clean sinusoidal grid inductor current can be obtained.

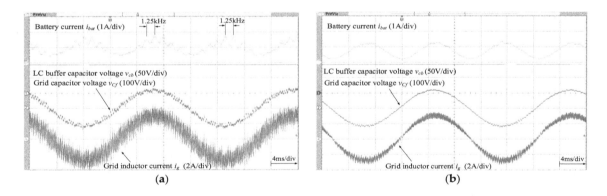

Figure 31. Comparison of experimental results between delta-sigma and zero-vector commutation. (**a**) Delta-sigma shows fluctuation in the battery current at 1.25 kHz. (**b**) Zero-vector commutation shows a clean sinusoidal waveform that can be obtained in the grid inductor current and battery current.

Figure 32a,b shows the fast Fourier transform (FFT) analysis of the grid current between the delta-sigma and the zero-vector commutation, respectively. Even the number of harmonic components contains the battery current. Then, it can be noticed that the 4th, 6th, and 8th harmonic components in the delta-sigma conversion is higher than that of the zero-vector commutation. The result can confirm that the zero-vector commutation could achieve better THD than the conventional ones.

Figure 33 shows the effectiveness of power decoupling with the zero-vector commutation. In Figure 33a, the power decoupling control was disabled and therefore the battery current contains a low frequency component. Then, after being applied to the power decoupling control as shown in

Figure 33b, the single-phase power fluctuation occurs in the capacitor voltage. The average capacitor voltage is constantly kept at half of the battery voltage, then Δv_{cb} of approximately 30 V is controlled at 100 Hz sinusoidal waveform to compensate the single-phase power fluctuation. This is also identical to the theoretical calculation which is explained in Figure 3, where a 400 μF capacitor with 30 V voltage difference is designed for the power decoupling. As a result, the single-phase power fluctuation can be reduced in the battery current. Note that the low frequency fluctuation occurrs in the battery current because of the DC bias effect in the ceramic capacitor.

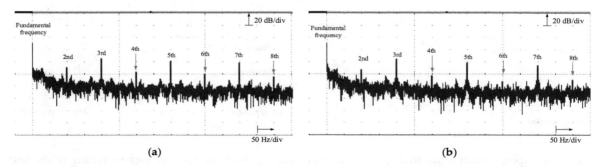

Figure 32. Comparison of the fast Fourier transform (FFT) analysis on grid current between delta-sigma and zero-vector commutation, the even harmonic component is lower in the zero-vector commutation. (**a**) Delta-sigma. (**b**) Zero-vector commutation.

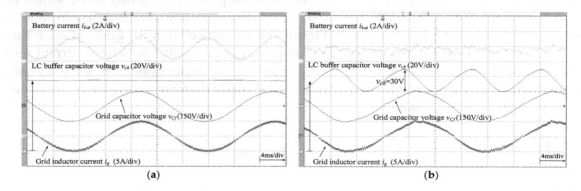

Figure 33. Experimental results to demonstrate the effectiveness of power decoupling control with the zero-vector commutation. (**a**) Without the decoupling control; capacitor voltage is constantly controlled at the average of battery voltage. (**b**) With the decoupling control; the capacitor voltage difference is 30 V with the decoupling control to compensate the single-phase power fluctuation.

Figure 34 shows the experimental measurement efficiency of the prototype. The prototype achieves the highest efficiency 91.5% at 1 kW. Optimization of the losses will be considered in the future work to improve the efficiency.

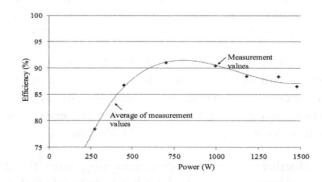

Figure 34. Measurement of efficiency, the prototype achieves the highest efficiency of 91.5%.

6. Conclusions

The comparisons among the three methods and along with other literature reviews are summarized in Table 4. Power decoupling is obviously not considered in the past studies due to the difference of circuit structure. The control also shows a difficult level due to complex commutation rules. This paper describes and compares three methods of modulation of MC for the power decoupling with the transformer integration. Zero-vector communication is introduced in this paper. The effectiveness of these modulations have been demonstrated in simulation and experimental.

Table 4. Comparison results among the three methods of modulation.

Modulation Methods	Power Decoupling	Control Complexity	Quality Waveform (Grid Current THD, Battery Current Ripple)		LCL Filter Sizing
			Low-Power	High-Power	
SPWM synchronous rec. [12]	×	△	○	○	50 kHz L_f: 10 µH C_f:1 µF
PWM four-step comm. [13]	×	×	○	○	5 kHz L_f: 300 µH C_f:4.7 µF
PWM comm. [14]	×	×	△	○	N/A
D-FF [15]	○	○	×	△	× 3 kHz L_f: 100 µH C_f:22 µF
Delta-sigma [16]	○	△	△	○	△ 2 kHz L_f: 100 µH C_f: 45 µF
Zero-vector commutation	○	△	○	○	○ 5 kHz L_f: 50 µH C_f:22 µF

○ = good, △ = average, × = poor.

Modulations that consider the power decoupling are summarized as follows. The D-FF is simple in terms of control but the quality waveform is poor due to the voltage error. Delta-sigma achieves average among the three, in order to improve the THD during low-power, a bigger size of LCL filter is required. The zero-vector commutation can produce a better quality waveform but the control complexity requires a high-bandwidth controller. If the design level is only concerned for high-power, delta-sigma with a lower bandwidth controller is another option of choice.

Author Contributions: Conceptualization, G.T.C. and T.S.; methodology, G.T.C.; software, G.T.C.; validation, G.T.C. and T.S.; formal analysis, G.T.C.; investigation, G.T.C.; resources, G.T.C.; data curation, G.T.C.; writing—original draft preparation, G.T.C.; writing—review and editing, G.T.C. and T.S.; visualization, G.T.C.; supervision, T.S.; project administration, T.S.; funding acquisition, T.S. All authors have read and agreed to the published version of the manuscript.

References

1. Raggl, K.; Nussbaumer, T.; Doerig, G.; Biela, J.; Kolar, J.W. Comprehensive design and optimization of a high-power density single-phase PFC. *IEEE Trans. Ind. Electron.* **2009**, *56*, 2574–2587. [CrossRef]
2. Choi, W.; Rho, K.-M.; Cho, B.-H. Fundamental duty modulation of dual-active bridge converter for wide-range operation. *IEEE Trans. Power Electron.* **2016**, *31*, 4048–4606. [CrossRef]
3. Jovanovic, M.M.; Jang, Y. State-of-the art, single-phase, active power factor correction techniques for high power applications. *IEEE Trans. Ind. Electron.* **2009**, *56*, 2574–2587. [CrossRef]
4. Musavi, F.; Eberle, W.; Dunford, W.G. A high-performance single-phase bridgeless interleaved PFC converter for plug-in hybrid electric vehicle battery chargers. *IEEE Trans. Ind. Appl.* **2011**, *47*, 1833–1843. [CrossRef]

5.   Xue, L.; Shen, Z.; Boroyevich, D.; Mattavelli, P. GaN-based high frequency totem-pole bridgeless PFC design with digital implementation. In Proceedings of the IEEE Applied Power Electronics Conference and Exposition (APEC), Charlotte, NC, USA, 15–19 March 2015; pp. 759–766.

6.   Kolar, J.W.; Friedli, T.; Rodriguez, J.; Wheeler, P.W. Review of three-phase PWM AC-AC converter topologies. *IEEE Trans. Ind. Electron.* **2011**, *58*, 11. [CrossRef]

7.   Empringham, L.; Kolar, J.W.; Rodrigues, J.; Wheeler, P.W.; Clare, J.C. Technological issues and industrial application of matrix converters: A review. *IEEE Trans. Ind. Electron.* **2013**, *60*, 10. [CrossRef]

8.   Sun, Y.; Liu, Y.; Su, M.; Xiong, W.; Yang, J. Review of active power decoupling topologies in single-phase systems. *IEEE Trans. Power Electron.* **2016**, *31*, 4778–4794. [CrossRef]

9.   Komeda, S.; Fujita, H. A power decoupling control method for an isolated sing-phase AC-to-DC converter based on direct AC-to-AC converter topology. *IEEE Trans. Power Electron.* **2018**, *33*, 9691–9698. [CrossRef]

10. Itoh, J.-I.; Hayashi, F. ripple current reduction of a fuel cell for a single-phase isolated converter using a DC active filter with a center tap. *IEEE Trans. Power Electron.* **2009**, *25*, 550–556. [CrossRef]

11. Takaoka, N.; Takahashi, H.; Itoh, J.-I. Isolated single-phase matrix converter using center-tapped transformer for power decoupling capability. *IEEE Trans. Ind. Appl.* **2018**, *54*, 1523–1531. [CrossRef]

12. Wang, M.; Huang, Q.; Yu, W.; Huang, A.Q. An isolated bi-directional soft-switched DC-AC converter using wide-band-gap devices with novel carrier-based unipolar modulation technique under synchronous rectification. In Proceedings of the IEEE Applied Power Electronics Conference and Exposition (APEC), Charlotte, NC, USA, 15–19 March 2015; pp. 2317–2324.

13. Varajao, D.; Rui, E.A.; Miranda, L.M.; Lopes, J.A.P.; Weise, N. Control of an isolated sing-phase bidirectional AC-DC matric converter for V2G applications. *Electr. Power Syst. Res.* **2017**, *149*, 19–29. [CrossRef]

14. Norrga, S. Experimental study of a soft-switched isolated bidirectional AC-DC converter without auxiliary circuit. *IEEE Trans. Power Electron.* **2006**, *21*, 6. [CrossRef]

15. Nakata, Y.; Orikawa, K.; Itoh, J.-I. Several-hundred-kHz single-phase to commercial frequency three-phase matrix converter using Delta-sigma modulation with space vector. In Proceedings of the IEEE Energy Conversion Congress and Exposition (ECCE), Pittsburg, PA, USA, 14–18 September 2014; pp. 571–578.

16. Takaoka, N.; Takahashi, H.; Itoh, J.-I.; Chiang, G.T.; Sugiyama, T.; Sugai, M. Power decoupling method comparison of isolated single-phase matrix converters using center-tapped transformer with PDM. In Proceedings of the IEEE Energy Conversion Congress and Exposition (ECCE), Montreal, QC, Canada, 20–24 September 2015.

17. Chiang, G.T.; Takahide, S.; Masaru, S. Optimal design of a matrix converter with a LC active buffer to onboard vehicle battery charger in single phase grid structure. In Proceedings of the 18th European Conference on Power Electronics and Applications, Karlsruhe, Germany, 5–9 September 2016.

18. He, J.; Li, Y.W. Hybrid voltage and current control approach for DG grid interfacing converters with LCL filters. *IEEE Trans. Ind. Electron.* **2013**, *60*, 1797–1809. [CrossRef]

19. She, H.; Lin, H.; He, B.; Wang, X.; Yue, L.; An, X. Implementation of voltage-based commutation in space-vector modulated matrix converter. *IEEE Trans. Ind. Electron.* **2012**, *59*, 154–166.

20. Afsharian, J.; Xu, D.; Wu, B.; Gong, B.; Yang, Z. A new PWM and commutation scheme for one phase loss operation of three-phase isolated buck matrix-type rectifier. *IEEE Trans. Power Electron.* **2018**, *33*, 9854–9865. [CrossRef]

9

Challenges and Design Requirements for Industrial Applications of AC/AC Power Converters without DC-Link

Pawel Szczesniak

Institute of Electrical Eng., University of Zielona Góra, 65-516 Zielona Góra, Poland;
P.Szczesniak@iee.uz.zgora.pl

Abstract: AC/AC converters that do not have a DC energy storage element, such as a matrix chopper and a matrix converter, are increasingly becoming alternatives to conventional two-stage AC/DC/AC converters and thyristor choppers. In such systems, the main DC-link capacitor does not exist, so the system provides more reliable operation and makes it possible to reduce the financial costs of its construction. It should be noted that AC/AC converters without an energy storage element in a form of DC-link capacitors have not been implemented on an industrial scale. The reasons involve technical aspects and cost components. The main aim of this paper is to present some of the challenges and selected design requirements for industrial applications of AC/AC high reliability power converters.

Keywords: power electronic converter; AC/AC converter; matrix converter; reliability

1. Introduction

The development of power electronic converters has led to them being applied in various areas of life, such as: industrial and household applications, renewable energies, Flexible AC transmission systems and micro grids, as well as automotive and transport applications. In the development of power converter topologies, much more attention is being paid to the, reliability [1,2] efficiency and robustness of power electronics converters [3–6]. Automotive industry and transport have very strict reliability requirements in power electronics systems due to safety requirements. Moreover, the industrial and energy sectors are striving for improvement in the efficiency and robustness of power electronics systems. Likewise, home solutions are often designed as economical and sustainable eco-friendly devices. Additionally, some novel devices have a smaller size, and have achieved higher power density and robustness for very high loads. This is a general trend in all areas of technology. High power density, higher switching frequency, and reduced overall dimensions of passive components and power electronic converters are possible due to the use of power transistors made using silicon carbide (SiC) or gallium nitride (GaN) technology [7–10].

In low-voltage industrial systems, the most commonly used topology is the unidirectional AC/DC/AC converter. The bidirectional (four-quadrant) structure of such a converter is a system called a back-to-back converter (B2B) (Figure 1) [11]. The power stage of converters contains power transistors with a heatsink and heat dissipation system, a DC-link capacitor, an input filter to provide a unit input power factor, and measurement and control units. In medium-voltage industrial systems, multilevel converters with DC-link are commonly used [12,13]. Multilevel converters offer numerous advantages compared with the two-level converter, e.g., better power quality and lower switching losses, low transistor voltage stress, and high voltage capability. It is evident that the multilevel converters will be used for medium voltage or high power electronic systems. However, in general terms, it can be concluded that the two-level converters will be used in low-voltage systems with lower power. In addition, the specialized design of low voltage power electronics converters using

SiC or GaN semiconductors, with the optimization of cooling systems, enables the construction of power plants with high efficiency or power densities. In the last decade, apart from the development of power electronics devices, there has also been a significant development of capacitors used in DC circuits. Aluminum electrolytic capacitors are probably the most common capacitors used in the DC-link circuits of modern power electronic converters. In addition, other technologies employed in the production of such capacitors are the multi-layer ceramic structure or the metallized-polypropylene structure [11]. Despite the significant reduction in the dimensions of modern capacitors used in DC circuits, these capacitors still constitute the main component increasing the dimensions, weight, and price of power converters. Furthermore, the capacitor is the component that is most often damaged due to disturbances from the power grid and as a result of improper operation. Figure 2 shows the elements that are most often damaged in circuits of power electronic converters [1,2,14]. For this reason, the most common electrolytic capacitors are the cause of the premature shortening of the lifespan of the converter device [11]. The ageing of DC-link capacitors is manifested by the increase in their equivalent series resistance R_{ESR}. The heating up of the electrolyte and its consequent evaporation and deterioration of electrical parameters is the most significant factor in the degradation of the electrolytic capacitor. One of the main reasons for the increase in capacitor temperature is the ripple of the capacitor current. In addition, the second most critical parameter for the failure of the capacitor is its voltage capacity, usually determined by the rated voltage or operating voltage, surge voltage, or allowed short-term maximum voltage. The voltage parameter is so critical that exceeding its nominal values for a few tenths of a second can cause an immediate failure or significantly accelerate the degradation of capacitor nominal parameters. For the presented reasons, in power converters, especially high power ones, particular attention is paid to environmental and operational conditions affecting their lifespan, with special emphasis on DC-link capacitors.

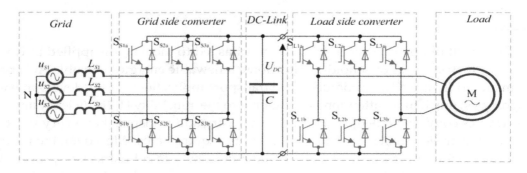

Figure 1. Two-level back-to-back converter with DC-link.

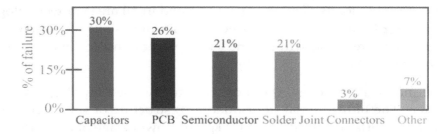

Figure 2. Distribution of faults in power electronic converters [14].

The development of power converters that do not have large capacitors in their structure (DC-link) is an interesting solution to reduce their cost and size. In addition, this can achieve greater reliability and thus extend their operating time [15–18]. Such converters are often considered an alternative solution for atypical applications and are not found in many industrial applications. One of the main reasons concerns technical aspects related to the small number of available and dedicated semiconductor power devices, as well as DSP processors with a sufficient number of PWM outputs and A/D converters.

An additional difficulty is undoubtedly the complex algorithms for modulation of the switch control function and transistor current commutation.

AC/AC converters without DC-link capacitors have several topologies with different functionalities. One of the main functionalities is the possibility to change the frequency of the output current/voltage (f_L = const [16], f_L = var [15–18]). In addition, different structures of AC/AC converters are distinguished, such as direct and indirect (with a fictitious DC-link) topology. An example of an AC/AC converter without an intermediate circuit capacitor and f_L = const is a matrix controller (matrix chopper), whereas f_L = var is a matrix converter.

This paper will present a review of current commercial applications of the AC/AC converters whose topologies do not have a DC energy storage element (capacitor or inductor). Particular attention will be paid to the technological issues and barriers concerning the design of such converters. In addition, possible construction solutions as well as potential applications will be indicated. The originality of this research is in the presentation of requirements and challenges for practical or future-proof applications of AC/AC high-reliability power converters. The paper includes (1) a review of semiconductor power elements dedicated to the discussed converters, (2) an indication of expectations regarding integrated intelligent modules and new SiC technologies, and (3) an indication of the development of modern control techniques (e.g., model predictive control (MPC)). In addition, new application possibilities of the discussed converters in AC power systems, (e.g., compensators for voltage changes) are indicated and their beneficial properties are discussed.

2. Description of Selected Topologies

The analysis presented in the article will be based on topologies of two converters: a matrix controller (chopper) and a matrix converter. Based on these, the main design problems and limitations related to the available components and the industrial application will be indicated.

One of the basic applications of AC/AC choppers, as an alternative to thyristor choppers, is in electric drive systems used as a soft-start for an AC motor. Issues with industrial temperature control and lighting intensity control are the further areas of AC/AC chopper application. A schematic diagram of the three phase AC/AC matrix controller power circuit is presented in Figure 3a [16]. The system consists of six bidirectional switches. Three switches are connected to the input terminals, while the other three are connected in parallel to the load. Control of the switches is carried out by means of the PWM signal (Figure 3b), and a duty cycle determines the RMS value of the output voltage. In addition, the choppers have an input low-pass filter to eliminate the higher harmonics components of the current drawn by the converter, which result from the frequency of the PWM signal. The second kind of structure based on the chopper shown in Figure 3a is the matrix-reactance chopper, which also enables the increase in amplitude of the output voltage [19] as well as various other types of compensators for AC voltage fluctuations in the power grid [20,21].

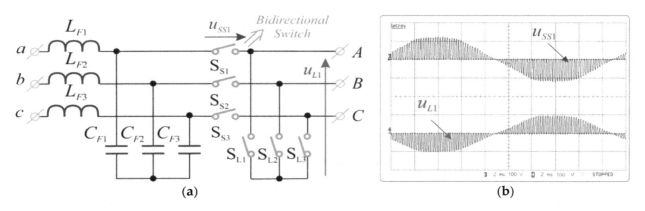

(a)   (b)

Figure 3. AC/AC controller: (a) main circuit; (b) voltage time waveforms.

The main advantages of the AC/AC chopper compared to the classic thyristor chopper are derived from the sinusoidal current drawn from the mains, and the use of LC output filters to obtain sinusoidal output voltages. Such properties allow one to work with the unit input power factor and use these choppers in systems of electricity conditioners [20,21]. The high switching frequency of the power transistors makes it possible to minimize the sizes for the input and output LC passive filters. The disadvantage of chopper devices is the increased internal losses resulting from the large number of power transistors and the necessity of using additional snubber devices on each of the power electronics switches or of using overvoltage protection of bidirectional switches with a clamp circuit [16,22].

The most widely known of frequency converters (f_L = var) without DC-link is the matrix converter (MC) in its direct topology shown in Figure 4a [15–18,22–24]. Generally, in three-phase systems, the MC consists of nine bidirectional switches connected in a matrix—each input is connected to each output. Similar to the matrix controller, there is an input low pass filter in the MC topology that fulfills identical functions. MC output voltages are formed from pieces of input voltages, shown in Figure 4b, as exemplary experimental time waveforms.

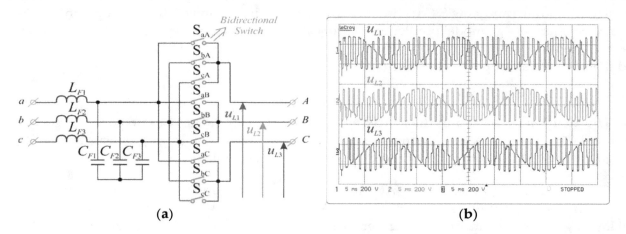

Figure 4. Matrix converter: (**a**) main circuit; (**b**) voltage time waveforms.

The MC is a four-quadrant topology with adjustable input power factor and a sinusoidal shape of input and output currents. In addition, the MC is a fully semiconductor device without large capacitor circuit capacitors. Similarly to chopper devices, there is a necessity to use additional overvoltage protection of bidirectional switches with a clamp circuit, where small capacitors accumulating commutation voltage spikes should be used. The disadvantages of the MC are the large number of power transistors and the associated complex control and switching strategies. In addition, the voltage gain of MCs does not exceed 0.866, which, for applications in variable speed drives, requires the use of motors with lower rated voltages.

In both MC and AC/AC chopper topologies, the main elements are bi-directional power electronic switches. These switches allow switching of both the positive and negative sine wave of the supply voltage and, in the case of the MC, the bi-directional power flow. Bidirectional switches, as will be discussed in the next section, have a configuration of transistors and diodes other than those used in classical, commonly used voltage inverters. Therefore, the commercial development of available bidirectional switches in various configurations is important for the development of AC/AC converters without DC-link circuits.

3. Design, Construction, and Implementation Barriers

The development of AC/AC converters encounters problems related to misunderstanding the specificity of their operation, especially in the context of MCs. In this chapter, basic construction problems and development barriers related to technological differences in relation to commonly used frequency converters will be indicated. Barriers hindering development include, among others, a lack of semiconductor components in more complex modules than single transistors, a specific distribution of control signals, a complex transistor switching strategy, and a large number of measuring sensors. However, in spite of these problems, this chapter will indicate the first commercial applications of AC/AC converters without DC-link and provide guidelines for the modular construction of intelligent power modules that would significantly contribute to the development of this technology. An important development impulse may also be the implementation of control methods that are developed for other topologies such as model predictive control. The number of topics discussed in this chapter results from the need to indicate both the application potential and difficulties in designing the discussed converters.

3.1. Power Semiconductors

As already mentioned in the previous chapter, bi-directional switches are used in the discussed converters. Basic configurations of such bidirectional switches made in Si technology are presented in Figure 5. Such configurations of Insulated Gate Bipolar Transistor (IGBT) transistors and diodes allow conducting currents in both directions and blocking voltages for positive and negative polarity.

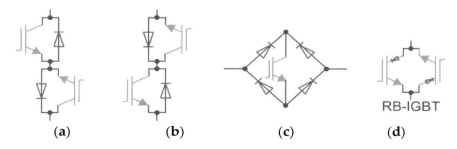

(a)   (b)   (c)   (d)

Figure 5. Bi-directional switches: (**a**) Common emitter IGBT, (**b**) common collector IGBT, (**c**) IGBT with a diode bridge, and (**d**) RB-IGBT reverse blocking IGBTs.

3.1.1. Si Semiconductors

AC/AC converters are full semiconductor systems and must be constructed of bi-directional switches. For power converters commonly used in industry, there are many developed modules containing complex structures on the market for the transistors or diode connections. However, for AC/AC converters, there are very few commercially available semiconductor modules that have bi-directional switches in their structure. As shown in Figure 5, the most commonly used configurations of bi-directional switches are IGBTs with anti-parallel diodes (two topologies, a common collector, and a common emitter). In addition, there are switches with a diode bridge and a single IGBT transistor, and reverse blocking IGBT (RB-IGBT), where anti-parallel diodes can be eliminated [21]. It is possible to build bi-directional switches with a combination of various discrete elements, but the creation of advanced power electronic devices in the form of integrated power modules facilitates a reduction in dimensions of the converter, allows higher power densities, and produces high reliability. Few examples of power modules for the bidirectional switches or complex structures are currently available on the market, though some of these modules are now commercially available. Table 1 lists the selected of manufacturers that have commercially available power modules with bi-directional switches.

The first example of the use of bi-directional Dynex Semiconductor switches with rated parameters of 1.2 kV and 200 A is illustrated in the structure of the matrix converter shown in Figure 6. An example of a structure containing a single housing topology of a matrix converter with RB-IGBT transistors is the module presented in Figure 7. The module is manufactured by the FUJI Electric company and its nominal parameters are 1.2 kV/50 A. As generally known, in addition to the power element, a dedicated transistor gate driver is also needed. In this case, designers can use existing drivers used in other converter topologies. However, there are dedicated drivers for available bi-directional modules. Examples of such dedicated drivers include Concept [14] drivers for Dynex switches (Figure 8a) and Semicron drivers and transistors presented in Figure 8b,c. These are examples of bi-directional switches. In the catalogues of other semiconductor companies, such solutions are found often and with increasing frequency. Based on the above review, it can be said that the number of dedicated components is gradually increasing over time. It should be noted that, without systematic growth of proposals for new, dedicated components for the design and construction of AC/AC converters, there will be no significant development of these topologies. The main direction of development should be to reduce component costs and increase their reliability.

Figure 6. MC construction: (**a**) Dynex 1.2 kV/200 A IGBT; (**b**) a built-in MC structure in the control cabinet.

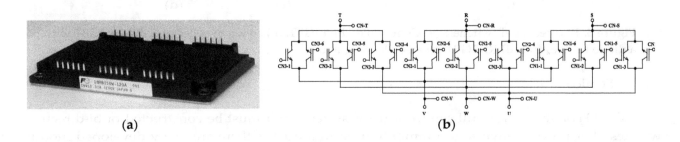

Figure 7. Matrix-connected RB-IGBTs module: (**a**) photograph; (**b**) topology structure.

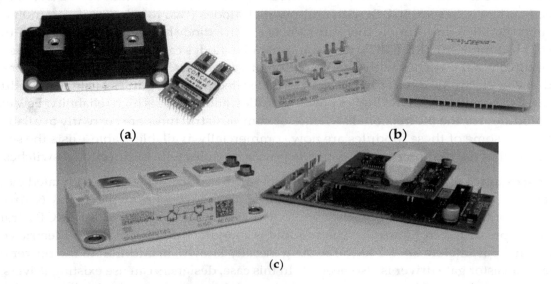

Figure 8. A photograph of the IGBTs with dedicated drivers: (**a**) Dynex IGBT with Concept driver; (**b,c**) Semicron drivers and IGBTs.

Table 1. Available, selected commercial modules with bidirectional switches.

Switch Model	Characteristic	Manufacturer	Number of Bidirectional Switches
DIM200MBS12-A	1.2 kV/200 A	DYNEX	1
DIM400PBM17-A	1.7 kV/400 A	DYNEX	1
DIM600EZM17-E	1.7 kV/600 A	DYNEX	1
SK 60GM123	1.7 kV/60 A	SEMICRON	1
SKM 150GM12T4G	1.2 kV/60 A	SEMICRON	1
SML150MAT12	1.2 kV/150 A	SEMELAB	3
SML300MAT06	0.6 kV/300 A	SEMELAB	3
18MBI50W-120A	1.2 kV/ 50A	FUJI	9
18MBI100W-060A	0.6 kV/100 A	FUJI	9
18MBI100W-120A	1.2 kV/100 A	FUJI	9
18MBI200W-060A	0.6 kV/200 A	FUJI	9

3.1.2. SiC Semiconductors

Thanks to the development of semiconductor power electronics with silicon carbide (SiC), which has certain advantageous features such as high temperature operation, low losses, and higher switching speeds, a full power electronic converter structure has also been developed. Moreover, thanks to the semi-conductor being made using SiC technology, a further improvement in the converters power density can be achieved. Because this is a technology that has not fully achieved technological maturity, there are some problems in the implementation of power converters with SiC devices due to the high speed of switching devices and the design of devices for the gate drive.

The most promising SiC devices are the normally-off SiC JFET, SiC MOSFET, and SiC BJTs. The bi-directional switch topologies in SiC technology are shown in Figure 9. Potential improvements regarding the efficiency and performance of SiC components in the context of bi-directional power electronics switches are reported in [24–27]. As demonstrated by existing research, the use of transistors made in SiC technology enables a significant increase in the efficiency of converters for switching frequencies of several dozen kHz. Sample results illustrating the beneficial properties of SiC transistors, taken from [24], are presented in Figure 10.

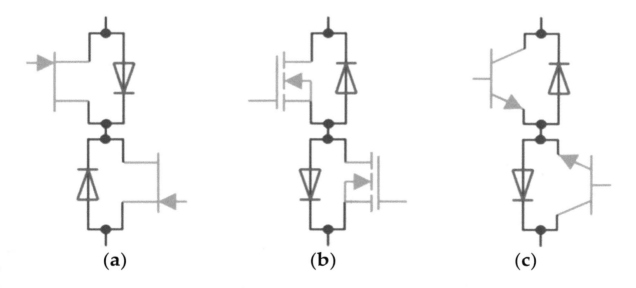

(a)   **(b)**   **(c)**

Figure 9. Main topologies for SiC bi-directional switches: (**a**) common drain anti-paralleled JFET, (**b**) common source anti-paralleled SiC MOSFET, and (**c**) common emitter anti-paralleled SiC BJT.

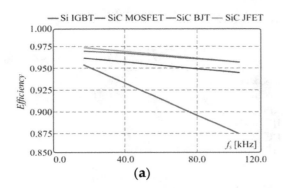 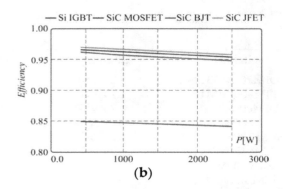

(a)   (b)

Figure 10. Illustration of the beneficial properties of the use of SiC transistors, taken from [24], presents the efficiency of a two-phase to single-phase 2.5 kW MC in Si and SiC technology (**a**) as a function of switching frequency at $T_c = 125\,°C$ and (**b**) as a function of power at $f_s = 100$ kHz, $T_c = 125\,°C$.

3.2. Control Units

The MC power stage circuit is made up from a table of nine bi-directional switches which provides a total of 18 transistors. The specificity of the MC construction also requires the generation of a PWM switching sequence. In addition, special attention should be paid to maintaining a safe current commutation strategy when the transistors are switched. An additional difficulty in the modulation process of the transistors control functions is the distribution of control pulses in a single period of the control sequence T_{Seq}. Since there are three bidirectional switches connected in one output branch (six transistors) and two or three of them cannot be switched on simultaneously, the switching sequence will be responsible for three switching operations during one period T_{Seq} (Figure 11a). In the classic bridge voltage inverter, for a single branch, there are two transistors working alternately. A pair of complementary control signals are then generated and are negated in relation to each other (Figure 11b). In addition, additional dead times (dead-band) are used to ensure correct current commutation. Modern DSP processors and microcontrollers have built-in PWM modulators that generate complementary control signals that occur in classic bridge voltage inverters with built-in dead time generation functions [28]. The control implementation for the matrix converter using built-in PWM modulators is not simple and requires either multiple-core systems (at least two) or additional logic circuits. It is also possible to solve this problem using software I/O outputs instead of built-in PWM modulator procedures. These are, of course, more complicated solutions than in the classic branches of the inverter bridge.

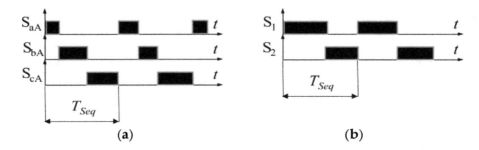

(a)   (b)

Figure 11. Distribution of control pulses in a single period of the control sequence T_{Seq}: (**a**) in the phase leg of a matrix converter, (**b**) in the phase leg of a classical inverter.

In addition to the algorithm for the modulation of switch control signals in the MC, it is necessary to use complex current switching algorithms for transistors. Several commutation strategies can be found in the scientific literature [17,21], where the most widely known is the four-step current commutation strategy. In this method, the current direction in the output line is used to determine switching sequences of transistors. The commutation process for one output phase is shown in

Figure 12a. An analogous switching pattern occurs between any transistors in each MC output stage. Examples of control sequences obtained at the FPGA output are shown in Figure 12b.

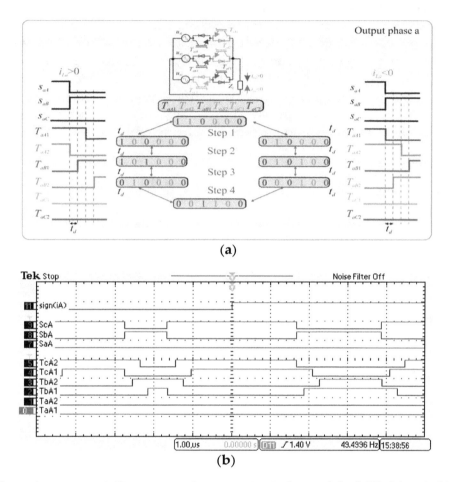

(a)

(b)

Figure 12. Four-step commutation process in one output phase of the MC: (**a**) switching diagram; (**b**) control signals from FPGA devices.

The digital implementation of the two presented algorithms of modulation and commutation requires the use of complex computing systems often combining DSP and FPGA devices. [29]. The control algorithms for a given application (electric drive, voltage compensator, etc.) and modulation algorithm are in most cases implemented in the DSP, while the FPGA, observing safe commutation rules, is used in the part related to the separation of control signals and transfer to individual transistors, so as to perform additional tasks to supplement the DSP function.

A significant limitation preventing faster development of AC/AC converters, such as MCs, is the large number of transistors and hence the need for DSPs with a large number of PWM outputs. The solution to this problem may be the rapid development of FPGAs, which has occurred in recent years. In the scientific literature, first articles related to the implementation of complete MC control algorithms implemented solely on FPGA have appeared [30,31]. Such an implementation is much more difficult when advanced mathematical functions such as trigonometric functions are used. Additionally, I/O systems in particular should be defined using a software solution, especially those related to cooperation with A/D and D/A converters. Nevertheless, parallelism and the speed of calculation means that FPGAs can replace DSP processors.

3.3. Measurement of Voltages and Currents

In order to correctly implement the control, modulation, and commutation algorithms in MCs, it is necessary to measure the values of voltages and currents [21,32]. Measured voltage and current signals are used, among others, in (1) semisoft commutation strategies, (2) fault detections, (3) modulation process (e.g., SVM), (4) phase-locked loop (PLL) units, (5) current loops in control algorithms, and (6) the clamp protection circuit. The measuring transducers should ensure galvanic isolation between the measured signals and the components of the control system and match the level of signals to the A/D converters. The most popular voltage and current measurement devices are LEM type transducers [33]. Of course, there are also many other manufacturers of this type of device, e.g., ABB, Honeywell, Allegro Microsystems, and Chen Young [34]. Transducer devices are quite expensive, so the measurement of voltages or currents can also be made on the basis of differential and isolated amplifiers, which are cheaper. Devices with galvanic isolation use optical, capacitive, or inductive insulation barriers. Table 2 presents examples of voltage and current measuring amplifiers as well as selected types of transducers [34].

Table 2. Selected available voltage and current measurement amplifiers and transducers.

Model	Isolation/Operation	Manufacturer	Type
HCPL-7860	Optic	Avago Technologies	Current
HCPL-786J	Optic	Agilent	Current
HCPL-7800	Optic	Avago Technologies	Current
ISO124	capacitive	Texas Instruments	Current/Voltage
ISO120/121	capacitive	Burr-Brown Corporation	Current/Voltage
INA270	difference amplifier	Texas Instruments	Current
LM358	difference amplifier	Texas Instruments	Current
AD8210	difference amplifier	Analog Devices	Current.
ACS752SCA-100	magnetic	Allegro MicroSystems	Current
ACS756xCB	magnetic	Allegro MicroSystems	Current
LV-25	magnetic	LEM	Voltage
LTS 25-NP	magnetic	LEM	Current
MP25P1	magnetic	ABB	Current
VS500B	magnetic	ABB	Voltage

One of the more precise issues related to the measurement of currents in AC/AC converters is the detection of its direction needed to ensure correct commutation of the tan resistors [21]. Due to the inductive nature of the load and the switched output voltage, the currents in the circuits of AC/AC converters contain higher harmonics. Therefore, accurate determination of the current direction is an issue requiring either advanced software algorithms or additional electronic circuits. The software solution for determining the current direction can be implemented using the network voltage synchronization algorithms (PLL) [35]. It is also possible to use an additional electric circuit that will perform the function of determining the current direction in a hardware manner. An example of such a solution based on a zero-crossing detector is shown in Figure 13 [36].

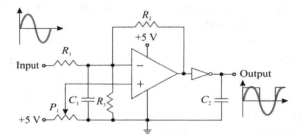

Figure 13. Circuit diagram of the zero-crossing detector.

3.4. Commercialization of Prototype Solutions

The first major challenge for AC/AC converters without a DC-link element is the commercialization of the semiconductor components and other elements necessary for their design and implementation.

The market situation of the semiconductor switches is discussed in Section 3.1.1. A growing number of dedicated semiconductor modules is visible. However, is this enough for faster development of power converters? This question is important because despite the growing interest in scientific research in this field, there are few AC/AC converters without DC-link available on the market. One of the first manufacturers of matrix converters is the Yaskawa Electric Corporation [32]. The Yaskawa products of an MC include low-voltage and medium-voltage solutions. The low-voltage MC solution is the classic matrix converter shown in Figure 7a. The classic MC product portfolio has two voltage levels: 200 V (from 9 to 63 kVA) and 400 V (from 10 to 114 kVA) [32]. In contrast, the solution for medium voltage is designed as a multi-level structure (Figure 14a). It uses a three-phase to single-phase MC structure as a basic power block (Figure 14b). By combining three blocks in series, a phase voltage three times higher than the voltage of a single module can be obtained (Figure 14a). The multilevel MC has similar features to the classical matrix converter. The medium voltage MC product portfolio has two voltage levels: a 3.3 kV level ranging from 200 to 3000 kVA and a 6.6 kV level from 400 to 6000 kVA [32].

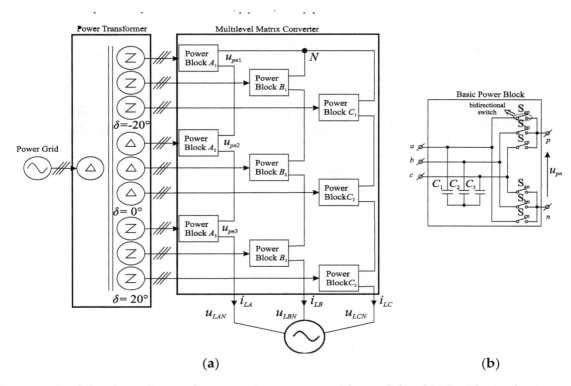

(a)   (b)

Figure 14. Multilevel, medium voltage matrix converters: (**a**) a multilevel MC with nine basic power switch blocks; (**b**) a three-phase to single-phase (3 × 2) MC power block.

3.5. Modular Construction

In commercially available modules dedicated to the discussed converters, there are no solutions for intelligent modules with power semiconductors, protection devices, drivers, or measurement sensors in their structure. This subsection indicates the expectations regarding this type of intelligent solution based on previous experience.

The power-electronic building block (PEBB) is the concept of building converters from basic modules. The effect of this approach is to increase the reliability and dimensions of power electronic converters [37,38]. The PEBB concept is used to integrate the following components into one module: power supply devices, gate drives, communication interfaces, measuring sensors, snubber protection circuits, and other components necessary for the proper operation of the converter. The concept of the PEBB offers a means for hardware standardization of power electronics systems.

In the analyzed AC/AC power converters, depending on the topology, there are several basic blocks presented in Figure 15. Figure 15a is the most widely used single bidirectional switch structure, which consists of two active switches and two diodes anti-parallel with them. Practical implementation of such a basic block can be realized by using power electronic switches with the dedicated driver systems that are shown in Figure 8. Figure 15b shows a basic structure for a single leg of the matrix converter structure. This basic block may be used for the construction of both the classical MC (Figure 4a) as well as the multilevel MC (Figure 14). Two prototypes of power module in such configurations as SML150MAT12 and SML300MAT06 are proposed and presented in Table 1. Another basic block from Figure 15c is the arrangement of a three-to-single phase MC, which can be used in a multilevel MC (Figure 14) or can be operated as a rectifier [39]. The most complex block power is the whole structure of the MC in a single device shown in Figure 15d. Manufacturers of semiconductor power devices offer several of these dedicated structures, as shown in Table 1.

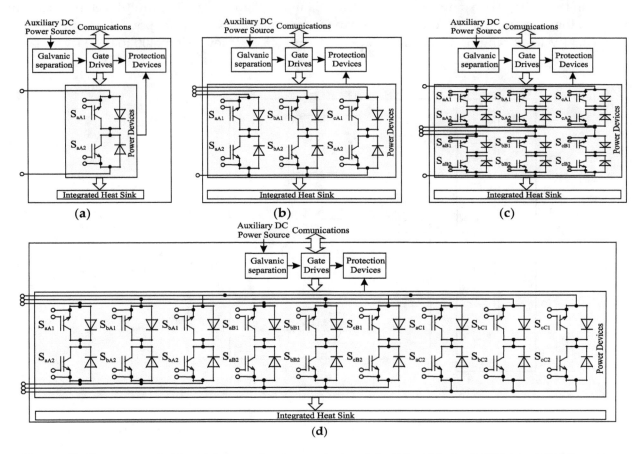

Figure 15. Common switching structures for power-electronic building blocks (PEBBs) in AC/AC power converters (**a**) single bidirectional switch structure, (**b**) single output phase of MC switch configuration, (**c**) two phase of MC switch configuration, (**d**) three phase of MC switch configuration.

3.6. The Development of Modulation Methods

The development of varied technologies is related to the possibility of following the evolving trends in technology. In this subsection, it is shown that, in the discussed topologies, it is possible to implement recent commonly developed methods of predictive control.

Modulation strategies in AC/AC converters without capacitors in the intermediate circuit are not a simple issue due to the high number of transistors. AC/AC choppers have much simpler modulation strategies. Most modulation techniques have been determined for classical, direct MCs, and they are presented in a review article [22]. In recent years, MPC has been the best developed method of modulation. At discrete time T_d, MPC examines a model of a controlled system and predicts its condition in the next step. The configuration of converter switches is selected to ensure a minimum value of the cost function [40]. Scientific publications with MPC of an MC show better achievements in the quality of current and voltage waveforms, torque ripple, and internal switching power losses [30,41,42]. An example MPC scheme for the MC is shown in Figure 16, and this example is one that minimizes the cost function related to the accuracy of shaping the output current and the input power factor. Development of predictive control methods in AC systems without DC-link capacitors is a future area of extensive conceptual and implementation research.

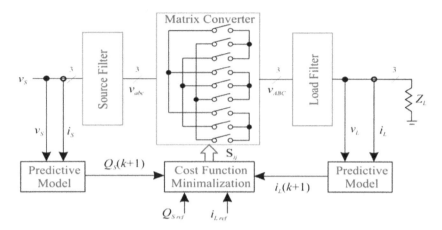

Figure 16. Model predictive control (MPC) of an MC with minimalization of the output current ripples and input power factor (reactive power).

The AC/AC chopper modulation algorithms are not complicated and are based only on the change in the control pulse duty factor. Research on control algorithms is mainly related to increasing efficiency, identifying faults, and optimizing their operation during component failures. New system structures are also being developed that produce significantly better properties regarding efficiency and power density [43,44].

4. Novel Applications of AC/AC Converters without DC-Link Capacitor

The classic application of frequency converters such as the matrix converter is to variable speed electric drives, while the AC/AC choppers are used in applications such as industrial process heating, the control of lighting intensity, or the soft-start of an induction AC motor. These applications are ubiquitously described in the literature and will not be analyzed in this article.

New potential areas for the application of choppers and AC/AC frequency converters without a DC-link circuit are in such devices as compensators for voltage fluctuations in the electric power grid and power flow controllers [19,45,46]. An example of such a compensator with a power flow control function is the hybrid transformer (HT) with a matrix converter or a matrix chopper [45,46], which is shown in Figure 17. The HT contains two main units: (1) a conventional three-phase transformer with two pairs of secondary windings and (2) a three-phase AC/AC converter. The first pair of output windings supplies the converter, whereas the second pair is connected in series with the output terminals of the converter. The compensating voltage is generated at the output of the converter. The output voltage of the whole HT is the sum of the power converter output voltage and the second secondary winding voltage of the transformer. The output terminals of the HT are connected with a sensitive load or a protected small industrial plant. The disadvantage of the HT is the necessity of using transformers with two pairs of secondary windings. Furthermore, under normal supply voltage conditions, the inverter can be switched off, resulting in this part of the transformer generating additional power losses.

A hybrid transformer with a matrix chopper allows compensation only for the amplitude of the supply voltage; using a matrix converter, it is possible to compensate for both symmetrical and asymmetrical voltage changes as well as harmonic distortions. As an example, to illustrate the beneficial properties of an HT with an MC, the results of the MPC algorithm (Figure 18) are presented in Figure 19 [47]. As can be seen from Figure 19, the output voltage of the compensator is kept at a constant amplitude with no harmonic distortion, despite large fluctuations in the main voltage.

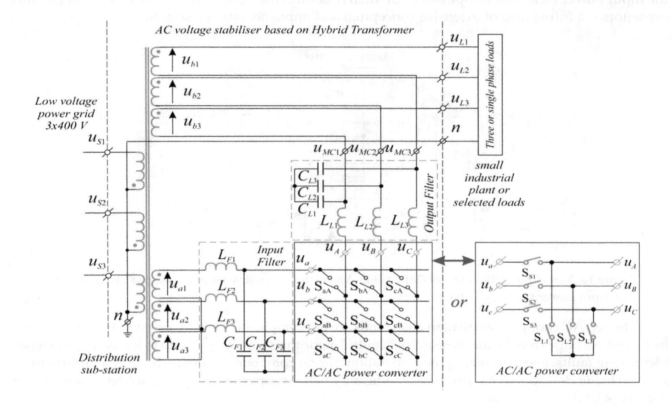

Figure 17. Compensator for voltage fluctuations, based on the HT with an MC or matrix chopper, installed at connection terminals of an industrial plant, a building, or selected industrial loads.

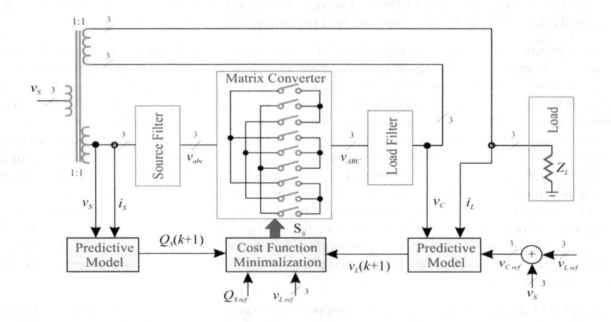

Figure 18. MPC of an HT voltage compensator with an MC and minimalization of the output voltage tracking error and input power factor.

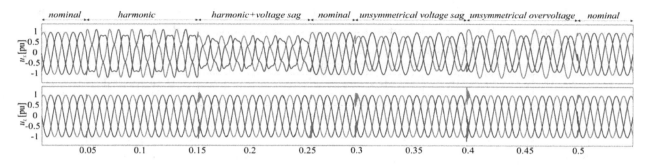

Figure 19. Results of compensations of electric power grid voltage fluctuation using an HT with an MC and MPC control (author's simulation results); red – voltage phase 1, blue – voltage phase 2, green – voltage phase 3.

If such an HT were connected to a dual-source network, it would be possible to regulate the power flow in the system by changing the phase angle of the compensator output voltage [46]. As can be seen from the above applications, AC/AC converters without a DC-link capacitor can be used in non-standard applications in low voltage power grids.

The main requirements for devices installed in the power system are a high reliability and a high efficiency factor. For an AC/AC converter, the parts most sensitive to damage are the power electronic elements. The installed converters should permit continuous work even after the occurrence of a fault in the power electronic converter, ensuring greater reliability. The reliability is high if the load can work with source voltage parameters (or slightly modified) even when damage has occurred in the power electronic part [19]. In this case, the bypass circuits should be used to disconnect or short-circuit the defective converter. In Figure 20, the single phase equivalent models of example AC/AC converters without a DC-link capacitor installed in the power system are presented. In the case of damage to the power electronic unit (AC/AC), the topologies based on the configuration shown in Figure 20a,b, could still operate with source (v_S) parameters, after activation of the additional bypass switch. In the case of the configuration shown in Figure 20c, the reliability can be assessed as low. In the event of damage to the AC/AC converter, the whole circuit must be turned off, because it is unable to continue to operate. As a result, all receivers connected to the load side have to be shut off.

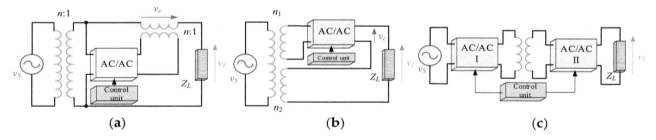

Figure 20. Single phase equivalent models of AC/AC converters installed in the power grid, (**a,b**) high reliability configurations, (**c**) low reliability configuration.

5. Conclusions

The presented converter offers a number of important benefits. However, such devices are still not widely used. Their application at the industrial level has been halted due to the fact that there are a number of challenges that must be overcome. This article has identified the challenges faced by the designers of such systems. The main aim of this article was to identify the problems that need to be addressed by research to enable the use of such converters without DC energy storage elements.

The main barrier to the development of AC/AC converters is due to the small number of available semiconductor power components. In addition, there are virtually no solutions in the form of intelligent modules that also include gate drivers and systems providing additional measuring and protection functions. It should be pointed out that minimizing the size of AC/AC converters without a DC-link circuit will become possible as a result of the development of modules which will be dedicated and optimized in terms of functionality, size, and performance of intelligent power modules. Another constraint is met in the complex algorithms for the modulation and commutation of transistors, which requires the construction of complex systems most often equipped with DSP and FPGA. In addition, the complexity of the above-mentioned algorithms also involves the use of additional electronic circuits, such as current flow detectors or a clamp protection circuit.

Finally, the article identifies the application possibilities of the analyzed topologies in modern power systems. The potential of modern control algorithms based on MPC has also been indicated. Comparative analysis shows that the use of such algorithms produces very good properties in voltage or current signal shapes.

References

1. Yang, Y.; Wang, H.; Sangwongwanich, A.; Blaabjerg, F. Design for Reliability of Power Electronic Systems. In *Power Electronics Handbook*, 4th ed.; Butterworth-Heinemann: Oxford, UK, 2018; pp. 1423–1440.
2. Chung, H.S.-H.; Wang, H.; Blaabjerg, F.; Pecht, M. *Reliability of Power Electronic Converter Systems*; The Institution of Engineering and Technology: London, UK, 2015.
3. Sang, Z.; Mao, C.; Lu, J.; Wang, D. Analysis and Simulation of Fault Characteristics of Power Switch Failures in Distribution Electronic Power Transformers. *Energies* **2013**, *6*, 4246–4268. [CrossRef]
4. Fischer, K.; Pelka, K.; Puls, S.; Poech, M.-H.; Mertens, A.; Bartschat, A.; Tegtmeier, B.; Broer, C.; Wenske, J. Exploring the Causes of Power-Converter Failure in Wind Turbines based on Comprehensive Field-Data and Damage Analysis. *Energies* **2019**, *12*, 593. [CrossRef]
5. Choi, U.M.; Blaabjerg, F.; Lee, K.B. Study and Handling Methods of Power IGBT Module Failures in Power Electronic Converter Systems. *IEEE Trans. Power Electron.* **2015**, *30*, 2517–2533. [CrossRef]
6. Kamel, T.; Biletskiy, Y.; Chang, L. Fault Diagnoses for Industrial Grid-Connected Converters in the Power Distribution Systems. *IEEE Trans. Ind. Electron.* **2015**, *62*, 6496–6507. [CrossRef]
7. Peftitsis, D.; Rabkowski, J. Gate and Base Drivers for Silicon Carbide Power Transistors: An Overview. *IEEE Trans. Power Electron.* **2016**, *31*, 7194–7213. [CrossRef]
8. Marzoughi, A.; Burgos, R.; Boroyevich, D. Characterization and Performance Evaluation of the State-of-the-Art 3.3 kV 30 A Full-SiC MOSFETs. *IEEE Trans. Ind. Appl.* **2019**, *55*, 575–583. [CrossRef]
9. Zhang, L.; Yuan, X.; Wu, X.; Shi, C.; Zhang, J.; Zhang, Y. Performance Evaluation of High-Power SiC MOSFET Modules in Comparison to Si IGBT Modules. *IEEE Trans. Power Electron.* **2019**, *34*, 1181–1196. [CrossRef]
10. Jones, E.A.; Wang, F.F.; Costinett, D. Review of Commercial GaN Power Devices and GaN-Based Converter Design Challenges. *IEEE J. Emerg. Sel. Top. Power Electron.* **2016**, *4*, 707–719. [CrossRef]
11. Wang, H.; Blaabjerg, F. Reliability of capacitors for DC-Link applications in power electronic converters—An overview. *IEEE Trans. Ind. Appl.* **2014**, *50*, 3569–3578. [CrossRef]
12. Yun, G.; Cho, Y. Active Hybrid Solid State Transformer Based on Multi-Level Converter Using SiC MOSFET. *Energies* **2019**, *12*, 66. [CrossRef]
13. Zhu, Q.; Dai, W.; Guan, L.; Tan, L.; Li, Z.; Xie, D. A Fault-Tolerant Control Strategy of Modular Multilevel Converter with Sub-Module Faults Based on Neutral Point Compound Shift. *Energies* **2019**, *12*, 876. [CrossRef]
14. Volosencu, C. *System Reliability*; Intech Open: Rijeka, Croatia, 2017.
15. Kolar, J.W.; Friedli, T.; Rodriguez, J.; Wheeler, P.W. Review of three-phase PWM AC–AC converter topologies. *IEEE Trans. Ind. Electron.* **2011**, *58*, 4988–5006. [CrossRef]
16. Szcześniak, P.; Kaniewski, J.; Jarnut, M. AC/AC power electronic converters without DC energy storage: A review. *Energy Convers. Manag.* **2015**, *92*, 483–497. [CrossRef]
17. Zhang, J.; Li, L.; Dorrell, D.G. Control and applications of direct matrix converters: A review. *Chin. J. Electr. Eng.* **2018**, *4*, 18–27.
18. Empringham, L.; Kolar, J.W.; Rodriguez, J.; Wheeler, P.W.; Clare, J.C. Technological issues and industrial application of matrix converters: A review. *IEEE Trans. Ind. Electron.* **2013**, *60*, 4260–4271. [CrossRef]
19. Kaniewski, J.; Fedyczak, Z.; Szcześniak, P. AC voltage transforming circuits in power systems. *Electr. Rev.* **2015**, *91*, 8–17.
20. Kaniewski, J.; Szcześniak, P.; Jarnut, M.; Benysek, G. Hybrid voltage sag/swell compensator. A review of AC/AC converters. *IEEE Ind. Electron. Mag.* **2015**, *9*, 37–48. [CrossRef]

21. Szcześniak, P.; Kaniewski, J. Power electronics converters without DC energy storage in the future electrical power network. *Electr. Power Syst. Res.* **2015**, *129*, 194–207. [CrossRef]

22. Andreu, J.; Kortabarria, I.; Ormaetxea, E.; Ibarra, E.; Martin, J.L.; Apiñaniz, S. A step forward towards the development of reliable matrix converters. *IEEE Trans. Ind. Electron.* **2012**, *59*, 167–183. [CrossRef]

23. Rodriguez, J.; Rivera, M.; Kolar, J.W.; Wheeler, P.W. A review of control and modulation methods for matrix converters. *IEEE Trans. Ind. Electron.* **2012**, *59*, 58–70. [CrossRef]

24. Safari, S.; Castellazzi, A.; Wheeler, P. Experimental and Analytical Performance Evaluation of SiC Power Devices in the Matrix Converter. *IEEE Trans. Power Electron.* **2014**, *29*, 2584–2596. [CrossRef]

25. Empringham, L.; de Lillo, L.; Schulz, M. Design Challenges in the Use of Silicon Carbide JFETs in Matrix Converter Applications. *IEEE Trans. Power Electron.* **2014**, *29*, 2563–2573. [CrossRef]

26. Trentin, A.; de Lillo, L.; Empringham, L.; Wheeler, P.; Clare, J. Experimental Comparison of a Direct Matrix Converter Using Si IGBT and SiC MOSFETs. *IEEE J. Emerg. Sel. Top. Power Electron.* **2015**, *3*, 542–554. [CrossRef]

27. Trentin, A.; Empringham, L.; de Lillo, L.; Zanchetta, P.; Wheeler, P.; Clare, J. Experimental Efficiency Comparison Between a Direct Matrix Converter and an Indirect Matrix Converter Using Both Si IGBTs and SiC mosfets. *IEEE Trans. Ind. Appl.* **2016**, *52*, 4135–4145. [CrossRef]

28. Buccella, C.; Cecati, C.; Latafat, H. Digital Control of Power Converters—A Survey. *IEEE Trans. Ind. Inform.* **2012**, *8*, 437–447. [CrossRef]

29. Kobravi, K.; Iravani, R.; Kojori, H.A. Three-Leg/Four-Leg Matrix Converter Generalized Modulation Strategy—Part II: Implementation and Verification. *IEEE Trans. Ind. Electron.* **2013**, *60*, 860–872. [CrossRef]

30. Gulbudak, O.; Santi, E. FPGA-Based Model Predictive Controller for Direct Matrix Converter. *IEEE Trans. Ind. Electron.* **2016**, *63*, 4560–4570. [CrossRef]

31. Wiśniewski, R.; Bazydło, G.; Szcześniak, P.; Wojnakowski, M. Petri Net-Based Specification of Cyber-Physical Systems Oriented to Control Direct Matrix Converters with Space Vector Modulation. *IEEE Access* **2019**, *7*, 23407–23420. [CrossRef]

32. Yamamoto, E.; Hara, H.; Uchino, T.; Kawaji, M.; Kume, T.J.; Kang, J.-K.; Krug, H.-P. Development of MCs and its Applications in Industry. *IEEE Ind. Electron. Mag.* **2011**, *5*, 4–12. [CrossRef]

33. Oancea, C.D.; Dinu, C. LEM transducers interface for voltage and current monitoring. In Proceedings of the 9th International Symposium on Advanced Topics in Electrical Engineering (ATEE'15), Bucharest, Romania, 7–9 May 2015; pp. 949–952.

34. Sozański, K. *Digital Signal Processing in Power Electronics Control Circuits*, 2nd ed.; Springer: London, UK, 2017.

35. Bobrowska-Rafal, M.; Rafal, K.; Jasinski, M.; Kazmierkowski, M.P. Grid synchronization and symmetrical components extraction with PLL algorithm for grid connected power electronic converters—A review. *Bull. Pol. Acad. Sci. Tech. Sci.* **2011**, *59*, 485–497. [CrossRef]

36. Irmak, E.; Colak, I.; Kaplan, O.; Guler, N. Design and application of a novel zero-crossing detector circuit. In Proceedings of the International Conference on Power Engineering, Energy and Electrical Drives, Malaga, Spain, 11–13 May 2011.

37. Iyer, A.R.; Kandula, R.P.; Moghe, R.; Hernandez, J.E.; Lambert, F.C.; Divan, D. Validation of the Plug-and-Play AC/AC Power Electronics Building Block (AC-PEBB) for Medium-Voltage Grid Control Applications. *IEEE Trans. Ind. Appl.* **2014**, *50*, 3549–3557. [CrossRef]

38. Laka, A.; Barrena, J.A.; Chivite-Zabalza, J.; Rodriguez, M.A. Analysis and Improved Operation of a PEBB-Based Voltage-Source Converter for FACTS Applications. *IEEE Trans. Power Deliv.* **2013**, *28*, 1330–1338. [CrossRef]

39. You, K.; Xiao, D.; Rahman, M.F.; Uddin, M.N. Applying Reduced General Direct Space Vector Modulation Approach of AC/AC Matrix Converter Theory to Achieve Direct Power Factor Controlled Three-Phase AC-DC Matrix Rectifier. *IEEE Trans. Ind. Appl.* **2014**, *50*, 2243–2257. [CrossRef]

40. Rodríguez, J.; Cortes, P. *Predictive Control of Power Converters and Electrical Drives*; John Wiley & Sons: Hoboken, NJ, USA, 2012.

41. Zhang, J.; Norambuena, M.; Li, L.; Dorrell, D.; Rodriguez, J. Sequential Model Predictive Control of Three-Phase Direct Matrix Converter. *Energies* **2019**, *12*, 214. [CrossRef]

42. Rivera, M.; Rojas, C.; Rodriguez, J.; Wheeler, P.W.; Wu, B.; Espinoza, J.R. Predictive Current Control with Input Filter Resonance Mitigation for a Direct Matrix Converter. *IEEE Trans. Power Electron.* **2011**, *26*, 2794–2803. [CrossRef]

43. Khan, A.A.; Cha, H.; Ahmed, H.F. High-Efficiency Single-Phase AC–AC Converters Without Commutation Problem. *IEEE Trans. Power Electron.* **2016**, *31*, 5655–5665. [CrossRef]

44. Khan, A.A.; Cha, H.; Kim, H. Three-Phase Three-Limb Coupled Inductor for Three-Phase Direct PWM AC–AC Converters Solving Commutation Problem. *IEEE Trans. Ind. Electron.* **2016**, *63*, 189–201. [CrossRef]

45. Kaniewski, J.; Fedyczak, Z.; Benysek, G. AC voltage sag/swell compensator based on three-phase hybrid transformer with buck-boost matrix-reactance chopper. *IEEE Trans. Ind. Electron.* **2014**, *61*, 3835–3846. [CrossRef]

46. Szcześniak, P.; Kaniewski, J. Hybrid transformer with matrix converter. *IEEE Trans. Power Deliv.* **2016**, *31*, 1388–1396. [CrossRef]

47. Szcześniak, P.; Tadra, T.; Kaniewski, J.; Fedyczak, Z. Model predictive control algorithm of AC voltage stabilizer based on hybrid transformer with a matrix converter. *Electr. Power Syst. Res.* **2019**, *170*, 222–228. [CrossRef]

10

Nonsingular Terminal Sliding Mode Control Based on Binary Particle Swarm Optimization for DC–AC Converters

En-Chih Chang [1], Chun-An Cheng [1] and Lung-Sheng Yang [2,*]

[1] Department of Electrical Engineering, I-Shou University, No.1, Sec. 1, Syuecheng Rd., Dashu District, Kaohsiung City 84001, Taiwan; enchihchang@isu.edu.tw (E.-C.C.); cacheng@isu.edu.tw (C.-A.C.)
[2] Department of Electrical Engineering, Far East University, No.49, Zhonghua Rd., Xinshi Dist., Tainan City 74448, Taiwan
* Correspondence: yanglungsheng@yahoo.com.tw

Abstract: This paper proposes an improved feedback algorithm by binary particle swarm optimization (BPSO)-based nonsingular terminal sliding mode control (NTSMC) for DC–AC converters. The NTSMC can create limited system state convergence time and allow singularity avoidance. The BPSO is capable of finding the global best solution in real-world application, thus optimizing NTSMC parameters during digital implementation. The association of NTSMC and BPSO extends the design of classical terminal sliding mode to converge to non-singular points more quickly and introduce optimal methodology to avoid falling into local extremum and low convergence precision. Simulation results show that the improved technique can achieve low total harmonic distortion (THD) and fast transients with both plant parameter variations and sudden step load changes. Experimental results of a DC–AC converter prototype controlled by an algorithm based on digital signal processing have been shown to confirm mathematical analysis and enhanced performance under transient and steady-state load conditions. Since the improved DC–AC converter system has significant advantages in tracking accuracy and solution quality over classical terminal sliding mode DC–AC converter systems, this paper will be applicable to designers of relevant robust control and optimal control technique.

Keywords: binary particle swarm optimization (BPSO); nonsingular terminal sliding mode control (NTSMC); global best solution; total harmonic distortion (THD); DC–AC converter

1. Introduction

DC–AC converters have been widely applied in renewable energy systems, such as solar photovoltaic (PV) energy systems, wind turbine generator systems and fuel cell power generation systems. For example, a solar PV energy system can convert sunlight into usable electrical energy. The simplified solar PV systems include PV panels, DC–DC converters, DC–AC converters and loads. Such system can be designed to yield maximum power delivered to the load. There are two power conversion stages in this structure, so it can be considered a two-stage system. The DC–DC converter is used to handle maximum power point tracking (MPPT) and regulate the DC load voltage. At the same time as the grid connection occurs, the power is generated by the PV panel and converted to AC power by the DC–AC converter. Furthermore, the operation of DC–AC conversion and maximum power point tracking (MPPT) can be combined into a single-stage system using only one DC–AC converter. In various power converter topologies for renewable energy applications, inductor capacitor (LC) filter DC–AC converters are often used as an interface between renewable energy and the grid. The converter DC link is connected to the PV panel either directly or through an intermediate DC–DC power conversion stage. The LC low-pass filter eliminates higher harmonics in the converter's pulse width modulation (PWM) output, enabling pure sine. Therefore, even under plant parametric variations and external load disturbances, the requirements of high-performance DC–AC converters must involve fast dynamic response and low total harmonic distortion (THD) of the output voltage. In order to meet these requirements, a proportional plus integral (PI) controller is frequently used; nevertheless, the controller may not be able to withstand severe disturbances, thereby degrading the performance of the system [1,2]. In order to obtain better tracking accuracy, the different control schemes are discussed in the research literature [3–7].

The repetitive control related to the H-infinity concept is proposed for the inductor-capacitor-inductor (LCL) grid-tied inverter to achieve near-zero steady-state error and reduce harmonic distortion of the output voltage caused by the nonlinear load. However, this method requires a complex control algorithm [3]. A simpler fractional repetitive method is developed for the voltage control of the microgrid to suppress the generation of harmonics. Although the structure is simple and exhibits a rapid dynamic response when the load suddenly changes, the steady-state response is not significantly improved [4]. A deadbeat control based on predictive model is proposed to control the grid-connected inverter. The proposed inverter with nonlinear load shows good steady state, but this method depends largely on the accuracy of the parameters. The transient performance is somewhat mediocre [5]. The improved direct deadbeat voltage control is applied to the closed-loop regulation of an island AC microgrid. This method is sensitive to changes in plant parameters, and even if the system exhibits a fast dynamic response, it may lead to non-zero steady-state errors [6]. Since the problem of parameter uncertainty and external disturbance can be reduced, it is recommended to combine mu-synthesis with the H-infinity method for island microgrid control; however, the mu-synthesis algorithm complicates the digital implementation and the resulting waveform has visible distortion, especially in strong nonlinear cases [7].

Sliding mode control (SMC) has inherent robustness to system uncertainty and is used as an effective technique in many different engineering fields [8–10]. The SMC system theory and its related sliding surfaces have been widely used in control design for the past 40 years, and the application fields are increasing (power system control, aerospace design problems, robot manipulator control) [11–14]. The primary purpose of the sliding behavior in the SMC direction allows the system state to tend to a predetermined desired hyperplane, i.e., a sliding surface or slip manifold defined in the state space. Once the state trajectory hits the sliding surface, it enters the sliding mode and stays there; after that, the system can achieve its control objectives and can suppress internal parameter changes and external load disturbances [15–18]. Of course, the controller of the DC–AC converter is also universally designed by SMC [19–21]. For the single-phase inverter, a fixed switching frequency sliding mode is proposed; the control design adopts the traditional sliding surface, which causes the output voltage distortion under non-linear load [19]. In order to retrieve the incomplete system dynamics of the grid-connected inverter, a sliding surface based on multi-resonance is designed. Although it can enhance the performance of steady state and transient, this algorithm is time-consuming calculation [20]. The improved SMC shows the ability to suppress uncertainty interference for voltage regulation in the microgrid, but it has complex hardware design and significant jitter [21]. As mentioned above, these classical SMC methods have problems with non-time-limited convergence and jitter.

In the case of a path tracking system, the invariant characteristics exhibited by the classical SMC are only maintained during the sliding phase, and the tracking trajectory may be affected by external load disturbances or changes in internal parameters of the arrival phase. Previous research efforts have attempted to reduce tracking errors and speed up arrival times. The observer can shorten the arrival time, but it uses a traditional reduced-order design, resulting in large jitter, which is not desirable in dynamic systems [22]. In order to eliminate the phase of arrival, a time varying sliding surface given by the constraint of zero error under initial conditions is applied to a rotary actuator system. However, it does not conform to the general situation because the initial conditions can be arbitrarily assigned in the actual system [23]. An indirect sliding mode power control method was developed to control the grid-connected power converter. Although this method allows the system to slide to the sliding surface in a suitable short time, there is a phenomenon of jitter around the sliding surface [24].

In recent years, an interesting series of SMC controllers, named nonsingular terminal sliding mode (NTSMC) has allowed finite time convergence, and overcome the singularity problem [25]. The NTSMC has been well applied in various fields [26,27]. Although the NTSMC can drive the system state to converge to the origin within a limited time while still retaining the robustness of the classical SMC, it has a jitter problem [28]. From a practical point of view, system parameter changes, external load disturbances, and unmodeled dynamics are difficult to know. If the system uncertainty limit is large or small, the jitter or steady-state error may occur, and the existence and invariance of the sliding mode cannot be guaranteed. Many studies have used adaptive control methods to tackle the effect of the jitter caused by boundary uncertainty. Such solutions effectively reduce the jitter and steady-state error, enhancing both transient and steady-state behavior [29–38]. In addition, the NTSMC has the difficulty in choosing optimal controller parameters particularly in face of large variations of model parameters and load changes.

In the upcoming era of artificial intelligence, the BPSO method has been widely applied in solving optimization problems due to its simplicity, fast execution and high-quality solution [39–43]. For this reason, the BPSO is used to find the optimal values of the NTSMC parameters, therefore significantly improving the control performance and avoiding the tedious trial and error tuning. This improved technique provides another option and potential recommendation as opposed to not adding optimal methodology in the classical terminal sliding mode or SMC. Although the final performance results of the improved system are not superior to the recent THD results of the previous work, it does improve the TSMC method and produces a systematic optimal tuning for the determination of the controller parameters. It may be noted that the association of the presented NTSMC and BPSO results in a closed loop feedback DC–AC converter system that has low THD under steady state load and a fast response under transient load. Finally, the efficacy of the improved technique is verified by the implementation of a digital signal processing (DSP)-based DC–AC converter system, and the improved system is also evaluated by MATLAB/Simulink software.

2. Modeling of DC–AC Converter

Figure 1 depicts a commonly used DC–AC converter consisting of a full-bridge switching element with MOSFETs (metal-oxide-semiconductor field-effect transistors), an LC filter and a resistive load. The DC bus voltage V_s is expressed by the output voltage v_c, and the load R is represented by the output current i_o. In the case of a situation $x_1 = v_c$ and $x_2 = \dot{v}_c$ defined as a state variable, the state equation of the system can be written as

$$\dot{x} = Ax + Bu \tag{1}$$

where $x = \begin{bmatrix} x_1 & x_2 \end{bmatrix}^T$, $A = \begin{bmatrix} 0 & 1 \\ 1/LC & 1/RC \end{bmatrix}$, $B = \begin{bmatrix} 0 & K_{PWM}/LC \end{bmatrix}^T$, and u is the control signal. When the switching frequency is much higher than the fundamental frequency of the AC output, it can be considered as the proportional gain K_{PWM} of the PWM full-bridge switching element and is consistent with $V_s/\hat{v}_{tr}$; $\hat{v}_{tr}$ represents the amplitude of the triangular wave v_{tr} in the PWM. According to (1), the output voltage v_c must follow a sinusoidal reference voltage v_r. Therefore, the tracking request $v_c(t) \rightarrow v_r(t)$ as $t \rightarrow \infty$ is maintained and the design problem of the DC–AC converter will be the trajectory tracking control problem. The tracking errors $e = \begin{bmatrix} e_1 & e_2 \end{bmatrix}^T$ is defined as

$$e = x - x_r \tag{2}$$

where $x_r = \begin{bmatrix} v_r & \dot{v}_r \end{bmatrix}^T$.

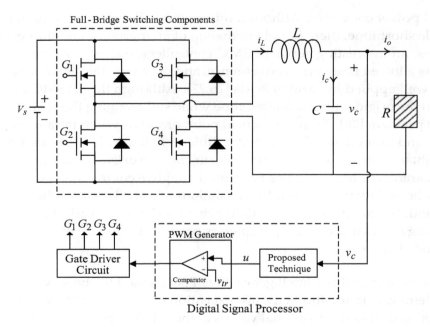

Figure 1. Structure of DC–AC converter.

It can be seen from (1) and (2) that the error state equation of the DC–AC converter can be expressed as

$$\begin{cases} \dot{e}_1 = e_2 \\ \dot{e}_2 = -\frac{1}{LC}e_1 - \frac{1}{RC}e_2 + \frac{K_{PWM}}{LC}u - N \end{cases} \tag{3}$$

where $N = \frac{1}{LC}v_r + \frac{1}{RC}\dot{v}_r + \ddot{v}_r$ is the disturbance.

From (3), the control signal u must be designed so well that the tracking error e can converge to zero. In fact, the NTSMC is a fine control method with nonsingular fast convergence characteristics. The system (3) using NTSMC will achieve fast finite time convergence, strong robustness and infinite stability. However, when the load on a DC–AC converter is a large step change or uncertainty or even a severe nonlinear environment, a global optimization algorithm for the systematic and optimal choice of NTSMC parameter values is becoming more important. Based on such a motivation, and the practical application of artificial intelligence method is rapidly becoming a hot topic in engineering and science, it is a good idea to introduce an optimal methodology in the NTSMC design, providing an alternative reference for researchers. Therefore, NTSMC with BPSO method is proposed to improve the transience and steady-state behaviors of the classical TSMC to provide more accurate tracking. A DC–AC converter using this improved control design can produce a higher performance AC output voltage.

3. Proposed Control Technique

3.1. Problem Statement

First, a brief summary of the problem statement for a nonlinear system using the classical TSMC, is summarized and an improved technique is then designed. Consider the second-order uncertain nonlinear dynamic systems below:

$$\begin{aligned} \dot{x}_1 &= x_2 \\ \dot{x}_2 &= f(x) + g(x) + b(x)u \end{aligned} \tag{4}$$

where the system state is $x = \begin{bmatrix} x_1 & x_2 \end{bmatrix}^T$, $f(x)$ and $b(x)$ stands for a smooth nonlinear function x, $g(x)$ denotes parameter uncertainty and external disturbance and u represents the control input.

To achieve finite time convergence of the system state, the following first-order terminal sliding variable can be defined as

$$s = x_2 + \mu x_1^{\gamma} \tag{5}$$

where $\mu > 0$ is the design constants and $0 < \gamma = \gamma_1/\gamma_2 < 1$ (γ_1 and γ_2 are positive odd integers).

An SMC law $u = u^+(x)$, $u^-(x)$ for $s > 0$, $s < 0$ can be used, which is expressed as driving s to the sliding mode $s = 0$ for a limited time. Therefore, system dynamics can be controlled by the following differential equations:

$$x_2 + \mu x_1^{\gamma} = \dot{x}_1 + \mu x_1^{\gamma} = 0 \tag{6}$$

The limited time t_s from the initial state $x_1(0)$ to zero can be determined by

$$t_s = \frac{|x_1(0)|^{1-\gamma}}{\mu(1-\gamma)} \tag{7}$$

This indicates the convergence of the two system states x_1 and x_2 the convergence of zero in a finite time in the NTSMC manifold. Considering the Jacobian matrix J, the system state converges to zero gain in a finite time around the equilibrium $x_1 = 0$.

$$J = \frac{\partial \dot{x}_1}{\partial x_1} = -\frac{\mu\gamma}{x_1^{(1-\gamma)/\gamma}} \tag{8}$$

From (8), we can obtain the eigenvalues of the first-order approximation matrix as follows.

$$J \to -\infty \quad \text{when } x_1 \to 0 \tag{9}$$

It is inferred that the eigenvalue has a tendency of negative infinity at the equilibrium point, that is, the speed of the system trajectory to the equilibrium becomes infinite, resulting in limited time accessibility.

Thus, for error dynamics (3), the finite-time terminal sliding function can be expressed as

$$s = \dot{e}_1 + \mu e_1^{\gamma} \tag{10}$$

Using the (10), the $s = 0$ and e_1 arrive within a limited time. The control law can be designed to ensure that TSM occurs as follows.

$$u = u_e + u_s \tag{11}$$

with

$$u_e(t) = b^{-1}[a_1 e_1 + a_2 e_2 - \mu(\gamma e_1^{\gamma-1} \cdot e_2)] \tag{12}$$

$$u_s(t) = -b^{-1}[K \operatorname{sgn}(s)], K > |N(t)| \tag{13}$$

where $a_1 = 1/LC$, $a_2 = 1/RC$, $b = K_{PWM}/LC$, and u_e called the equivalent control component, control undisturbed plants, such that $s = 0$ and $\dot{s} = 0$. The named sliding control component suppresses system uncertainty. Therefore, the state trajectory will reach the sliding mode $s = 0$ and perform a limited system state convergence time. However, it is worth noting that there are the following problems in the equivalent control component: (i) If $e_2 \neq 0$ when $e_1 = 0$ and $0 < \gamma < 1$, the $u_e(t)$ with $e_1^{\gamma-1} e_2$ may result in a singularity. This singularity causes the control law to produce an unbounded control signal, resulting in an unstable closed loop system. (ii) An imaginary number $e_1^{\gamma-1}$ can be generated in a given situation $0 < \gamma < 1$.

3.2. Control Design

To overcome the singularity problem, the (10) is reconstructed into

$$s = e_1 + \frac{1}{\lambda} e_2^{\frac{q}{p}} \tag{14}$$

where $\lambda > 0$ and p, q are positive odd numbers ($p < q < 2p$). Then, a sliding-mode reaching equation $\dot{s} = -\eta_1 s - \eta_2 |s|^{1-\gamma} \text{sgn}(s)$ is employed. The control law can be expressed as

$$u(t) = u_{nft}(t) + u_s(t) \tag{15}$$

with

$$u_{nft}(t) = b^{-1}[a_1 e_1 + a_2 e_2 - \lambda \frac{p}{q} e_2^{2-\frac{q}{p}}] \tag{16}$$

$$u_s = -b^{-1}\left[\eta_1 s + \eta_2 |s|^{1-\gamma} \text{sgn}(s)\right], \eta_1, \eta_2 > 0, 0 < \gamma < 1 \tag{17}$$

where there is no negative index equivalent control u_{nft}, which results in non-singularity, and u_s represents the sliding control for compensating the influence of the disturbance. Therefore, the system state will be forced to arrive $s = 0$ and converge in a limited time.

Proof. Choose Lyapunov candidate as

$$V = \frac{1}{2}s^2 \tag{18}$$

Along the dynamic system trajectory (3) and the control law (15), and use (14), the time derivative V is given as

$$\begin{aligned}
\dot{V} &= s\dot{s} \\
&= s\left(\dot{e}_1 + \frac{1}{\lambda}\frac{q}{p}e_2^{\frac{q}{p}-1}\dot{e}_2\right) \\
&\leq -s\left(\frac{1}{\lambda}\frac{q}{p}e_2^{\frac{q}{p}-1}(\eta_1 s + \eta_2 |s|^{1-\gamma})\text{sgn}(s)\right)
\end{aligned} \tag{19}$$

Since $e_2^{q/p-1} > 0$, $\dot{V} \leq 0$, the surface of the NTSMC in (19) is allowed to converge to equilibrium in a limited time. Once $s = e_1 + \lambda^{-1}e_2^{q/p}$, the state of system (3) will also converge to equilibrium within a finite time. However, the load may be a large load disturbance that provides inaccurate tracking performance in the system (3), i.e., the output voltage of the DC–AC converter is not exactly equal to the desired sinusoidal waveform. It is thus important to find out the optimal values of the NTSMC parameters in the (15) to maintain satisfactory performance of the DC–AC converter. To avoid tedious, time-consuming trial-and-error calculations, and obtain global best solutions, the BPSO method is employed to get the optimal value of NTSMC parameters. Finally, the combination of the DC–AC converter in (3) with BPSO method and NTSMC is asymptotically stable, and then achieves finite-time convergence to zero of tracking errors. The BPSO algorithm can be used illustrated in (20) and (21). The (20) and (21) show the evolution models of a particle. The speed and position of each particle can be renovated while flying toward aim.

$$v_i^{t+1} = \sigma_0 v_i^t + \sigma_1 k_1 (x_i^{pb} - \chi_i) + \sigma_2 k_2 (x_i^{gb} - \chi_i) \tag{20}$$

$$\chi_i^{t+1} = \begin{cases} 0, & \text{if rand} \geq Sig(v_i^{t+1}) \\ 1, & \text{if rand} < Sig(v_i^{t+1}) \end{cases} \tag{21}$$

where σ_0, σ_1 and σ_2 denote variables, and k_1, k_2 indicate random numbers, v_i is present flying speed, χ_i stands for present position, x_i^{pb} is local best position, x_i^{gb} represents global best position, and Sig symbolizes sigmoid function that converts the particle velocity into the probability $1/1 + e^{-v_i^{t+1}}$. $\square$

4. Simulation and Experimental Results

In order to test the performance and robustness of the improved technique, the simulation and experimental results of the improved technique were compared to those obtained using the classical TSMC. The system parameters are as follows: The system parameter are listed as follows: $V_s = 200$ V,

$v_c = 110\,V_{rms}$, $f_o = 60\,Hz$, $v_r(t) = \sqrt{2} \cdot 110 \cdot \sin(2\pi \cdot 60 \cdot t)$, $f_s = 15\,kHz$, $L = 0.12\,mH$, $C = 2\,\mu F$, rated load $= 12\,\Omega$. Under the abrupt load change from no load to 12 ohm, the simulated results obtained using the improved technique and classical TSMC are shown in Figures 2 and 3, respectively. Compared to classical TSMC, the improved technique exhibits a slight voltage drop and allows for fast output voltage recovery to verify its limited time accessibility. Due to the optimal tuning of the BPSO, after an instantaneous voltage dip (7 V_{rms}), the output voltage with the improved technique can be restored to the sinusoidal reference voltage, but the classical TSMC results in a large voltage dip (22 V_{rms}). Figure 4 shows that the improved technique can allow random variations in filter parameters L and C from 10% to 200% and 10% to 200% of the nominal value under a 12 ohm resistive load, respectively; however, the classical TSMC shown in Figure 5 yields the significant oscillations and leads to a descent in system robustness. Table 1 shows the simulated comparison of the voltage drop and %THD of the output voltage for step load and LC variation. A prototype of the DC–AC converter depicted as Figure 6 is constructed. Figure 7 illustrates the experimental waveform obtained using the improved technique under the step load from no load to 12 ohm at a 90 degree firing angle.

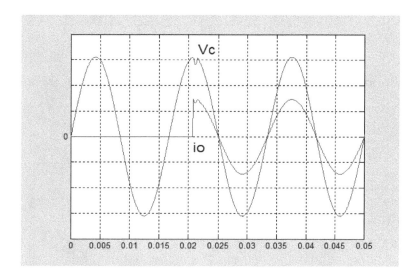

Figure 2. Simulated waveforms under step load change (load suddenly turn on) for the improved technique (50 V/div; 10 A/div).

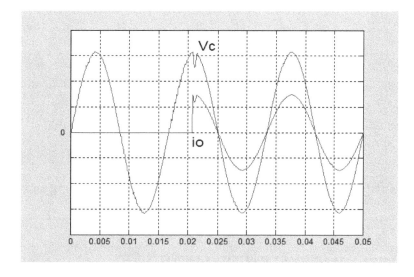

Figure 3. Simulated waveforms under step load change (load suddenly turn on) for the classical TSMC (50 V/div; 10 A/div).

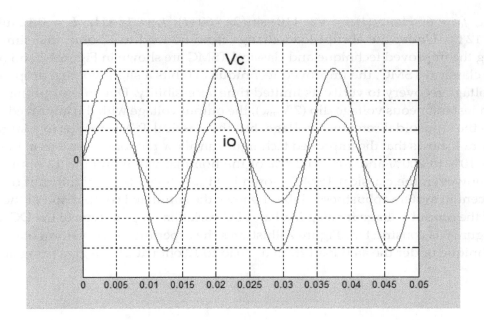

Figure 4. Simulated waveform under LC (inductor capacitor) variation for the improved technique (50 V/div; 10 A/div).

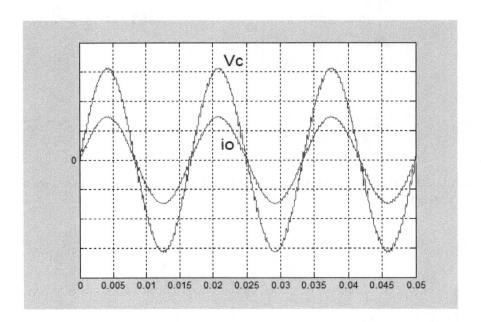

Figure 5. Simulated waveforms under LC variation for the classical terminal sliding mode control (TSMC) (50 V/div; 10 A/div).

Table 1. Simulated output-voltage slump and %THD under step loading and LC variation.

	Simulations	
Improved technique	Step loading (Voltage Slump)	7 V_{rms}
	LC variation (%THD)	0.23%
Classical TSMC	Step loading (Voltage Slump)	22 V_{rms}
	LC variation (%THD)	15.98%

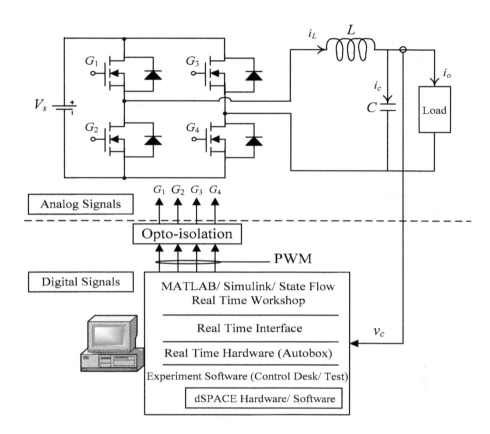

Figure 6. Improved algorithm for DSP (digital signal processing)-based implementation.

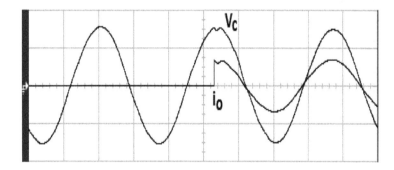

Figure 7. Experimental waveforms under step loading from no load to full load with the improved technique (100 V/div; 20 A/div; 5 ms/div).

After the fast transient response with a voltage slump, the voltage waveform still keeps high tracking precision. Reversely, the experimental waveform obtained using the classical TSMC plotted in Figure 8 appears a large voltage slump and has a slow recovery time. In other words, the output voltage of the improved system due to the action of the BPSO can reach the 110 V_{rms} reference sine wave after the 8 V_{rms} small voltage slump, but the output-voltage slump with the classical TSMC is close to 36 V_{rms}, yielding an unsatisfactory performance in transience. The experimental system performance under the value of filter parameter (L and C) be assumed to undergo a random variation, i.e., 10% ~ 200% of nominal value at a 12 ohm resistive load, is investigated. As can be seen, the output-voltage of the Figure 9 obtained using the improved technique provides the robust ability of greater parameter variation tolerance. However, it is worth noting that a long-time distortion with the sensitivity at the beginning of the waveform exists in the output-voltage with the classical TSMC as shown in Figure 10. Table 2 lists the experimental output-voltage slump and THD values under step load and LC variation.

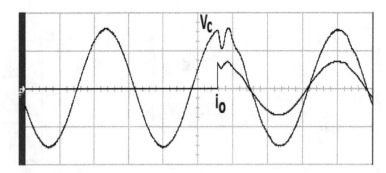

Figure 8. Experimental waveforms under step loading from no load to full load with the classical TSMC (100 V/div; 20 A/div; 5 ms/div).

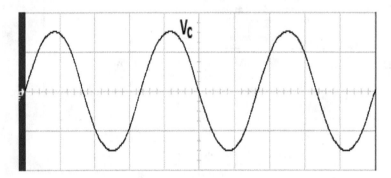

Figure 9. Experimental output-voltage under LC variation with the improved technique (100 V/div; 5 ms/div).

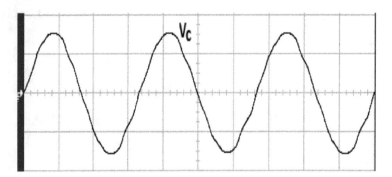

Figure 10. Experimental output-voltage under LC variation with the classical TSMC (100 V/div; 5 ms/div).

Table 2. Experimental output-voltage slump and %THD under step loading and LC variation.

	Experiments	
Improved technique	Step loading (Voltage Slump)	8 V_{rms}
	LC variation (%THD)	0.41%
Classical TSMC	Step loading (Voltage Slump)	36 V_{rms}
	LC variation (%THD)	11.43%

5. Discussion and Future Research

The improved technique has been proposed for reducing jitter, steady-state error attenuation and greater interference rejection resulting in good system performance. However, in order to advance future research, we have reviewed a large amount of literature on sliding modes, especially in high-order SMC (HOSMC) and adaptive SMC [30–33,38]. The adaptive SMC not only reduces the jitter and steady-state error but also avoids nominal knowledge requirement of the system. The HOSMC method [44,45] reported on a topic of recent interest in SMC theory. For example, in the rth-order HOSMC, the derivative of the $(r-1)$th control input becomes continuous, and both the sliding plane and its high-order derivative need to be zero. Therefore, the HOSMC retains the original features of the traditional SMC while producing less jitter and better convergence accuracy. The HOSMC method has been used to control DC–AC converter related systems [46–49]. The grid-connected wind energy conversion system is designed by multiple input multiple output HOSMC, so it can adjust active and reactive power, and even develop a switching control scheme based on voltage grid measurement. This approach yields attractive benefits such as robust robustness to system uncertainty, reduced jitter, and finite time convergence of system states to sliding surfaces [46]. A reverse-threshold HOSMC strategy is proposed for grid-connected distributed generation (DG) units. It provides excellent regulation of the inverter output current and provides a perfect sinusoidal balanced current for the grid, resulting in a distributed generator system with good performance against model uncertainties, parameter variations and unmodeled dynamics as well as external disturbances [47]. HOSM observers are effectively introduced into the control design of single-phase DC–AC inverter systems to suppress multiple sources of interference/uncertainty, including parametric perturbations, complex nonlinear dynamics, and external disturbances. Based on the Lyapunov function criterion, the stability of the whole system and the effectiveness of the high-order sliding mode observer are rigorously proved [48]. In addition, an improved HOSM observer is proposed for use in a proton exchange membrane fuel cell system based on excess oxygen ratio, which provides an observation of unmeasurable system conditions. Even under the influence of measurement noise, modeling error, parameter uncertainty and strong external interference, the method still has good robustness and fast convergence, and has limited time stability [49]. As mentioned above, HOSMC produces high-order derivative constraints on the sliding surface while preserving the main advantages of traditional SMC; it is undeniable that HOSMC eliminates jitter effects and produces more accurate control performance. Therefore, HOSMC will promote further research in this area of the DC–AC converter.

6. Conclusions

In this paper, a BPSO optimized NTSMC for a DC–AC converter is described that is capable of producing low THD and fast transients. The importance of the NTSMC is limited system state convergence time and no singularities. Also, the parameters of the NTSMC should be chosen well to obtain optimal performance. These parameters are traditionally determined by a trial and error method, which is very tedious, laborious to implement, and time-consuming. Therefore, the BPSO is used to optimize the NTSMC parameters, yielding better transient and steady-state response. The Lyapunov method is used to analyze the stability of the improved technique. The finite time accessibility of the sliding surface, the asymptotic stability of the closed-loop system and the finite time convergence of the tracking error are proved. Therefore, we believe that the improved technique will contribute to the control design of future artificial intelligence related systems. Simulations and experiments have been developed on prototypes of DC–AC converters using DSP to verify the applicability of the improved technique.

Author Contributions: E.-C.C. conceived, investigated and designed the circuit, and developed the methodology. C.-A.C. and L.-S.Y. prepared software resources, set up simulation software, and E.-C.C. performed circuit simulations. E.-C.C. carried out DC–AC converter prototype, and measured as well as analyzed experimental results. E.-C.C. wrote the paper and revised it for submission.

Acknowledgments: The authors gratefully acknowledge the financial support of the Ministry of Science and Technology of Taiwan, R.O.C., under project number MOST 107-2221-E-214-006.

References

1. Jeyraj, S.; Rahim, N.A. Multilevel inverter for grid-connected PV system employing digital PI controller. *IEEE Trans. Ind. Electron.* **2009**, *56*, 149–158.
2. Mirzaei, M.; Tibaldi, C.; Hansen, M.H. PI controller design of a wind turbine: Evaluation of the pole-placement method and tuning using constrained optimization. *J. Phys. Conf. Ser.* **2016**, *753*, 1–7. [CrossRef]

3.   Jin, W.; Li, Y.L.; Sun, G.Y.; Bu, L.Z. H-infinity repetitive control based on active damping with reduced computation delay for LCL-type grid-connected inverters. *Energies* **2017**, *10*, 586.

4.   Xie, C.; Zhao, X.; Savaghebi, M.; Meng, L.; Guerrero, J.M.; Quintero, J.C.V. Multirate fractional-order repetitive control of shunt active power filter suitable for microgrid applications. *IEEE J. Emerg. Sel. Top. Power Electron.* **2017**, *5*, 809–819. [CrossRef]

5.   Benyoucef, A.; Kara, K.; Chouder, A.; Silvestre, S. Prediction-based deadbeat control for grid-connected inverter with L-filter and LCL-filter. *Electr. Power Compon. Syst.* **2014**, *42*, 1266–1277. [CrossRef]

6.   Kim, J.; Hong, J.; Kim, H. Improved direct deadbeat voltage control with an actively damped inductor-capacitor plant model in an islanded AC microgrid. *Energies* **2016**, *42*, 978. [CrossRef]

7.   Bevrani, H.; Feizi, M.R. Robust frequency control in an islanded microgrid: H-infinity and mu-synthesis Approaches. *IEEE Trans. Smart Grid* **2016**, *7*, 706–717. [CrossRef]

8.   Ren, F.Y.; Lin, C.; Yin, X.H. Design a congestion controller based on sliding mode variable structure control. *Comput. Commun.* **2005**, *28*, 1050–1061. [CrossRef]

9.   Ignaciuk, P.; Bartoszewicz, A. *Congestion Control in Data Transmission Networks: Sliding Mode and Other Designs (Communications and Control Engineering)*; Springer: New York, NY, USA, 2013.

10. Tan, S.C.; Lai, Y.M.; Tse, C.K. *Sliding Mode Control of Switching Power Converters: Techniques and Implementation*; Taylor & Francis: Boca Raton, FL, USA, 2012.

11. Yu, X.; Kaynak, O. Sliding mode control with soft computing: A survey. *IEEE Trans. Ind. Electron.* **2009**, *56*, 3275–3285.

12. Kang, S.W.; Kim, K.H. Sliding mode harmonic compensation strategy for power quality improvement of a grid-connected inverter under distorted grid condition. *IET Power Electron.* **2015**, *8*, 1461–1472. [CrossRef]

13. Kaynak, A.B.; Utkin, V.I. Industrial applications of sliding mode: Control Part II. *IEEE Trans. Ind. Electron.* **2009**, *56*, 3271–3274. [CrossRef]

14. Kayacan, E.; Cigdem, O.; Kaynak, O. Sliding mode control approach for online learning as applied to type-2 fuzzy neural networks and its experimental evaluation. *IEEE Trans. Ind. Electron.* **2012**, *59*, 3510–3520. [CrossRef]

15. Vardan, M.; Ekaterina, A. *Sliding Mode in Intellectual Control and Communication: Emerging Research and Opportunities*; IGI Global: Hershey, PA, USA, 2017.

16. Derbel, N.; Ghommam, J.; Zhu, Q.M. *Applications of Sliding Mode Control*; Springer: Singapore, 2017.

17. Yan, X.G.; Spurgeon, S.K.; Edwards, C. *Variable Structure Control of Complex Systems: Analysis and Design*; Springer International Publishing: New York, NY, USA, 2017.

18. Andreas, R.; Luise, S. *Variable-Structure Approaches: Analysis, Simulation, Robust Control and Estimation of Uncertain Dynamic Processes*; Springer International Publishing: New York, NY, USA, 2016.

19. Abrishamifar, A.; Ahmad, A.A.; Mohamadian, M. Fixed switching frequency sliding mode control for single-phase unipolar inverters. *IEEE Trans. Power Electron.* **2012**, *27*, 2507–2514. [CrossRef]

20. Hao, X.; Yang, X.; Liu, T.; Huang, L.; Chen, W.J. A sliding-mode controller with multiresonant sliding surface for single-phase grid-connected VSI with an LCL filter. *IEEE Trans. Power Electron.* **2013**, *28*, 2259–2268. [CrossRef]

21. Aghatehrani, R.; Kavasseri, R. Sensitivity-analysis-based sliding mode control for voltage regulation in microgrids. *IEEE Trans. Sustain. Energy* **2013**, *4*, 50–57. [CrossRef]

22. Camila, L.C.; Tiago, R.O.; Jose, P.V.S.C. Output-feedback sliding-mode control of multivariable systems with uncertain time-varying state delays and unmatched non-linearities. *IET Control Theory Appl.* **2013**, *7*, 1616–1623.

23. Hung, L.C.; Lin, H.P.; Chung, H.Y. Design of self-tuning fuzzy sliding mode control for TORA system. *Expert Syst. Appl.* **2007**, *32*, 201–212. [CrossRef]

24. Hemdani, A.; Dagbagi, M.; Naouar, W.M.; Idkhajine, L.; Belkhodja, I.S.; Monmasson, E. Indirect sliding mode power control for three phase grid connected power converter. *IET Power Electron.* **2015**, *8*, 977–985. [CrossRef]

25. Liu, J.K.; Wang, X.H. Terminal Sliding Mode Control. In *Advanced Sliding Mode Control for Mechanical Systems*; Springer: Berlin, Germany, 2011; pp. 137–162.

26. Bhave, M.; Janardhanan, S.; Dewan, L. A finite-time convergent sliding mode control for rigid underactuated robotic manipulator. *Syst. Sci. Control Eng.* **2014**, *2*, 493–499. [CrossRef]

27. Hong, Y.G.; Yang, G.W.; Cheng, D.Z.; Spurgeon, S. Finite time convergent control using terminal sliding mode. *J. Control Theory Appl.* **2004**, *2*, 69–74. [CrossRef]

28. Yu, X.H.; Xu, J.X. Variable Structure Systems with Terminal Sliding Modes. In *Variable Structure Systems: Towards the 21st Century*; Springer: Berlin, Germany, 2002.
29. Roy, S.; Roy, S.B.; Kar, I.N. Adaptive–robust control of euler–lagrange systems with linearly parametrizable uncertainty bound. *IEEE Trans. Control Syst. Technol.* **2018**, *26*, 1842–1850. [CrossRef]
30. Roy, S.; Roy, S.B.; Kar, I.N. A new design methodology of adaptive sliding mode control for a class of nonlinear systems with state dependent uncertainty bound. In Proceedings of the 2018 15th International Workshop on Variable Structure Systems (VSS), Graz, Austria, 9–11 July 2018; pp. 414–419.
31. Roy, S.; Kar, I.N. Adaptive-robust control of uncertain euler-lagrange systems with past data: A time-delayed approach. In Proceedings of the 2016 IEEE International Conference on Robotics and Automation (ICRA), Stockholm, Sweden, 16–21 May 2016; pp. 5715–5720.
32. Roy, S.; Kar, I.N.; Lee, J.; Tsagarakis, N.G.; Caldwell, D.G. Adaptive-robust control of a class of el systems with parametric variations using artificially delayed input and position feedback. *IEEE Trans. Control Syst. Technol.* **2019**, *27*, 603–615. [CrossRef]
33. Roy, S.; Kar, I.N.; Lee, J.; Jin, M.L. Adaptive-robust time-delay control for a class of uncertain euler–lagrange systems. *IEEE Trans. Ind. Electron.* **2017**, *64*, 7109–7119. [CrossRef]
34. Plestan, F.; Shtessel, Y.; Brégeault, V.; Poznyak, A. Sliding mode control with gain adaptation—Application to an electropneumatic actuator. *Control Eng. Pract.* **2013**, *21*, 679–688. [CrossRef]
35. Plestan, F.; Shtessel, Y.; Brégeault, V.; Poznyak, A. New methodologies for adaptive sliding mode control. *Int. J. Control* **2010**, *83*, 1907–1919. [CrossRef]
36. Utkin, V.I.; Poznyak, A.S. Adaptive sliding mode control with application to super-twist algorithm: Equivalent control method. *Automatica* **2013**, *49*, 39–47. [CrossRef]
37. Moreno, J.A.; Negrete, D.Y.; González, V.T.; Fridman, L. Adaptive continuous twisting algorithm. *Int. J. Control* **2016**, *89*, 1798–1806. [CrossRef]
38. Roy, S.; Lee, J.N.; Baldi, S. A New Continuous-Time Stability Perspective of Time-Delay Control: Introducing a State-Dependent Upper Bound Structure. *IEEE Control Syst. Lett.* **2019**, *3*, 475–480. [CrossRef]
39. Alireza, S.; Mozhgan, A. Binary PSO-based dynamic multi-objective model for distributed generation planning under uncertainty. *IET Renew. Power Gener.* **2012**, *6*, 67–78.
40. Lin, C.J.; Chern, M.S.; Chih, M.C. A binary particle swarm optimization based on the surrogate information with proportional acceleration coefficients for the 0-1 multidimensional knapsack problem. *J. Ind. Prod. Eng.* **2016**, *33*, 77–102. [CrossRef]
41. Silva, S.A.; Sampaio, L.P.; Oliveira, F.M.; Durand, F.R. Feed-forward DC-bus control loop applied to a single-phase grid-connected PV system operating with PSO-based MPPT technique and active power-line conditioning. *IET Proc. Renew. Power Gener.* **2017**, *11*, 183–193. [CrossRef]
42. Babu, T.S.; Ram, J.P.; Dragicevic, T.; Miyatake, M.; Blaabjerg, F.; Rajasekar, N. Particle Swarm Optimization Based Solar PV Array Reconfiguration of the Maximum Power Extraction Under Partial Shading Conditions. *IEEE Trans. Sustain. Energy* **2018**, *9*, 74–85. [CrossRef]
43. Awadallah, M.A.; Venkatesh, B. Bacterial Foraging Algorithm Guided by Particle Swarm Optimization for Parameter Identification of Photovoltaic Modules. *Can. J. Electr. Comput. Eng.* **2016**, *39*, 150–157. [CrossRef]
44. Emel'yanov, S.V.; Korovin, S.K.; Levant, A. High order sliding modes in control systems. *Comput. Math. Model.* **1996**, *7*, 294–318.
45. Levant, A. Higher-order sliding modes, differentiation and output-feedback control. *Int. J. Control* **2003**, *76*, 924–941. [CrossRef]
46. Valenciaga, F.; Fernandez, R.D. Multiple-input–multiple-output high-order sliding mode control for a permanent magnet synchronous generator wind-based system with grid support capabilities. *IET Renew. Power Gener.* **2015**, *9*, 925–934. [CrossRef]
47. Dehkordi, N.M.; Sadati, N.; Hamzeh, M. A robust backstepping high-order sliding mode control strategy for grid-connected DG units with harmonic/interharmonic current compensation capability. *IEEE Trans. Sustain. Energy* **2017**, *8*, 561–572. [CrossRef]
48. Chen, D.; Jun, Y.; Wang, Z.; Li, S.H. Universal active disturbance rejection control for non-linear systems with multiple disturbances via a high-order sliding mode observer. *IET Control Theory Appl.* **2017**, *11*, 1194–1204.
49. Deng, H.W.; Li, Q.; Chen, W.R.; Zhang, G.R. High-order sliding mode observer based OER control for PEM fuel cell air-feed system. *IEEE Trans. Energy Convers.* **2018**, *33*, 232–244. [CrossRef]

11

Fault Investigation in Cascaded H-Bridge Multilevel Inverter through Fast Fourier Transform and Artificial Neural Network Approach

G. Kiran Kumar [1], E. Parimalasundar [2], D. Elangovan [1,*], P. Sanjeevikumar [3,*], Francesco Lannuzzo [4] and Jens Bo Holm-Nielsen [3]

[1] School of Electrical Engineering, VIT Vellore, Tamil Nadu 632014, India; kiran215vit@gmail.com
[2] Department of Electrical and Electronics Engineering, Sree Vidyanikethan Engineering College, Tirupati 517102, India; parimalpsg@gmail.com
[3] Center for Bioenergy and Green Engineering, Department of Energy Technology, Aalborg University, 6700 Esbjerg, Denmark; jhn@et.aau.dk
[4] Department of Energy Technology, Aalborg University, 9220 Aalborg, Denmark; fia@et.aau.dk
* Correspondence: elangovan.devaraj@vit.ac.in (D.E.); san@et.aau.dk (P.S.)

Abstract: In recent times, multilevel inverters are used as a high priority in many sizeable industrial drive applications. However, the reliability and performance of multilevel inverters are affected by the failure of power electronic switches. In this paper, the failure of power electronic switches of multilevel inverters is identified with the help of a high-performance diagnostic system during the open switch and low condition. Experimental and simulation analysis was carried out on five levels cascaded h-bridge multilevel inverter, and its output voltage waveforms were synthesized at different switch fault cases and different modulation index parameter values. Salient frequency-domain features of the output voltage signal were extracted using a Fast Fourier Transform decomposition technique. The real-time work of the proposed fault diagnostic system was implemented through the LabVIEW software. The Offline Artificial neural network was trained using the MATLAB software, and the overall system parameters were transferred to the LabVIEW real-time system. With the proposed method, it is possible to identify the individual faulty switch of multilevel inverters successfully.

Keywords: Artificial Neural Networks (ANN); fault diagnosis; Fast Fourier Transform (FFT); Multilevel Inverter (MLI); LabVIEW

1. Introduction

In recent years, multilevel inverters are drawing intense interest in the research of solid industrial electric drives organizes in the direction of attaining the high power demands necessary with them. The foremost merits of Multilevel Inverters (MLIs) are minimization of harmonic deformation of the output voltage waveform by way of raising incapacity of levels as well as litheness for the usage of battery sets or fuel for in-between periods [1–3].

Although MLIs are effectively used in engineering applications employing a confirmed technology, the collapse of power electronic switches and its fault investigation is until now a recent research issue for researchers. In engineering applications, it is implemented to examine the state of power switches which is available in inverters. The extent of levels in the inverter changes, the number of power also switches varied, which can raise the chance of collapse of any one of the switches; hence, any such fault should be acknowledged at the initial stage so that the process of drive and motor during anomalous conditions is not affected [4–10]. The different modes of the collapse of power semiconductor switches, an open-switch as well as the short-switch fault, directs to current harmonics and generates troubles in the gate driving circuits. Therefore, it reduces the system's concert. Several researchers used the inverter output current and voltage for constructing that fault identification arrangement [11,12]. Surin Khomfoi et al. [13], created an open-switch fault analytic coordination of an MLI created based on that output voltage with Fast Fourier Transform (FFT) model as well as five parallel neural networks employing 40 contribution neurons for every system. While the range of that neural network is extremely complex because of 40 input neurons, inside of a new paper, Surin Khomfoi et al. projected a different method that contains a mixture of FFT principal constituent analysis, genetic algorithm as well as neural network technology for identifying the fault category as well as fault location in an inverter [14].

Recognition of faulty switches of MLIs is still an emerging research area, and numerous researchers are steadily working on the way to identify the faults precisely. On the other hand, information about the actual instance implementation of high concert fault investigation system for cascaded H-bridge multilevel inverter is inadequate. With that consideration, Multilevel Inverter (MLI) output voltages are measured as a significant constraint to identifying faulty switches. Real-time implementation of overall fault analytic schemes has been executed in National Instruments (NI)—LabVIEW software with a version of DAQmx 19.1; it has been a complicated apparatus designed for rising as well as operating actual instant submission. LabVIEW makes use of graphical encoding language formed by National Instruments and is also successfully applied for data attaining, instrumentation control, as well as in automotive industries. ANN is a useful tool in the classification of patterns in the course of learning as well as nonlinear mapping. Together, LabVIEW and ANN are arranged to measure. It is essential as an additional test for obtaining research work. Therefore, the actual instant fault detection method is programmed at some stage in LabVIEW with the help of ANN [15–17].

2. Structure of H-Bridge Multilevel Inverter

The H-Bridge inverter modules with separate DC sources are connected in series to form a cascaded multilevel inverter. Figure 1 illustrates a typical three-phase cascaded MLI using 3 H-bridge modules within every phase associated with a three-phase asynchronous motor load. The number of output voltage levels can be calculated with the formula 2S+1; here, 'S' is the number of H-Bridge modules utilized for it. The 3 phase MLI structure is generally used for industrialized applications. In current work, the single-phase MLI is used because that projected fault analytic scheme has the capability of expanding in favour of three-phase applications.

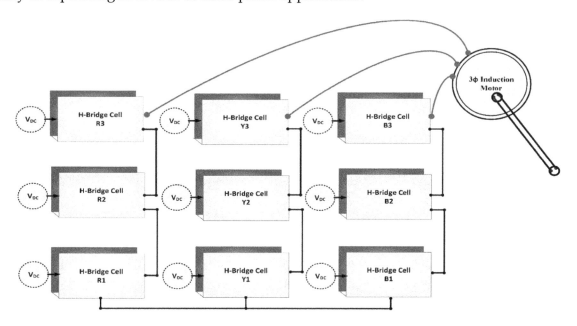

Figure 1. Representation of a three-phase H-bridge cascaded multilevel inverter fed induction motor.

Figure 2 illustrates the representation of the single-phase five levels cascaded H-bridge inverter utilized; it contains 2 H-Bridge modules, as well as 8 number of powers, switches Insulated-Gate Bipolar Transistors (IGBTs). Every IGBT switch is named following its module location like S1A, S2A, S3B, S4B, etc. The cascaded inverter has been linked along with a dynamic load like 1ψ, 0.5 HP, 50 Hz asynchronous motor. The Sinusoidal Pulse Width Modulation (SPWM) technique has been applied for generating the necessary switching signals for IGBTs. In this SPWM, higher frequency contained triangular carrier waves are used along with a sinusoidal reference wave. Figure 3 shows that production of switching sequence corresponding to module A at a carrier wave frequency value (fc) of 3 kHz as well as a modulation index value (m) of sinusoidal wave 0.85. Now in this work, the modulation index is deferred in-between sort of values 0.8 to 0.95.

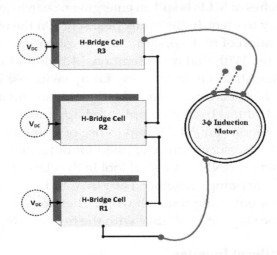

Figure 2. Simulink model of 1-Φ cascaded H-Bridge 5-level inverter fed with Induction Motor load.

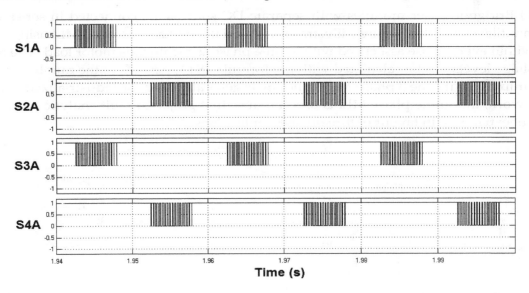

Figure 3. Sinusoidal Pulse Width Modulation cell A switching patterns created with a carrier wave frequency of 3 kHz and modulation index value of 0.85.

3. Fault Analysis of Output Voltage and Current

Simulation fault analysis was carried out with the help of Matlab/Simulink (R2019b) to realize the output load voltage, as well as output, load current waveform earlier than as well as later than the fault commencement of the MLI. Foremost, the open circuit fault occurs at 1 s on switch S1A of module A. Figure 4 shows the actual load voltage of the inverter, its exaggerated outlook as well as current wave earlier and later of that faulty condition of single-phase cascaded H-Bridge MLI associated employing the induction motor load. The voltage, as well as current patterns, were exampled for 20 kHz. Likewise, Figure 5 shows the output load voltage, as well as the output, load current waveforms later than the fault creation of open-circuit fault in switch S2A of Module A.

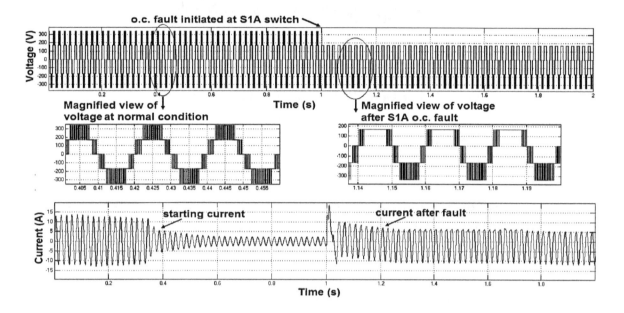

Figure 4. Load current and voltage waveform analysis of Bridge—A of S1A during an open-circuit fault.

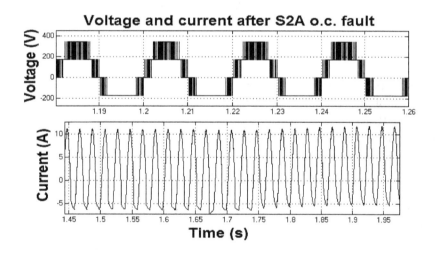

Figure 5. Load current and voltage waveform analysis of Bridge—A of S2A during an open-circuit fault.

4. Concept of Fast Fourier Transform and Feature Extraction Process

In the Fast Fourier Transform method, the output voltage waveform features are extracted. In order to generate a powerful fault analysis method, it is essential on the way to execute frequency domain analyses for the output load voltage patterns. The FFT method has been used for pulling out different parameters of the output voltage signal. As seen, signals of output parameters are not easy to be measured as a significant attribute to categorizing a faulty assumption. Consequently, the signal conversion method is required. A proper choice of that characteristic extractor is to give sufficient important information about the neural network in the example set; thus, that was the maximum amount of precision within that neural network concert that was attained. Single probable method of execution with Digital Signal Processing (DSP) microchip is executed with the help of FFT [9]. Initiating through Discrete Fourier Transform in Equation (1), after that FFT technique is with a combination of decimation within time decomposition algorithm has been represented in Equations (2) and (3):

$$F_k = \sum_{n=0}^{N-1} f_n W_N^{nk} \quad \text{for} \quad k = 0, \ldots, N-1 \tag{1}$$

where $W_N = e^{-j\frac{2\pi}{N}}$

$$F_k = G_k + W_N^k H_k \quad \text{for} \quad k = 0, , \frac{N}{2} - 1 \tag{2}$$

$$F_{k+\frac{N}{2}} = G_k - W_N^k H_k \quad \text{for} \quad k = 0, , \frac{N}{2} - 1 \tag{3}$$

G_k is to be on behalf of even-numbered essentials of f_n, while H_k is to be on behalf of odd-number essentials of fn. G_k, Z as well as Hk, are to be evaluated as exposed in Equations (4) and (5).

$$G_k = \sum_{n=0}^{\frac{N}{2}-1} f_{2n} W_{\frac{N}{2}}^{nk} \tag{4}$$

$$H_k = \sum_{n=0}^{\frac{N}{2}-1} f_{2n+1} W_{\frac{N}{2}}^{nk} \tag{5}$$

Figure 6 shows the schematic diagram representation of that fault analysis system implemented on behalf of the recognition of the failures in power electronic switches using those features extracted from the FFT technique. The inverter load voltage wave has been deliberated and using FFT technique, essential functions of the voltage signal, i.e., Total Harmonic Distortion (THD) (%), harmonic/fundamental ratio (%) values up to 11th harmonics are extracted. Hence, the RMS value analysis is also simultaneously carried out in the output voltage signal, and the extracted 12 features are specified as input for that the Artificial Neural Network. Figure 6 illustrates the feature 100/50 representing the ratio of the second harmonic to fundamental; feature 150/50 represents the ratio of third harmonic/fundamental and so on.

The output of the trained patterns of the ANN with LabVIEW software identifies the faulty switch of the multilevel inverter. Therefore, the FFT method is useful for every load voltage waveform underneath the open circuit fault condition as well as short circuit fault conditions with corresponding harmonic level variations, Vrms, and THD values. Figure 7 illustrates the FFT frequency plot of output voltage during a healthy and open-circuit fault condition. Figure 8 illustrates different features of the load voltage wave obtained from the FFT technique on various open switch faulty cases from second harmonic ratios to 11th Harmonic ratios. Figure 9 illustrates the features of the load voltage wave obtained from the FFT technique on various short switch faulty cases for different harmonic ratio values. Figure 10 shows THD of the output voltage waveform obtained from the FFT technique on various faulty cases as well as on various values of modulation index at different switches. Figure 11 illustrates the RMS values of the output voltage waveform obtained on various faulty cases as well as on various values of modulation index at different switches. Extracting the distinct features from the waveform of load voltage will automate the fault analysis process.

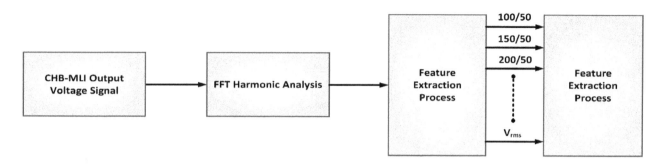

Figure 6. LabVIEW based fault investigative system using FFT features.

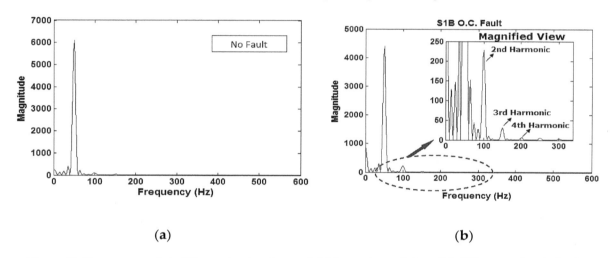

(a)   (b)

Figure 7. Frequency plot of the output voltage. (a) Normal condition; (b) S1B open circuit fault.

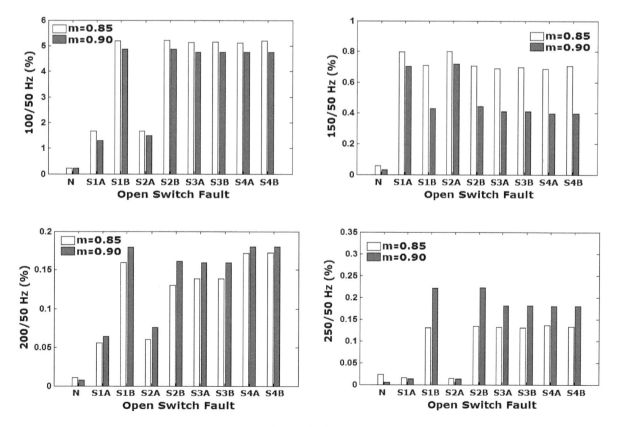

Figure 8. *Cont.*

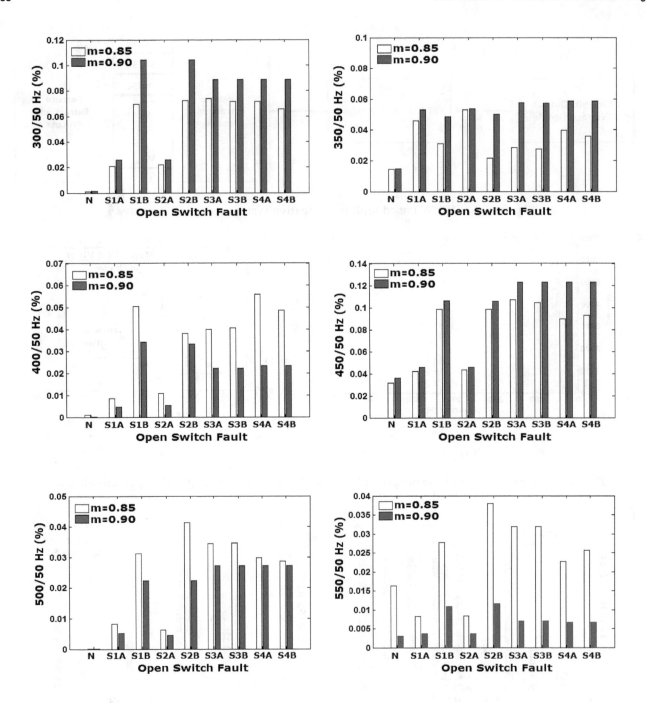

Figure 8. Features of the output voltage waveform acquired from FFT technique at distinct open switch fault cases from 2nd harmonic ratios to 11th harmonic ratios.

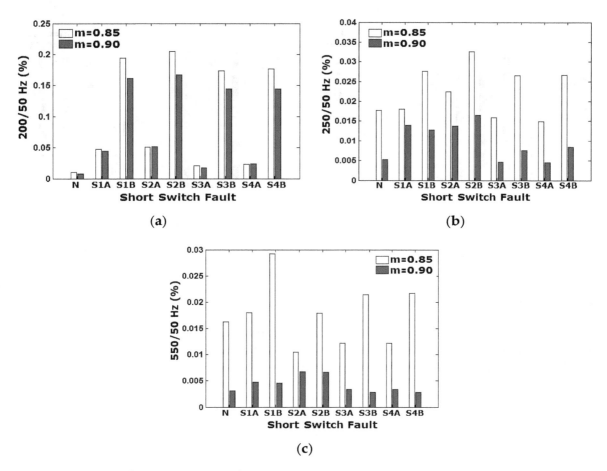

Figure 9. Features of the voltage waveform acquired from the FFT technique at distinct short switch fault cases (**a**) 4th harmonic ratio, (**b**) 5th harmonic ratio, (**c**) 11th harmonic ratio.

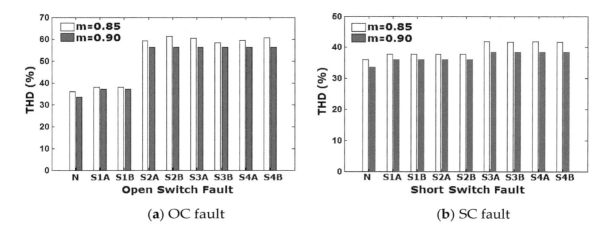

Figure 10. THD value of the output voltage waveform obtained from the FFT technique at different fault cases and different modulation index values.

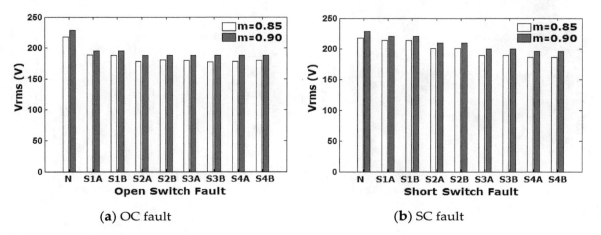

(a) OC fault   **(b)** SC fault

Figure 11. Output voltage waveforms attained at distinct modulation index values at various fault conditions.

5. Structure of Fault Diagnostic System

Figure [12] shows the schematic diagram of the overall fault diagnosis scheme built to recognize the power electronic switch failure in MLI. The hardware arrangement contains a DC Source, MLI, Induction Motor, as well as A-Data Fetching System.

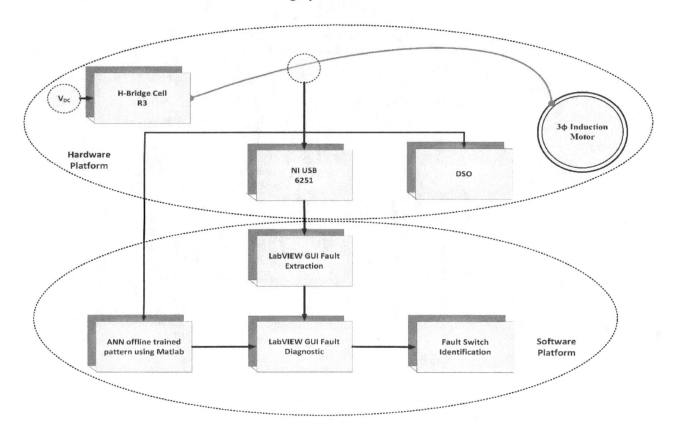

Figure 12. Implementation of the overall fault analytical system.

Primarily, extracting the voltage wave features as well as various modulation index values using FFT harmonic analysis in Matlab/Simulink (R2019b), the various fault conditions have been created in both practical setting and simulation. The obtained features were given to ANN offline training with Matlab. This trained model contains both weight values, as well as bias values of ANN, are fed on the way to graphical programming language LabVIEW used for fault analysis. For actual instant submission, outputs of voltage sensors are specified toward NI Universal Serial Bus data acquisition system that has been attached with PC. Voltage information is operated in LabVIEW FFT attribute, taking out an examination as well as compared by way of that offline trained example. After that, the LabVIEW GUI indicates fault switches in the MLI that helps maintain the system's reliability.

Figure 13 shows the laboratory experiment arrangement worn to gather load voltage waveforms of MLI on behalf of various switch faulty situation. To obtain five-level output voltages, the cascaded arrangement of PWM modules triggered 2 H-bridge inverters must be created. IGBTs having the specifications of 600 V, 25 A are chosen as switches. PWM module containing the reference as well as carrier signal selecting, modulation index as well as switching frequency modification are worn for sending necessary gate signals to IGBTs.

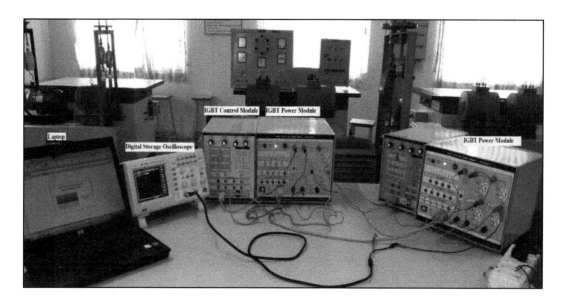

Figure 13. Laboratory experimental arrangement.

Figure 14a,b show the data acquisition system interfacing to LabVIEW software in the computer on behalf of actual instant submissions. National Instruments (NI) USB-6251 (1.25 MSa/Sec) are worn as data fetching arrangement that are interfaced with the computer to record as well as facilitate the new process for obtaining signals.

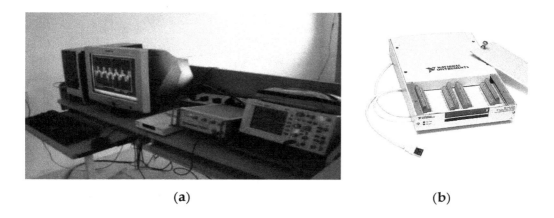

(a)   (b)

Figure 14.   (a) Data acquisition system interfaced with LabVIEW software in PC (**b**) NI USB-6251 hardware.

The total arrangement can measure the 16 analogue signals as well as 16 bits. The Digital Storage Oscilloscope (DSO) and Agilent (1 GSa/Sec), are worn as well for visualizing load voltage waves. Open circuit and short circuit faulty conditions are formed on every switching device, and consequent output voltage waves are stored. A voltage sensor has been utilized for collecting the signals at the output and is directly interfaced to NI USB-6251. The output signals are extracted from inverter for variety of indexed modularity values, and FFT harmonic analysis is also conceded. A significant characteristic of that voltage sensor is to extract the energy substance of that signal at various levels of dissolution.

5.1. Simulation Results at open circuit Fault

Firstly, an open circuit faulty condition is subjected to switch S1A, and then the resulting load voltage waveform recorded. Likewise, open circuit faulty condition is formed on the remaining switches of H-Bridge A as well as H-Bridge B. The resulting output voltage waveforms are recorded for an advanced characteristic of pulling out progression. Figure 15 shows the distinctive output waveforms attained when open circuit switch faulty condition on H-Bridge A. For assessment, that output voltage waveform when in the no-fault situation has also been included in Figure 15. These output voltage waveform patterns illustrate to find variations between healthy as well as faulty conditions effortlessly.

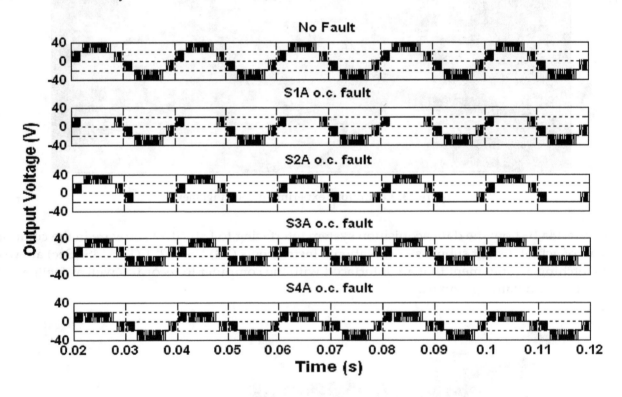

Figure 15. Fault condition output voltage waveforms.

5.2. Simulation Results at short circuit Fault

During this condition, short circuit fault has been formed on every one switch of H-Bridge A and H-Bridge B, one after one as well as that resulting output voltage waveforms are also recorded on behalf of supplementary characteristic pulling out method. Figure 16 shows that characteristic output voltage waves attained when normal conditions as well as short- switch faulty circumstances of H-Bridge B. by observing that the countable dissimilarity has been there in every output voltage waves of short circuit unsatisfactory situation while comparing with the reasonable condition. FFT method has been used for every output voltage wave underneath short circuit fault circumstance, and then consequent FFT harmonic analysis is evaluated.

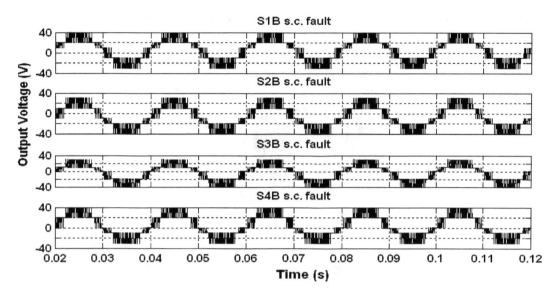

Figure 16. Short Circuit fault condition output voltage waveforms.

6. Experimental Validation

6.1. Feature Extraction Analysis Using LabVIEW

The whole faulty diagnosis system has implemented with National Instruments (NI) - LabVIEW software with a version of DAQmx 19.1. Figure 17 shows the implemented LabVIEW frontage panels containing the output voltage imprison as well as examination unit before identifying the fault switch. It shows the initial Data Acquisition capture settings in terms of max and min value, sampling frequency, number of samples per channel, multi-factor value, the magnitude of RMS voltage in max and min and time scale parameters. The GUI improved within LabVIEW displayed the fetched output waveform of voltage have frontage on the side panel of that programming algorithm. Their front side panel behaves similar to a UI someplace the user be able to set up as well as pull out information. Formerly that NI USB gadget has correctly interfacing to that LabVIEW front side panel; the piece of equipment contact indicates in green colour blinking. The front side panel has had power over parameters of output signal like the scale of time as well as magnitude, sampling frequency as well as no. of samples of every indication.

Moreover, with the help of acquisition setting the no. of sample signal capturing is also controlled for a particular time. Separate sub VI has power over frequency domain investigation of that output voltage signal. Because of the NI Data Acquisition System (DAS) as well as LabVIEW software's has the capability to capturing as well as analyzing the data set on a precise instant. Signals are capturing endlessly as well as recorded in the computer on behalf of supplementary dispensation. This front panel also shows the variations in the RMS value of the output voltage signal concerning time which helps in understanding the trend analysis of the Vrms parameter. Figure 18 shows that front end side panel of LabVIEW intended to examination frequency domain of the output voltage signals using FFT technique. This unit contains the have power over choosing the signals for FFT analysis purpose. Within this unit, control also provided for tracking of individual FFT plot of the voltage signal. Peak values of the harmonic frequencies are used to evaluate the harmonic ratios concerning fundamental frequencies. The Fast Fourier Transform frequency domain harmonic examination of the front panel is improved within the research work as well as it shows in Figure 19. This has been deliberated on the way toward the observation of various harmonic/fundamental ratios, and THD value of the output voltage signal of multilevel inverter and Figure 20 shows that VI front panel of MLI faulty switch analysis. In that screen, that is shown the possibility of tracking the harmonic scheme for each separate output voltage signals. This software module is developed in such a way to evaluate up to

11th harmonic ratio. Trend analysis of THD value and harmonic ratios is possible in this front panel and the faulty.

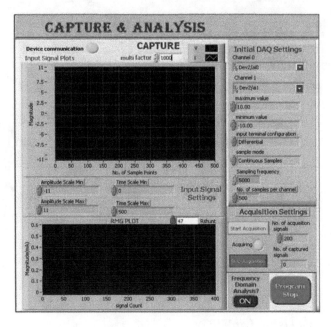

Figure 17. LabVIEW output Voltage imprisons in the front panel and analysis of multilevel inverter.

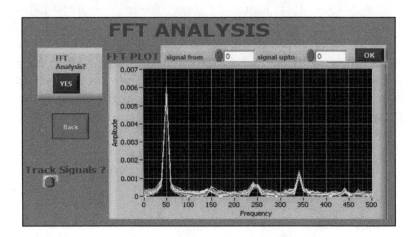

Figure 18. LabVIEW FFT based frequency domain analysis in the front panel.

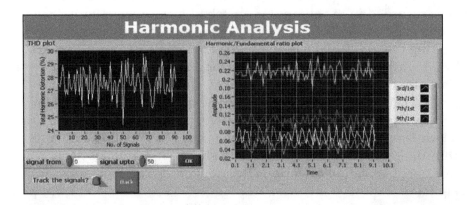

Figure 19. The front panel of FFT based THD and Harmonic analysis.

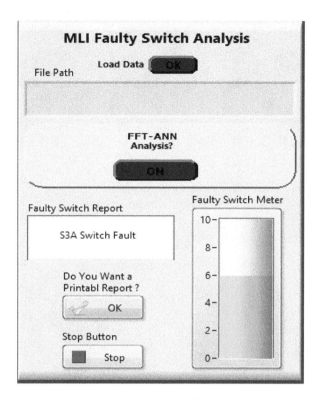

Figure 20. The front panel of MLI faulty switch analysis.

The switch must be identified and Figure 21 illustrates the output voltage pattern that relates to MLI under various faulty switch conditions at real-time implementation.

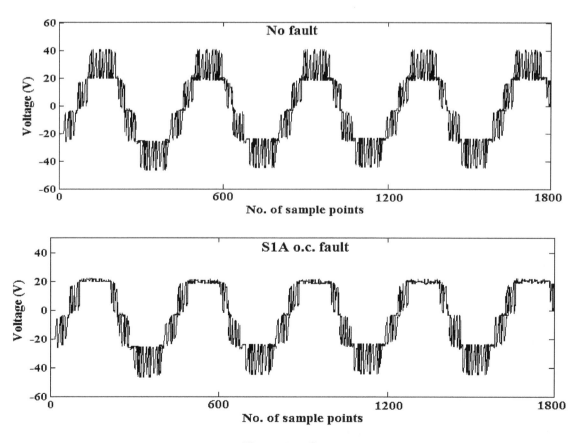

Figure 21. *Cont.*

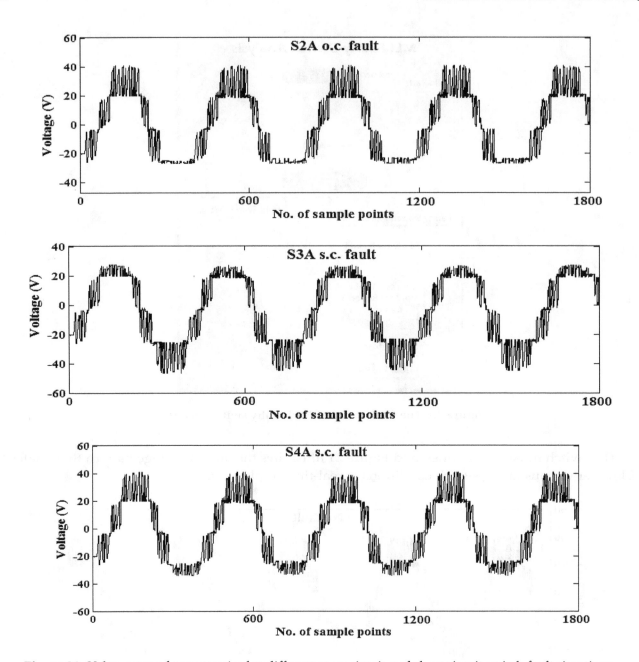

Figure 21. Voltage waveforms acquired at different open circuit and short circuit switch fault situations.

6.2. Real-Time Fault Diagnosis Results from LabVIEW-ANN Approach

On the way towards the mechanize the development that of fault analysis in MLI, a multilayer feed-forward network, as well as a backpropagation learning algorithm was utilized [15].

Figure 22 shows the ANN schematic diagram. It has a structure containing an input layer, one forbidden layer, and one output layer. The targeted and input vectors are primarily fed with some values for a training network. Exercising the arrangement is completed by altering the weight as well as the bias of the unit depends among the significant fault. Backpropagation training algorithm contains a frontward pass as well as toward the back pass is conceded out in anticipation of the Mean Square Error (MSE) has been evaluated up to the lowest value. The collected data is achieved at what time error between them calculated as well as the preferred amount produced, which is a lesser amount of that set value.

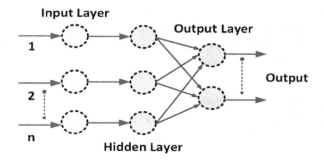

Figure 22. Schematic diagram of an Artificial Neural Network.

Here MSE, that has been meaning of the number of errors for every set of input as well as resulting output, been evaluated with Equation (6):

$$MSE = \frac{1}{m}\sum_{k}^{m}(S_k - Y_k)^2 \qquad (6)$$

Here S_k, as well as Y_k are, correspondingly, the preferred, as well as measured output on behalf of kth input set and m, is the whole quantity of their output parameters [15]. Here information of the revised neural network is utilized and is displayed in Table 1.

Table 1. Specifications of FFT—ANN-Lab VIEW Approach.

No. of Inputs	12
No. of Neurons in Hidden Layer	24
No. of Neurons in Output Layer	9
Learning Rate (η)	0.1
No. of Iterations	3800
No. of Training Sets	200
No. of Test Input Sets	150
Convergence Criteria	0.01

Within the proposed work, 12 parameters (10 harmonic ratios, THD and Vrms) obtained as features from the FFT technique of a faulty conditioned output voltage signals are fed to the input of the neural network. There is a total of 9 produced neurons for classifying that fault as that of no-fault, S1A fault up to S4B fault are shown in Table 2.

Table 2. Training pattern of Neural Network.

Classification of Fault	Position of Neuron	Output Pattern
No fault	1	[1 0 0 0 0 0 0 0 0]
S1A fault	2	[0 1 0 0 0 0 0 0 0]
S1B fault	3	[0 0 1 0 0 0 0 0 0]
S2A fault	4	[0 0 0 1 0 0 0 0 0]
S2B fault	5	[0 0 0 0 1 0 0 0 0]
S3A fault	6	[0 0 0 0 0 1 0 0 0]
S3B fault	7	[0 0 0 0 0 0 1 0 0]
S4A fault	8	[0 0 0 0 0 0 0 1 0]
S4B fault	9	[0 0 0 0 0 0 0 0 1]

It is shown that the ANN training sequence approaches various switches faults in MLI. In the exercise sequence, each neuron in that output layer of the neural network is assigned to particular faults and then trained for a binary value of 1 or 0, as shown in Table 2. For example, in the case of the no-fault condition, the first neuron in the output layer has been assigned a value of 1, and all other neurons are trained for a value of 0. Similarly, for different fault cases, the output layer neurons are trained for different binary training patterns. For the offline training of the neural network, 200 training sets were used, and the weight matrix of the trained pattern is given as an input to the LabVIEW GUI module for testing purposes with 150 test inputs.

Figure 23 shows the performance of the network at various iterations. The training of the current system reaches the junction criterion after close to 3800 iterations. It is proved that 3800 iterations were enough to know the successful training of that revised neural network. Consequently, a concert of that back PNN is known by way of 24 forbidden layer neurons maintained that worth of that learning rate is 0.1, and the number of iterations is 3800.

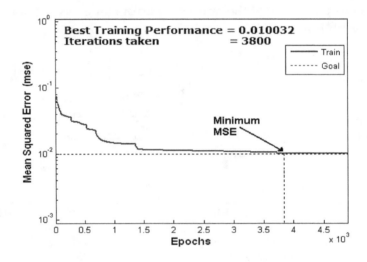

Figure 23. Variations occurred in MSE of the ANN during training pattern concerning an increase in the number of iterations.

In general, it is noticed that the neural network accurately predicts the no-fault case at all tested numbers of forbidden layer neurons. Since that network convergence has not been reached within the specified revised neural network parameters in such cases of 15 or 20 forbidden layer neurons, then the accuracy of identification tempo is affected. It has identified their concert of the neural network been enhanced on behalf of 24 forbidden layer neurons while comparing employing further situations. The average recognition tempo on behalf of every faulty case is 100% in the considered case, as well as the neural network, has capable of discovering that fault in all faulty cases effectively. Table 3 gives a detailed analysis of the identification rate, and Figure 24 illustrates the evaluation of neural network means the square error of various numbers of forbidden layer neurons.

Table 3. Overall Identification rates of FFT-ANN-LabVIEW approach.

Classification of Fault	Identification Rate (%) at Different Number of Hidden Layer Neurons		
	15	20	24
No fault	100	100	100
S1A fault	91	92	100
S1B fault	93	95	100
S2A fault	92	95	100
S2B fault	91	92	100
S3A fault	95	95	100
S3B fault	92	96	100
S4A fault	93	95	100
S4B fault	92	94	100

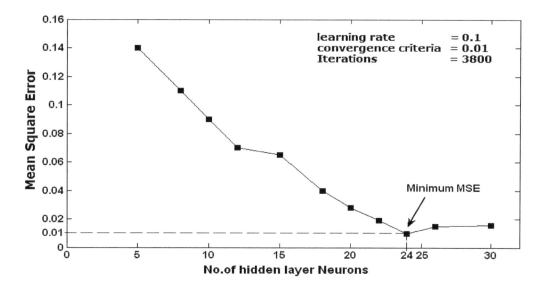

Figure 24. Evaluation of the mean square error of the neural network at different numbers of hidden layer neurons.

Real-time implementation of LabVIEW is dependent on the trail of the fault analysis of MLI with load voltage characteristics like FFT harmonic analysis showing the possibilities of determining the switch failure in a particular position in an MLI. When comparing with other techniques mentioned in previous publications [13,14], the projected technique considerably decreases the number of inputs to ANN network as well as analyzes the occurrence of faulty conditions in 10 msec, and it can find the faulty condition of an exact switch (OC fault or SC fault) in the MLI. Also, the projected method gives a 100% detection rate among no-fault as well as fault conditions of a switch. Therefore, on one occasion, the switch fault is analyzed that has helped operate and carry out precautionary safeguarding.

7. Conclusions

This paper's proposed H-bridge cascaded five level multilevel inverters associated through a load of induction motor with faulty switch analysis is carried out. First, the significant attributes of load voltage output waveform are examined with simulation as well as experimental analysis on dissimilar open-switch and short-switch faulty conditions. Meanwhile, the existing problems are further examined. Furthermore, essential features like harmonic analysis with FFT technique are specified to be input to that backpropagation trained ANN along with an evaluation of the mean square error at the different number of hidden layer neurons; to make the neural network offline the Matlab software is utilized. The actual instance function of that anticipated fault diagnosis scheme is executed with the LabVIEW software. This projected fault diagnosis scheme has the potential to accurately recognize each separate fault switch of the cascaded multilevel inverter. It can categorize 100% precisely typical as well as fault situations. Therefore, in this instance, the faulty switch conditions were recognized immediately, and implementing this system will help the operator in performing protective maintaining work.

Author Contributions: G.K.K., and E.P., have developed the proposed research concept, and they both are involved in studying the execution and implementation with statistical software by collecting information from the real environment and developed the simulation model for the same. D.E., P.S., F.L., J.B.H.-N. shared their expertise and validation examinations to confirm the concept theoretically with the obtained numerical results for its validation of the proposal. All authors are to frame the final version of the manuscript as a full. Moreover, all authors involved in validating and to make the article error-free technical outcome for the set investigation work. All authors contributed to the research investigation equally and presented in the current version of the full article. All authors have read and agreed to the published version of the manuscript.

References

1. Zheng, Z.; Wang, K.; Xu, L.; Li, Y. A hybrid cascaded multilevel converter for battery energy management applied in electric vehicles. *IEEE Trans. Power Electron.* **2014**, *29*, 3537–3546. [CrossRef]
2. Javad, G.; Reza, N. Analysis of Cascaded H-Bridge Multilevel Inverter in DTC-SVM Induction Motor Drive for FCEV. *J. Electr. Eng. Technol.* **2013**, *8*, 304–315.
3. Banaei, M.R.; Salary, E. A New Family of Cascaded Transformer Six Switch Sub-Multilevel Inverter with Several Advantages. *J. Electr. Eng. Technol.* **2013**, *8*, 1078–1085. [CrossRef]
4. Ui-Min, C.; Lee, K.-B.; Frede, B. Diagnosis and tolerant strategy of an open-switch fault for T-type three-level inverter systems. *IEEE Trans. Ind. Appl.* **2014**, *50*, 495–508. [CrossRef]
5. Chen, A.; Hu, L.; Chen, L.; Deng, Y.; He, X. A multilevel converter topology with fault-tolerant ability. *IEEE Trans. Power Electron.* **2005**, *20*, 405–415. [CrossRef]
6. Pablo, L.; Josep, P.A.; Thierry, M.; Jose, R.; Salvador, C.; Frédéric, R. Survey on fault operation on multilevel inverter. *IEEE Trans. Ind. Electron.* **2010**, *57*, 2207–2218.
7. Ma, M.; Hu, L.; Chen, A.; He, X. Reconfiguration of carrier-based modulation strategy for fault-tolerant multilevel inverters. *IEEE Trans. Power Electron.* **2007**, *22*, 2050–2060. [CrossRef]
8. Diallo, D.; Benbouzid, M.H.; Hamad, D.; Pierre, X. Fault detection and diagnosis in an induction machine drive—A pattern recognition approach based on concordia stator mean current vector. *IEEE Trans. Energy Conv.* **2005**, *20*, 512–519. [CrossRef]
9. Estima, J.; Cardoso, A.M. A new algorithm for real-time multiple open-circuit fault diagnosis in voltage-fed PWM motor drives by the reference current errors. *IEEE Trans. Ind. Electron.* **2013**, *60*, 3496–3505. [CrossRef]
10. Khan, M.A.S.K.; Rahman, M.A. Development and implementation of a novel fault diagnostic and protection technique for IPM motor drives. *IEEE Trans. Ind. Electron.* **2009**, *56*, 85–92. [CrossRef]
11. Lezana, P.; Aguilera, R.; Rodriguez, J. Fault detection on multicell converter based on output voltage frequency analysis. *IEEE Trans. Ind. Electron.* **2009**, *56*, 2275–2283. [CrossRef]
12. Masrur, M.A.; Chen, Z.; Murphey, Y. Intelligent diagnosis of open and short circuit faults in electric drive inverters for real-time applications. *IET Power Electron.* **2010**, *3*, 279–291. [CrossRef]
13. Surin Khomfoi, S.; Tolbert, L.M. Fault diagnostic system for a multilevel inverter using a neural network. *IEEE Trans. Power Electron.* **2007**, *22*, 1062–1069. [CrossRef]
14. Surin Khomfoi, S.; Tolbert, L.M. Fault diagnosis and reconfiguration for multilevel inverter drive using AI-based techniques. *IEEE Trans. Ind. Electron.* **2007**, *54*, 2954–2968. [CrossRef]
15. Sivakumar, M.; Parvathi, R.M.S. Diagnostic Study of Short-Switch Fault of Cascaded H-Bridge Multilevel Inverter using Discrete Wavelet Transform and Neural Networks. *Int. J. Appl. Eng. Res.* **2014**, *9*, 10087–10106.
16. Hochgraf, C.; Lasseter, R.; Divan, D.; Lipo, T.A. Comparison of multilevel inverters for static VAR compensation. In Proceedings of the Conference Record of the IEEE Industry Application Society Annual Meeting, Denver, CO, USA, 2–6 October 1994; pp. 921–928.
17. Kastha, D.K.; Bose, B.K. Investigation of fault modes of voltage-fed inverter system for induction motor drive. *IEEE Trans. Ind. Appl.* **1994**, *30*, 1028–1038. [CrossRef]

12

Analysis and Control of Electrolytic Capacitor-Less LED Driver Based on Harmonic Injection Technique

Mahmoud Nassary [1,*], **Mohamed Orabi** [1], **Manuel Arias** [2], **Emad M. Ahmed** [1] and **El-Sayed Hasaneen** [1]

[1] APAERC, Faculty of Engineering, Aswan University, 81542, Aswan, Egypt; morabi@apearc.aswu.edu.eg (M.O.); eelbakoury@apearc.aswu.edu.eg (E.M.A.); hasaneen@aswu.edu.eg (E.-S.H.)

[2] Departamento de Ingeniería Eléctrica, Electrónica, de Computadores y Sistemas University of Oviedo, 33204 Gijón, Spain; ariasmanuel@uniovi.es

* Correspondence: mnassary@ieee.org

Abstract: AC-DC LED drivers may have a lifespan shorter than the lifespan of LED chips if electrolytic capacitors are used in their construction. Using film capacitors solves this problem but, as their capacitance is considerably lower, the low-frequency ripple will increase. Solving this problem by limiting the output ripple to safe values is possible by distorting the input current using harmonic injection technique, as long as these harmonics still complies with Power Factor Regulations (Energy Star). This harmonic injection alleviates the requirements imposed to the output capacitor in order to limit the low-frequency ripple in the output. This idea is based on the fact that LEDs can be driven by pulsating current with a limited Peak-To-Average Ratio (PTAR) without affecting their performance. By considering the accurate model of LEDs, instead of the typical equivalent resistance, this paper presents an improved and more reliable calculation of the intended harmonic injection. Wherein, its orders and values can be determined for each input/output voltage to obtain the specified PTAR and Power Factor (PF). Also, this harmonic injection can be simply implemented using a single feedback loop, its control circuit has features of wide bandwidth, simple, single-loop and lower cost. A 21W AC-DC buck converter is built to validate the proposed circuit and the derived mathematical model and it complies with IEC61000 3-2 class D standard.

Keywords: pulsating output current; light emitting diode (LED); peak to average ratio (PTAR); power factor correction; harmonic injection; modelling; feedback loop control

1. Introduction

LED technology has several merits over conventional lamps such as: high-efficiency, very long lifespan (approximately 100,000 h [1]), lower power consumption, low maintenance cost and instantaneous switch-on [2,3]. Besides, the LEDs are environmentally friendly. Regarding their efficiency, CREE claimed to be the first company to break the 300 lumens per watt barriers (still being the highest level achieved) [4]. Moreover, lighting consumes 20% of the electrical energy in the industrial countries pushing forward the replacement of conventional lighting with LED lighting.

The critical part that defines the LED lamp lifespan is the driver. One-stage AC-DC LED drivers normally use a bulky electrolytic capacitor to balance power between the pulsating input and the constant output, minimizing the double line frequency current ripple [5]. This capacitor limits the lifespan of the LED lamp to its own lifespan, typically between 1000 and 10,000 h, considerably lower than the lifespan of LEDs [6]. In addition, the lifespan of the E-Caps follows the 10-degree-law that states that it decreases by a factor of 2 for each +10 °C temperature increase. Even assuming this, operation at 85 °C only pushes its lifespan to 20,000 h. Consequently, eliminating the E-Cap is mandatory and many research efforts have been made in this direction.

Wound and soft winding film capacitors can be used instead of E-Caps due to their long lifespan [7]. However, their energy density is low and that increases the output voltage and the output current ripple of the LED driver. This causes a depraved effect on the LED chip. The light perceived by humans' eyes is proportional to its average value because the light ripple is filtered as long as its frequency is higher than a few hundreds of hertz. Nonetheless, increasing LED peak current (as a consequence of the ripple) results in a change in the chromaticity coordinates, color correlated temperature (CCT), color rendering index (CRI), flux and efficacy degradation, so LED light is perceived as bluish-white [8,9].

Several studies proposed different topologies and control circuits to enable using low-density capacitors with limited LED peak current under a defined ratio called PTAR. The most effective way is harmonic injection technique. The idea is to inject predefined harmonics into the input current to limit the PTAR of the output LED current while observing PF regulations. Figure 1a–c shows different harmonics combinations for the LED current [10,11] with the result of a lower PTAR if those harmonics are wisely selected. First, Figure 1a shows the double line frequency output current in dashed line and the third harmonic order in dashed-dot line. The combination result is shown as a solid line and, as can be seen, its PTAR is lower than in the case of the first harmonic. Second, Figure 1b shows the double line frequency output current combined with the fifth harmonics. In this case, the resulting PTAR is higher than in the case of the first harmonic alone. Finally, Figure 1c shows the double line frequency output current combined with the third harmonics and the fifth harmonics. In this case, the PTAR is the lowest of the three cases. As more harmonic orders are injected, the combination has lower PTAR. However, it is very important to define the amplitudes that lead to the lowest PTAR while keeping the input PF within the regulation limits.

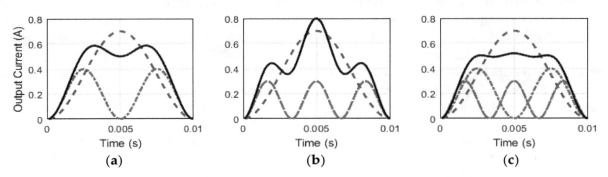

Figure 1. Different cases of the output current waveforms and its circuit diagram under harmonic injection [10,11]. (a) Combination of double and third line harmonics orders; (b) Combination of double and fifth line harmonics orders; (c) Combination of double, third and fifth line harmonics orders.

In [12], a two-stage LED driver is introduced. The first stage is a high frequency boost converter that operates at discontinuous conduction mode (DCM) as a pre-regulator PF correction (PFC) and injecting the 3rd harmonics into the input current. The second stage is a Flyback converter for regulating the LEDs current. The design replaced the storage E-Cap with film one; however, three 47-μF capacitors connected in parallel are still needed.

More harmonics are injected in the input current in [10,11]. The developed circuit is a single stage AC-DC Flyback converter. It has only a 0.47-μF film capacitor in its output. However, the model analysis introduced in [10] was based only on a resistive load model without considering the accurate LED chip model. As a consequence, the theoretical analysis and the experimental results were not well matched. Practical results show a PTAR of 1.43 for the output LED current instead of the designed one of only 1.34. Also, the presented controller included a feedforward loop with a multiplier and a divider that increase the complexity and the cost. In [13], limiting the PTAR of the LED current is achieved by using two loops, feedback and feedforward. Results recorded a PTAR of less than 1.34 by injecting more harmonics orders. Again, the LED model consists of the equivalent resistor, which results in inaccuracy. In [14,15] a feedforward circuit was proposed to generate a distorted sinusoidal reference signal that contains the required injected harmonics to limit the PTAR. The implemented circuit included one microcontroller in the current loop with two voltage sensors and one current sensor in addition to the multiplier circuit. It uses a look-up table with a normalized value to generate the reference signal for the duty cycle. Moreover, injecting only the third harmonic resulted in a higher PTAR compared to the previous approaches. Similar results can be found in [16], wherein a feedforward loop for harmonic injection in addition to PLL, multiplier and divider circuits are used. A similar strategy was proposed in [17], which reduces the electrolytic output capacitor to almost 24.2% by using an Active Ripple Compensation (ARC). In [18] the same methodology was used to reduce the electrolytic capacitance by 46.3% by using a different ARC technique.

Harmonics injection is not only method to eliminate the E-cap. Ripple cancelation method can be also used to maintain the same purpose. This technique is based on adding a bidirectional converter connected in parallel or series to the output capacitor in order to cancel the double line frequency of the current ripple produced from decreasing the output capacitance as shown in Figure 2 [19]. By controlling the bidirectional converter to absorb the double line harmonics, the output inductor L_o and output capacitor C_o are used only to filter the high frequency harmonics in the output current [20,21]. Consequently, E-cap is eliminated and can be replaced with small capacitance.

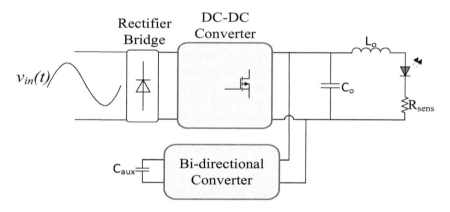

Figure 2. Parallel type of ripple cancellation circuit diagram [19].

Thus, many of these methods are using double stage with more than two semiconductors switches. Cost-wise is still not considered in this technique by using many components counts and using complex control circuits as well.

In [22], a boost converter was used as a pre-regulator to act as a PFC and it was cascaded with LLC resonant converter. Wherein, the frequency modulation technique is used in the control circuit. The behavior of the control circuit acts to vary the frequency when the line voltage changes. It generates the highest frequency at the peak of the line voltage. Therefore, the output current peak is expected to decrease. This kind of technique is suffering from many components counts in power stage and the implementation of the complex controller circuit.

Ripple cancellation technique can maintain the lowest LED current peak. However, the main disadvantage is using an extra converter to absorb the low frequency ripple current and using extra capacitor as shown in Figure 2. Therefore, it increases the component counts and circuit complexity.

Passive LED driver is considered one of the E-cap less topologies. In [23,24], four different passive LED driver topologies were proposed. The first type used valley-fill circuit to keep a high PF. It used a high inductance value of 1.47 H at the input, another bulky inductor value of 1.9 H at the output stage and two output capacitors in the valley fill circuit with a value of 20 µF polypropylene capacitors. The second type was done by making a modification in the valley-fill circuit. It has the same structure of the first type, in addition to the two output polypropylene capacitors in the valley-fill circuit with a value of 20 μF. The third type was proposed without using the valley-fill circuit. This circuit was considered the most cost-effective one among all passive LED driver types as it used less component counts than the previous two types. The fourth type used a bulky inductor for the input stage with the same value of the first type. Also, it used a coupled inductor and one capacitor to filter the output current ripples.

Passive LED drive is only considered a cost-effective solution in a high-power application where size and cost of this passive elements are not considerable. Wherein, the usage of the bulky inductor instead of the bulky electrolytic capacitor increases the LED driver footprint.

In conclusion, prior researches have the following limitations:

- Using additional control loop circuit to inject harmonics increases the component count. Consequently, the LED lamp cost will increase, limiting its penetration to the market, especially in the case of single-stage solution.
- The best practical result is a PTAR of 1.43 in nominal conditions which is still a high value.
- Using an inaccurate model of the LED chip (i.e., replacing it by a simple resistor), leads to a deviation between the practical results and mathematical ones.

This paper introduces a step forward to modify the harmonic injection techniques to be simpler and more economic through using just one feedback loop, that can achieve the target PTAR and PF by means of using an accurate model for LED chip.

2. Modelling of E-Cap-Less Converter for LED Applications

AC-DC buck converter operating in DCM without electrolytic capacitor is the simplest topology for implementing an AC-DC LED driver because high PF can be achieved with the converter operating as a resistor emulator. Figure 3 shows the LED driver circuit of the converter under study. The LED is modeled as a series branch of three components: a small dynamic resistor r_e, a DC source representing the knee voltage V_{knee} and an ideal diode. Although this is the standard model of a diode, it has not been normally used in AC-DC LED driver design in favor of just the typical and simple equivalent resistance. As will be shown, more accurate results will be obtained in the implementation of the harmonic injection technique for reducing the PTAR if the complex model is considered.

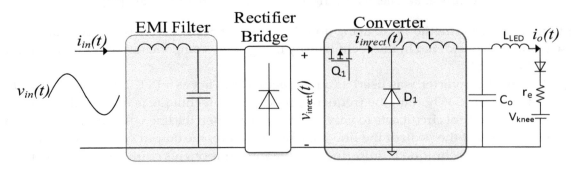

Figure 3. AC-DC E-Cap-less buck converter.

To simplify the analysis of the converter operation, elements such as diode D_1, MOSFET (switch Q_1), capacitor C_o, and inductor L are considered ideal. The buck converter operates with an input voltage equal to a rectified line voltage (whose pulsation is $2\omega_L$). The input current follows the input voltage as long as the input voltage is higher than the output one, defined by the required LED forward voltage (V_f). Figure 4a shows the rectified input voltage, inductor current and output current which starts increasing at T_c and reaches zero again at T_e. Their values can be expressed as follows:

$$T_c = \frac{\left|\sin^{-1}\left(\frac{V_o}{V_{inp}}\right)\right|}{\omega_L} = \frac{\left|\sin^{-1}(M)\right|}{\omega_L}, \tag{1}$$

$$T_e = \frac{T_L}{2} - T_c. \tag{2}$$

where V_{inp} is the peak value of the input voltage, V_o is the output voltage (LED voltage), M is the conversion ratio of the converter and T_L is the period of the input voltage. It should be mentioned that the output voltage is assumed to be constant to simplify the calculation.

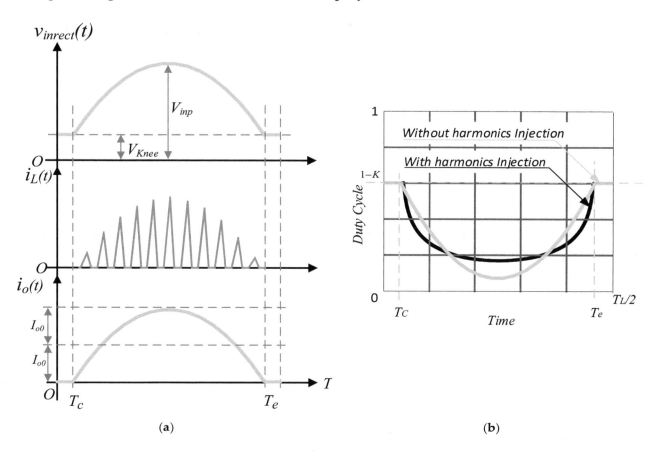

(a)   (b)

Figure 4. LED driver based buck converter for E-Cap less waveforms with and without harmonics injections. (**a**) Sketch for the rectified input voltage ($v_{inrect}(t)$) and inductor current (i_L) without harmonics injection. (**b**) The Duty cycle shape with and without harmonics injection.

2.1. Period 1: ($T_c \leq t \leq T_e$)

For period 1, the rectified voltage can be expressed as:

$$v_{inrect}(t) = V_{inp}|\sin(\omega_L t)|. \tag{3}$$

By referring to Figure 3, the inductor voltage during the on state of Q_1 can be expressed as:

$$v_L = L\frac{di_L}{dt} = v_{inrect}(t) - V_o. \tag{4}$$

The instantaneous peak inductor current, as shown in Figure 4a, is equal to Δi_L as the converter operates at DCM, and can be expressed as:

$$i_{PK}(t) = \frac{v_{inrect}(t) - V_o}{LF_s}d(t). \tag{5}$$

where $d(t)$ is constant in each switching period, and F_s is the switching frequency.

Also, the maximum peak inductor current over the line cycle can be calculated at the peak line input voltage V_{inp} as:

$$I_{PK} = \frac{V_{inp} - V_o}{LF_s} D_m. \tag{6}$$

where D_m is the maximum duty cycle.

The instantaneous input current, averaged over each switching period, can be expressed as:

$$i_{inrect}(t) = \frac{1}{2} i_{PK}(t) d(t) = \frac{v_{inrect}(t) - V_o}{2LF_s} d(t)^2. \tag{7}$$

where $i_{inrect}(t)$ is the instantaneous rectified current averaged over the switching period.

By ignoring the converter losses, the instantaneous input and output power are equal (over a switching period):

$$p_o(t) = p_{in}(t), \tag{8}$$

$$V_o i_o(t) = v_{inrect}(t) i_{inrect}(t). \tag{9}$$

where $p_o(t)$ is the output power and $p_{in}(t)$ is the input power. By referring to (7) and substituting it into (9) the expression for the output current can be expressed as:

$$i_o(t) = \frac{v_{inrect}(t) - V_o}{\frac{2LF_s V_o}{d^2}} v_{inrect}(t). \tag{10}$$

From (10), the instantaneous duty can be expressed as:

$$d(t) = \sqrt{\frac{2LF_s V_o i_o(t)}{v_{inrect}(t)(v_{inrect}(t) - V_o)}}. \tag{11}$$

Referring to (11), the duty cycle in DCM depends on the power stage inductance L, switching frequency F_s, output current $i_o(t)$, the input voltage $v_{inrect}(t)$ and the output voltage V_o.

2.2. Period 2: ($0 \leq t < T_c$ and $T_e < t \leq T_L/2$)

As explained in the section above, Equation (11) represents the derived control duty cycle for the system under study during the main conduction period ($T_c \leq t \leq T_e$). Here the duty cycle equation for the remaining part of the period will be derived.

The converter operates in DCM, where the conversion ratio M can be given as:

$$Conversion\ ratio\ (M) = \frac{2}{1 + \sqrt{1 + \frac{4k}{d^2}}}. \tag{12}$$

where $k = \frac{2L}{RT_s}$ and R is the equivalent load seen by the converter.

By replacing M by d in (12) to obtain the boundary between the DCM and CCM, so the duty cycle equals $1 - K$.

Combing period 1 and 2, using (11) and (12), the duty cycle is derived over the full range as:

$$d(t) = \begin{cases} 1 - K & 0 \leq t < T_c \\ \sqrt{\frac{2LF_s V_o i_o(t)}{v_{inrect}(t)(v_{inrect}(t) - V_o)}} & T_c \leq t \leq T_e \\ 1 - K & T_e < t \leq \frac{T_L}{2} \end{cases} \tag{13}$$

3. Modeling of Proposed E-Cap Less Converter under Harmonic Injection

Harmonic injection will be used to limit the PTAR of the output current. Equation (13) gives the duty cycle of the LED system under study. It is clear that the duty cycle is a function of the instantaneous output LED current and the input rectified voltage. If the output LED current is considered a dc value, the duty cycle shape will be as sketched in Figure 4b with orange line. This results in a PTAR of 2 for pure restive loads and even higher for real LEDs. Injecting harmonics into the current will imply modifying the duty cycle shape as shechted in Figure 4b with black line. In this section, a detailed study for the required harmonic injection components to limit the PTAR while keeping the target PF will be presented.

3.1. LED Output/Input Currents' Harmonics Relations

Figure 4a shows that the LED current during the positive half-line cycle can be expressed as:

$$i_o(t) = \begin{cases} 0 & 0 \le t < T_c \\ I_{o0} - \sum\limits_{n=1}^{\infty} I_{o(2n)} \cos 2n\omega_L t & T_c \le t \le T_e \\ 0 & T_e < t \le T_L/2 \end{cases} \tag{14}$$

where I_{o0} is the DC component, $I_{o(2n)}$ is the peak value of the harmonic component with order n. It is worth to note that if $n = 1$, this refers to conventional system without harmonic injection.

$I_{o(2n)}$ is a function of the conversion ratio M, resulting from the integration over the period $T_c \le t \le T_e$, where T_c is a function of M as derived in (1). Therefore, from (3), (7), (9) and (14) the input current can be derived as:

$$i_{in}(t) = \frac{V_o}{V_{inp}\sin(\omega_L t)} \left(I_{o0} - \sum\limits_{n=1}^{\infty} I_{o(2n)} \cos 2n\omega_L t \right). \tag{15}$$

Using a trigonometric expansion and Chebyshev polynomials to simplify the input current equation [25]:

$$i_{in}(t) = \frac{2V_o}{V_{inp}} \left\{ I_{o0}\sin\omega_L t + \sum\limits_{n=1}^{\infty} \left(\left(I_{o0} - \sum\limits_{x=1}^{n} I_{o(2x)} \right) \sin(2n+1)\omega_L t \right) \right\} \tag{16}$$

The Fourier series expansion of the input current is a sum of sinusoidal waveforms with different amplitudes and different frequencies:

$$i_{in}(t) = \sum\limits_{n=1}^{\infty} I_{i(2n-1)}\sin(2n-1)\omega_L t. \tag{17}$$

where $I_{i(2n-1)}$ is the peak value for the nth order harmonic component. By comparing (16) and (17), the relation between the input and output currents harmonics can be obtained:

$$I_{i(2n-1)} = 2\frac{V_o}{V_{inp}} \left(I_{o0} - \sum\limits_{x=2}^{n} I_{o(2(n-1))} \right). \tag{18}$$

3.2. Designed Operating Regions under Target PF and PTAR

From (14), there are multiple combinations of harmonics that can be injected to limit the PTAR while keeping the same PF and the same LED average current (I_{o0}). Therefore, a MATLAB script to sweep all possible combinations for the output current harmonics can be used. The target are those harmonic combinations that lead to minimum PTAR while keeping a PF greater than 0.9.

PF consists of two factors, distortion and displacement. In this case, the displacement factor is considered unity due to the converter operation in DCM. The distortion factor is then the significant one [26]:

$$PF = \frac{I_1}{\sqrt{I_1{}^2 + I_3{}^2 + I_5{}^2 + \ldots}}. \tag{19}$$

Substituting (18) into (19) to determine PF expression as a function of the LED current harmonics:

$$PF = \frac{I_{o0}}{\sqrt{I_{o0}^2 + \sum_{n=1}^{\infty}\left(I_{o0} - \sum_{x=1}^{n} I_{o(2x)}\right)^2}}. \tag{20}$$

The PTAR can be derived from (14) by normalizing the LED current with the DC component. Therefore, the PTAR value will be the peak value of the periodic signal, which can be expressed as:

$$PTAR = Max\left(1 - \sum_{n=1}^{\infty} I_{o(2n)}^* \cos 2n\omega_L t\right). \tag{21}$$

Regarding the MATLAB script, there are a many harmonics values which can be combined to implement the LED current considering the boundaries given in Equations (20) and (21), which determine the relation between the PTAR, the PF and the conversion ratio M. Therefore, the MATLAB script increases the amplitude of each harmonic component (I_2, I_4, I_6, etc.) and calculates the PTAR and the PF in each case. Valid combinations are those that satisfy the constraint of 0.9 as minimum PF value. Among them, the optimum one will be that with the lowest value of PTAR. Five cases are presented here. Case I is a double the frequency ($2\omega_L$). Case II is a combination of harmonics from the second ($2\omega_L$) to the fourth order ($4\omega_L$). Case III is a combination of harmonics from the second ($2\omega_L$) to the sixth order ($6\omega_L$). Case IV is a combination of harmonics from the second ($2\omega_L$) to the eighth order ($8\omega_L$). Finally, Case V is a combination from the second ($2\omega_L$) to the tenth order ($10\omega_L$).

The proposed analysis follows the flow chart presented in Figure 5. Adding more harmonics decreases the PTAR but also the PF. It is worth mentioning that the cases with higher harmonic orders (above tenth order) have been tested, but they are not considered here as they have a significant impact on PTAR but have a low PF (lower than 0.9) for any harmonic combination.

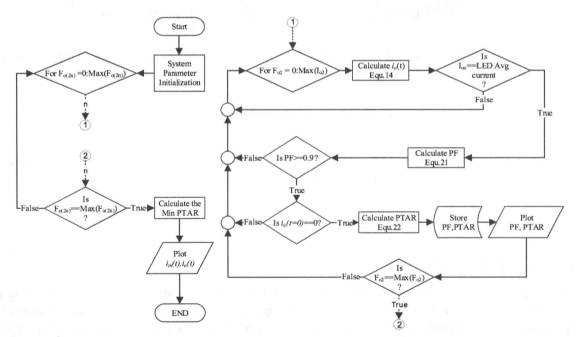

Figure 5. Flow chart of the MATLAB script.

It is in this script where using the real model of the LED rather than the typical equivalent resistance makes the conversion ratio to have a significant impact on PTAR and PF. Figure 6 shows the PTAR under Case V as a function of M while keeping the PF equals to 0.9. It should be mentioned that each point of the graph is obtained by the MATLAB script, which means that the minimum PTAR is shown. As illustrated in this graph, for a PTAR of less than 1.43, the accepted conversion ratio is limited to 0.2. This explain the mismatch in the results obtained by [10,11], where 1.43 was obtained experimentally, while 1.34 was targeted in the theoretical model. Also, this result clarifies the limitation of the conversion ratio in harmonic injection techniques. Using curve fitting, a relation between the PTAR and the conversion ratio M is found:

$$PTAR = -373.3M^4 + 265.3\ M^3 - 59.47\ M^2 + 5.677\ M + 1.142. \tag{22}$$

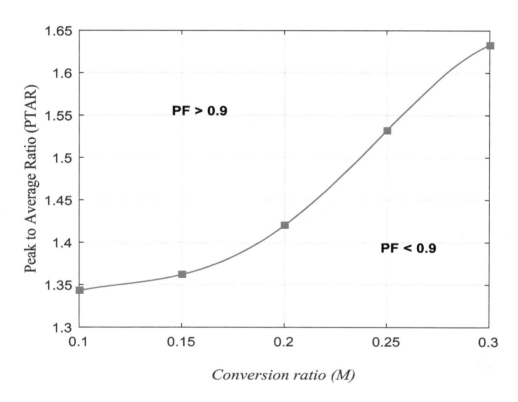

Figure 6. Curve fitted for different values for conversion ratio (M) and Peak to Average Ratio (PTAR).

4. Proposed Control Circuit and Simple Implementation

Taking the benefits described in the previous section, the control loop that reshapes the required LED current will be derived in this section using a straightforward control block. The harmonic combinations obtained from the MATLAB script for the output LED current. The duty cycle can be expressed as a Fourier series expansion using the curve fitting as:

$$d(t) = D_0 - \sum_{n=1}^{\infty} D_{(2n)}\cos 2n\omega_L t. \tag{23}$$

The duty cycle defined in (23) can be obtained by means of the proposed feedforward loop, which allows the chosen harmonic components to pass through. This reshapes the input and output currents as desired. Figure 7 shows the proposed closed-loop control for the LED driver. As can be seen, only a single feedback loop is implemented. The loop compensator, taking advantage of the previously explained harmonic injections optimization, should adjust the duty cycle according to the harmonics obtained from the MATLAB script.

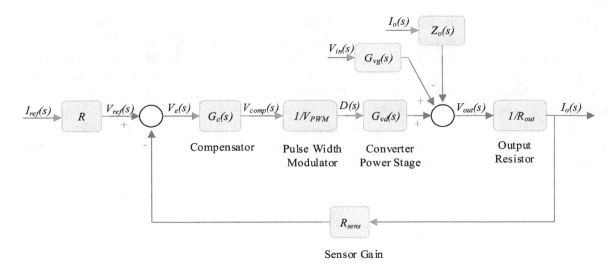

Figure 7. Closed system block configuration of DCM converter.

The loop compensator gain can be computed by dividing each order value of duty cycle harmonics, (24) by the output LED current harmonics, (14). The gain plot of the compensator output can be expressed in logarithmic scale as:

$$G_c = \frac{V_{comp}}{V_e} = \frac{V_{comp}}{V_{ref} - R_{sens}i_o} = \frac{V_{PWM}D(s)}{V_{ref} - R_{sens}I_o(s)}. \tag{24}$$

The results obtained with (25) states that this proposed reshaping block (compensation) is based on the division of the duty cycle by the output LED current harmonics.

5. Case Study

To validate the derived mathematical model and the proposed control, a case study will be analyzed in this section. The system parameters will be as follows:

- Nominal input voltage $V_{in} = 220$ VAC $\pm 10\%$.
- Line frequency $f_L = 50$ Hz.
- Output voltage $V_o = 60$ V.
- Average output power $p_o = 21$ W.
- Output current $I_O = 350$ mA.
- Switching frequency $f_s = 100$ kHz.
- Input filter capacitor $C_{in} = 47$ nF.
- Input filter inductor $L_{in} = 560$ µH.
- Output filter capacitor $C_{out} = 0.47$ µF.
- Output filter inductor $L = 270$ µH.
- Series inductor with LED $L_{LED} = 100$ µH.

It should be noted that the extra series inductor with LED is used to act as a low pass filter to attenuate the switching frequency harmonics and preventing it to flow through the LED chips.

5.1. Determination of the Targeted Injected Harmonics' Order Values

Firstly, using (23), the PTAR is determined for the selected conversion ratio M (in this case $M = 0.27$) For $V_{in} = 220$VAC $\pm 10\%$ (198 V, 220 V, 244 V) and 60 V output voltage, the PTAR is found to be 1.44, 1.41, and 1.38, respectively. It is important to highlight the advantage of this model where the designer can decide from the beginning of the design process if this resulting PTAR is acceptable or not and if another acceptable conversion ratio M has to be chosen.

Secondly, the MATLAB script is used to calculate the values of the required harmonics to be injected in the output current so that the predefined PTAR is obtained and the PF regulations are observed. Figure 8 shows the output LED current for different harmonic combination cases, as explained in the previous section. It can be shown that the peak value of the output current decreases gradually as the injected harmonic orders increases, while having the same LED current average. Figure 9 shows the input current under the same conditions. Table 1 illustrates the PTAR and the PF for each case. It is clear that Case V (with harmonics up to the tenth order) is the best one as the lowest PTAR is obtained for nominal input voltage (220 V).

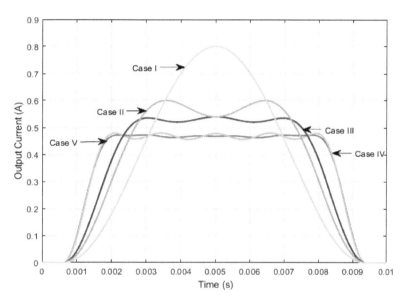

Figure 8. The LED current waveforms under different harmonics order combinations at $220V_{in}$.

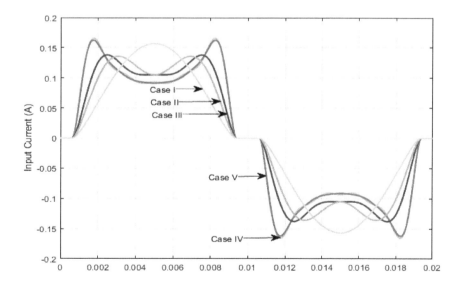

Figure 9. The input current waveform under different harmonics order combinations at $220V_{in}$.

Table 1. PTAR and PF values at different cases, $V_{in} = 220$ V.

Harmonics	Case I	Case II	Case III	Case IV	Case V
PF	0.983	0.979	0.963	0.9163	0.90
PTAR	2.28	1.713	1.541	1.431	1.41

5.2. Design Control Values for Single Multifunction Block (SMFB)

Once the LED current harmonics are obtained from MATLAB script, their values can be substituted in (14) in order to have the output LED current with its injected harmonic. Then, this output LED current is substituted into the duty cycle Equation (13). The resulting duty cycle is drawn and then the curve fitted as shown in Figure 10. Through this curve fitting, the Fourier components of the duty cycle waveform are obtained. Using (24), the results for the obtained gain, from the division of the duty cycle harmonics to the output LED harmonics, is plotted and fitted in Figure 11a.

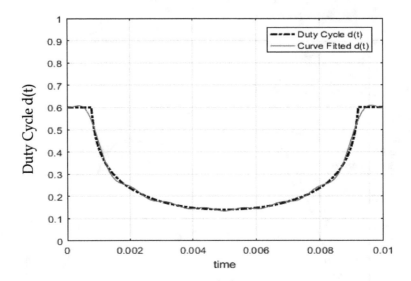

Figure 10. Duty cycle curve fitting.

It is possible to study the required features for this reshape SMFB with the results shown in Figure 11a. The required gain for the loop compensator is almost flat during the frequency range of 100 Hz–500 Hz, which is the frequency range for harmonic injection. As this analysis is concerned with the frequency range of the harmonic injection, it does not discuss the frequency range above this 500 Hz. However, this can follow regular design rules (i.e., having a low-pass filter with bandwidth in the range of one tenth of the switching frequency).

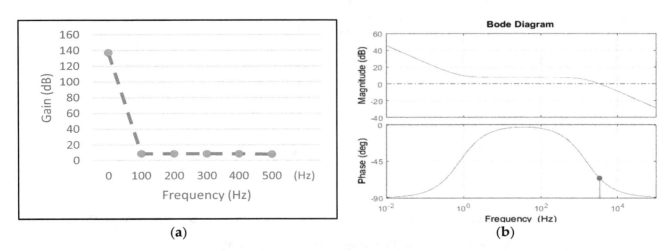

Figure 11. (a) Model of the bode plot. (b) Bode diagram of the system transformation function.

A simple implementation for this proposed SMFB can be an integrator and a pole before the double line frequency, in addition to an extra pole below one tenth of the switching frequency to filter undesired harmonics component. One critical point is that the required phase should be zero within the 100 Hz–500 Hz range to keep the line current and line voltage in phase (displacement PF requirement).

To summarize, the compensator parameters for this system can be summarized as:

1. Integrator for regulation purpose.
2. One zero at a frequency lower than one tenth of twice the line frequency to flat the gain and keep the phase angle to zero.
3. One pole between the last injected harmonic (>500 Hz, in this case) and one tenth of the switching frequency to filter undesired high frequency harmonics.

The transfer function of the described system can be presented as:

$$G_c = \frac{K\left(1 + \frac{s}{W_z}\right)}{s\left(1 + \frac{s}{W_p}\right)} = \frac{C_1 R_2 s + 1}{C_1 C_2 R_1 R_2 s^2 + (C_1 R_1 + C_2 R_2)s}. \tag{25}$$

Figure 11b shows the described system bode plot with the integrator, one zero at one hertz, and one pole at 4 kHz. One advantage of the proposed system is that it has a higher bandwidth, which introduces better dynamics than conventional systems with usually 10 Hz bandwidth.

The SMFB functions are summarized in Figure 12 regulation for the LED average current, duty cycle reshaping, harmonics injection and output current reshaping.

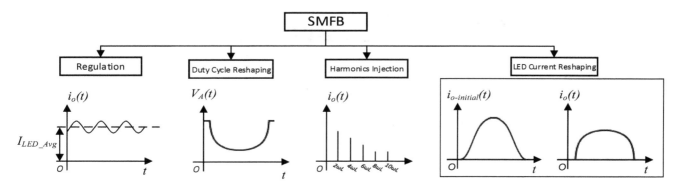

Figure 12. Single Multifunction Block (SMFB) functions.

6. Simulation & Experimental Results

To validate the proposed idea, simulation and experimentation are carried out. Figure 13a describes the circuit diagram used in both simulation and practice. It consists of the EMI filter, the rectifier bridge, and the buck converter. The buck converter switch is AOT22N50 500 V, 22 A N-Channel MOSFET and the diode is ES1J which is 1 A, 200 V Surface Mount Super-Fast Rectifier. The converter has a small output film capacitor to increase the lifetime of the LED driver. Small series resistor Surface Mount $R_{sems} = 0.5\ \Omega$ 1% 0.5 W is inserted to sense the output current for the control circuit.

The compensator is a Type II, as shown in Figure 13b Due to the simplicity of the control circuit, a generic PWM controller can be used, the UC3825a IC from Texas Instrument with a few surface mount technology (SMT) components. The compensator capacitors and resistors can be calculated using (25). Table 2 shows the compensator values.

Table 2. The compensator value.

Component	R1	R2	C1	C2
Value	100 k Ω	200 k Ω	1 uF	220 pF

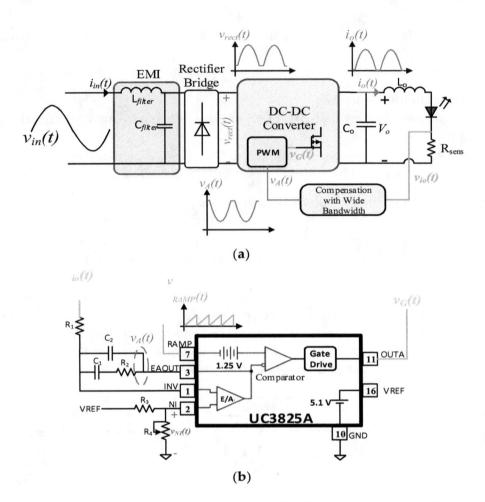

Figure 13. (**a**) Proposed circuit forharmonics injection. (**b**) Compensator circuit diagram for E-Cap less converter.

Figure 14 shows the simulation results obtained with PSIM (left column) as well as the experimental measurements on the prototype (right column). All the figures present the input voltage v_{in}, input current i_{in}, output voltage V_o and output current i_o in AC-DC electrolytic-capacitor-less buck converter. As shown, different line voltages were tested. Figure 14a,b show the results for $V_{in} = 198$ VAC which implies a conversion ratio of $M = 0.21$. The recorded PF is 0.9 and the PTAR is 1.43. Similarly, the system under study is tested for $V_{in} = 220$ VAC (nominal value), which represents $M = 0.19$, and its results are shown in Figure 14c,d. The recorded PF is 0.9 and the PTAR is 1.41. For $V_{in} = 244$ VAC, the results are shown in Figure 14e,f. The recorded PF is 0.9 and the PTAR is 1.39 ($M = 0.17$). Also, no phase shift happens between input current and line voltage as shown in the figures.

As discussed in introduction about the finding in [10], the circuit aimed to inject 3rd and 5th harmonics into the input current to eliminate the E-Cap. This was proposed using complex control technique which increase the component counts and leads to increase the LED driver technology as shown in Table 3 in compare to the proposed single-feedback loop in this paper. In addition, the mathematical model in [2] showed that the PTAR is 1.34, however the experimental results have a mismatch where a PTAR of 1.43 is reported. On the other hand, the proposed mathematical model in this paper has a good correlation with experimental results as shown in Figure 15. This graph shows different values for the PTAR and conversion ratio M under simulation, experimental and mathematical model.

Table 3. comparison between the proposed circuit and [10].

Circuit	PTAR Mathematical	PTAR Experimental	Op-Amp	Multiplier	Divider	No. Control Loops
[10]	1.34	1.43	4	1	1	2
Proposed	1.41	1.41	2	-	-	1

(Columns Op-Amp, Multiplier, Divider under "Control Circuit Component Count")

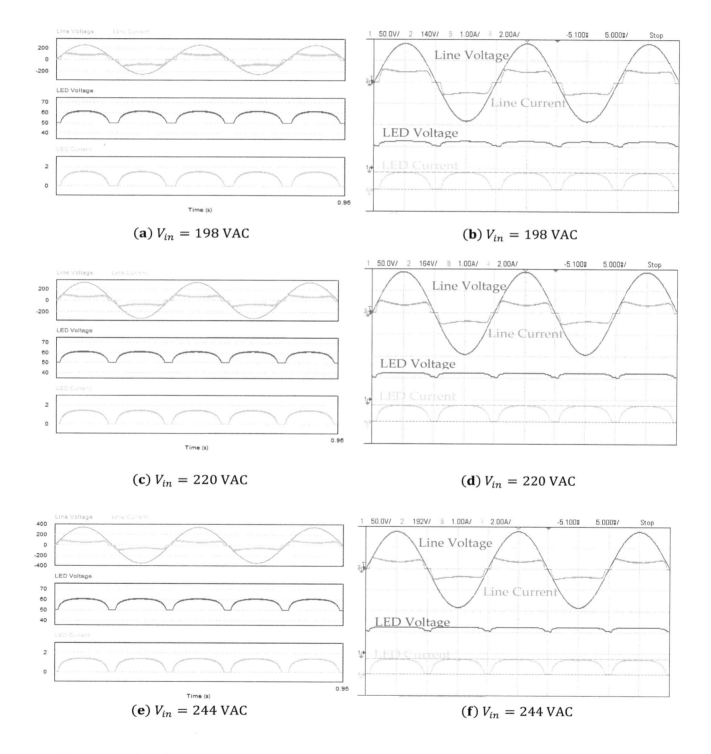

Figure 14. Simulation and prototype waveform results of input voltage v_{in}, input current i_{in}, LED output voltage V_o and output LED current i_o in buck converter.

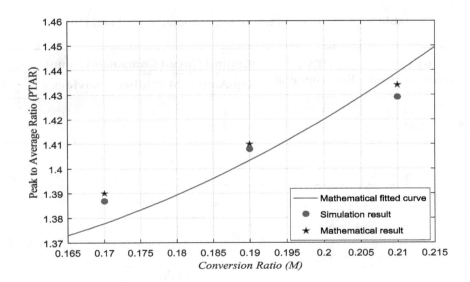

Figure 15. Obtained PTAR at different values for conversion ratio (*M*) from mathematical model, practical and simulation.

An AC-DC power converter should be complied with EN 61000-3-2 standard. This standard for limiting the harmonics current level of the electronics equipment which is injected by different loads back to the grid. Figure 16 shows a comparison between the input current values for the proposed circuit and the maximum values for EN 61000-3-2 Class D standard. This class should be complied for the lighting equipment with an input power smaller or equal than 25 W. The figure verifies that the proposed circuit is compliance with EN 61000-3-2 standard [27].

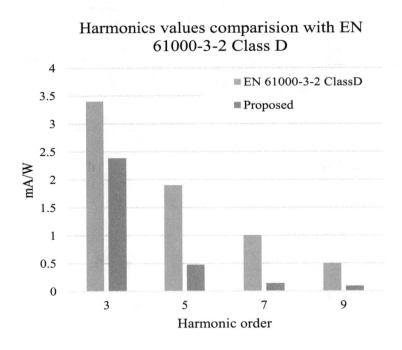

Figure 16. Comparison between the proposed Input current values and maximum level for EN 61000-3-2 Class D standard.

The proposed solution has the features of single feedback loop with low component counts for E-cap less LED solution under limited PTAR, by its turn lower cost solution. Within test, there was no record for visible flicker. However, flicker issue requires more investigation due to the increase in the LED current modulation percentage [28].

7. Conclusions

This paper has proposed the use of the complete LED model instead of the simple equivalent resistance. Its relevance and influence on the design, on the PF and on the PTAR are explored. The proposed model is derived under harmonic injection technique and it determines the required harmonic to limit the LED current PTAR and keep the input PF higher than 0.9. Based on the derived model, a reshape control block is proposed and implemented using a second-order compensator with a single feedback loop. This circuitry can be applied to different converters topologies such as Flyback converter. Results show a good agreement between simulation and experimentation. Results conclude that there is a specified range for the conversion ratio so the target PTAR can be achieved with while complying with ENERGY STAR and EN 61000-3-2 Class D standard.

Author Contributions: M.N. conceived and designed the experiments, performed the experiments, analyzed the data, and wrote the paper. M.O. define the concept, review results, supervision and manage the project. M.A. review results, revised and edited the manuscript. E.M.A., E.-S.H. co-supervised the work.

References

1. Machine, W. *Lifetime of White LEDs*; US Department of Energy: Washington, DC, USA, 2009.
2. Wang, Y.; Alonso, J.M.; Ruan, X. A Review of LED Drivers and Related Technologies. *IEEE Trans. Ind. Electron.* **2017**, *64*, 5754–5765. [CrossRef]
3. Gebreel, A.; Nassary, M.; Bakr, A. *Low Cost Low Voltage Low Power Integrated LED Driver*; LAP LAMBERT Academic Publishing: Saarbruggen, Germany, 2016.
4. Cree First to Break 300 Lumens-Per-Watt Barrier. CREE, 26 March 2014. Available online: http://www.cree.com/ (accessed on 2 November 2018).
5. Cheng, C.A.; Cheng, H.L.; Chung, T.Y. A Novel Single-Stage High-Power-Factor LED Street-Lighting Driver With Coupled Inductors. *IEEE Trans. Ind. Appl.* **2014**, *50*, 3037–3045. [CrossRef]
6. Fortunato, M. Ensure Long Lifetimes from Electrolytic Capacitors: A Case Study in LED Light Bulbs. Maximintegrated, 29 May 2013. Available online: www.maximintegrated.com (accessed on 2 November 2018).
7. Montanari, D.; Saarinen, K.; Scagliarini, F.; Zeidler, D.; Niskala, M.; Nender, C. Film Capacitors for Automotive and Industrial Applications. October 2008. Available online: www.kemet.com (accessed on 2 November 2018).
8. Almeida, P.S.; Nogueira, F.J.; Guedes, L.F.A.; Braga, H.A.C. An experimental study on the photometrical impacts of several current waveforms on power white LEDs. In Proceedings of the XI Brazilian Power Electronics Conference, Praiamar, Brazil, 11–15 September 2011; pp. 728–733.
9. Almeida, P.S.; Bender, V.C.; Braga, H.A.C.; Costa, M.A.D.; Marchesan, T.B.; Alonso, J.M. Static and Dynamic Photoelectrothermal Modeling of LED Lamps Including Low-Frequency Current Ripple Effects. *IEEE Trans. Power Electron.* **2015**, *30*, 3841–3851. [CrossRef]
10. Wang, B.; Ruan, X.; Yao, K.; Xu, M. A Method of Reducing the Peak-to-Average Ratio of LED Current for Electrolytic Capacitor-Less AC–DC Drivers. *IEEE Trans. Power Electron.* **2010**, *25*, 592–601. [CrossRef]
11. Ruan, X.; Yao, B.W.K.; Wang, S. Optimum Injected Current Harmonics to Minimize Peak-to-Average Ratio of LED Current for Electrolytic Capacitor-Less AC–DC Drivers. *IEEE Trans. Power Electron.* **2011**, *26*, 1820–1825. [CrossRef]
12. Gu, L.; Ruan, X.; Xu, M.; Yao, K. Means of Eliminating Electrolytic Capacitor in AC/DC Power Supplies for LED Lightings. *IEEE Trans. Power Electron.* **2009**, *24*, 1399–1408. [CrossRef]
13. Nassary, M.; Orabi, M.; Ahmed, E.; Hasaneen, El.; Gaafar, M. Control Circuit Optimization for Electrolytic Capacitor-Less LED Driver. In Proceedings of the IEEE MEPCON Conference, Cairo, Egypt, 19–21 December 2017.
14. Lamar, D.G.; Sebastian, J.; Arias, M.; Fernandez, A. On the Limit of the Output Capacitor Reduction in Power-Factor Correctors by Distorting the Line Input Current. *IEEE Trans. Power Electron.* **2012**, *27*, 1168–1176. [CrossRef]
15. Lamar, D.G.; Sebastián, J.; Arias, M.; Fernández, A. Reduction of the Output Capacitor in Power Factor Correctors by Distorting the Line Input Current. In Proceedings of the 2010 Twenty-Fifth Annual IEEE

Applied Power Electronics Conference and Exposition (APEC), Palm Springs, CA, USA, 21–25 February 2010; pp. 196–202.

16.   Rezaei, K.; Golbon, N.; Moschopoulos, G. A New Control Scheme for an AC-DC Single-stage Buck-Boost PFC Converter with Improved Output Ripple Reduction and Transient Response. In Proceedings of the APEC 2014, Beijing, China, 10–12 November 2014; pp. 1866–1873.

17.   Soares, G.M.; Alonso, J.M.; Braga, H.A.C. Investigation of the Active Ripple Compensation Technique to Reduce Bulk Capacitance in Off-line Flyback-Based LED Drivers. *IEEE Trans. Power Electron.* **2018**, *33*, 5206–5214. [CrossRef]

18.   Soares, G.M.; Almeida, P.S.; Alonso, J.M.; Braga, H.A.C. Capacitance Minimization in Offline LED Drivers Using an Active-Ripple-Compensation Technique. *IEEE Trans. Power Electron.* **2017**, *32*, 3022–3033. [CrossRef]

19.   Yang, Y.; Ruan, X.; Zhang, L.; He, J.; Ye, Z. Feed-Forward Scheme for an Electrolytic Capacitor-Less AC/DC LED Driver to Reduce Output Current Ripple. *IEEE Trans. Power Electron.* **2014**, *29*, 5508–5517. [CrossRef]

20.   Wang, S.; Ruan, X.; Yao, K.; Tan, S.C.; Yang, Y.; Ye, Z. A Flicker-Free Electrolytic Capacitor-Less AC–DC LED Driver. *IEEE Trans. Power Electron.* **2012**, *27*, 4540–4548. [CrossRef]

21.   Lee, K.W.; Hsieh, Y.H.; Liang, T.J. A Current Ripple Cancellation Circuit for Electrolytic Capacitor-less AC-DC LED Driver. In Proceedings of the 2013 Twenty-Eighth Annual IEEE Applied Power Electronics Conference and Exposition (APEC), Long Beach, CA, USA, 17–21 March 2013; pp. 1058–1061.

22.   Shen, Y.C.; Liang, T.J.; Tseng, W.J.; Chang, H.H.; Chen, K.H.; Lu, Y.J.; Li, J.S. Non-Electrolytic Capacitor LED Driver with Feedforward Control. In Proceedings of the IEEE Energy Conversion Congress and Exposition (ECCE), Montreal, QC, Canada, 20–24 September 2015; pp. 3223–3230.

23.   Hui, S.Y.; Li, S.N.; Tao, X.H.; Chen, W.; Ng, W.M. A Novel Passive Offline LED Driver With Long Lifetime. *IEEE Trans. Power Electron.* **2010**, *25*, 2665–2672. [CrossRef]

24.   Chen, W.; Li, S.N.; Hui, S.Y.R. A Comparative Study on the Circuit Topologies for Offline Passive LightEmitting Diode (LED) Drivers with Long Lifetime & High Efficiency. In Proceedings of the IEEE Energy Conversion Congress and Exposition, Atlanta, GA, USA, 12–16 September 2010; pp. 724–730.

25.   Kreyszig, E. *Advanced Engineering Mathematics*, 7th ed.; Wiley: Hoboken, NJ, USA, 1992.

26.   Robert, W. *Erickson and Dragan Maksimovic, Fundamentals of Power Electronics*, 2nd ed.; Springer Science+Business Media, LLC: New York, NY, USA, 2001.

27.   European Power Supply Manufacturers Association. *Harmonic Current Emissions: Guidelines to the Standard EN 61000-3-2*; EPSMA: Winchester, UK, 2010.

28.   *IEEE Recommended Practices for Modulating Current in HighBrightness LEDs for Mitigating Health Risks to Viewers*; IEEE Std: Piscataway, NJ, USA, 2015; pp. 1–80.

13

A Universal Mathematical Model of Modular Multilevel Converter with Half-Bridge

Ming Liu [1,2], Zetao Li [1,3,*] and Xiaoliu Yang [1]

[1] College of Electrical Engineering, Guizhou University, Guiyang 550025, China;
    weiminxiaohai@163.com (M.L.); xlyang1@gzu.edu.cn (X.Y.)
[2] Department of Mechanical Engineering, Guizhou College of Electronic Science and Technology,
    Guian 550003, China
[3] Guizhou Provincial Key Laboratory of Internet + Intelligent Manufacturing, Guiyang 550025, China
* Correspondence: gzulzt@163.com

Abstract: Modular multilevel converters (MMCs) play an important role in the power electronics industry due to their many advantages, such as modularity and reliability. In the current research, the simulation method is used to study the system. However, with the increasing number of sub-modules (SMs), it is difficult to model and simulate the system. In order to overcome these difficulties, this paper presents a universal mathematical model (UMM) of MMC using half-bridge cells as SMs. The UMM is a full-scale model with switching state, capacitance, inductance, and resistance characteristics. This method can calculate any number of SMs, and it does not need to build a simulation model (SIM) of physical MMC—in particular, parametric design can be realized. Compared with the SIM, the accuracy of the proposed UMM is verified, and the computational efficiency of the UMM is 8.7 times higher than the simulation method. Finally, by utilizing the proposed UMM method, the influence of the parameters of MMCs is studied, including the arm induction, SM capacitance, SM number, and output current/voltage total harmonic distortion (THD) based on the UMM in the paper. The results offer an engineering insight to optimize the design of MMCs.

Keywords: modular multilevel converter (MMC); total harmonic distortion (THD); universal mathematical model (UMM); switching state; nearest level modulation (NLM)

1. Introduction

With the rapid development of offshore wind farms, the demand for a high power, high-quality transmission system becomes more urgent. Modular multilevel converter (MMC)-based high voltage direct voltage (HVDC) technology provides a promising solution, due to its advantages of modularization, scalability, high efficiency, excellent harmonic performance, fault blocking ability, small filter size, high efficiency, and low redundancy cost [1–3]. MMC has been applied to many industries, such as energy storage systems, medium-voltage and high-power motor drive systems, distribution systems, etc. [4–8].

However, it is difficult to formulate an explicit expression of MMC, because it is a hybrid system of discrete and continuous models. The main feature of MMCs is the cascaded connection of a large number of sub-modules (SMs). These SMs are arranged in groups called arms or branches. The low-frequency voltage or current at the AC side is controlled by high-frequency switching values to manage SMs on/off. Therefore, the interaction between the arm and line quantities (variables) generates low- and high-frequency components on the AC and DC side of the SMs in an MMC [9]. In other words, MMC has strong coupling nonlinear multi-input and multi-output dynamic features [10]. The simulation studies were utilized to analyze the behavior of an MMC. However, the simulation process consumes time and computer resources to create a large number of SMs (up to 400 per arm) [11].

For example, a traditional detailed model (TDM) of MMC requires hundreds and thousands of Insulated Gate Bipolar Transistors(IGBTs) with antiparallel diodes and capacitors to be built and electrically connected in the simulation package's graphical user interface, resulting in a large admittance matrix.

To simplify the simulation model, the conventional switching models/detailed models with full capabilities of replicating the conduction of power electronic devices such as IGBTs and their anti-parallel diodes are inefficient for the modeling of MMC-HVDC, as the simulation time is prohibitively long [1]. To simplify the calculation, it was assumed that the SM capacitor voltages are well balanced at their reference values [12,13]. The SM terminal voltage in each arm was modeled as a single equivalent voltage source [14]. In [15], the equivalent model was used and a small-signal analysis was carried out. Alternatively, each arm of MMC was modeled as a nonlinear capacitor with a time-variant sinusoidal capacitance [16]. Moreover, to simplify the analysis, the average value models (AVMs) are presented in [17–20]. However, the methods above do not reflect the switching state and the transient process of the SM capacitor voltage.

To address this problem, an efficient model was proposed by Udana and Gole in [21], which is referred to as the detailed equivalent model (DEM) in this paper; yet, a drawback of the DEM is that the individual converter components are invisible to the user. A new model, referred to as the accelerated model (AM), was proposed by Xu et al. in [22], but a full and objective comparison could not be completed because different researchers built the models on different computers. In [23], an enhanced accelerated model (EAM) with improved simulation speed was proposed by Antony et al., which further improved the computational efficiency of one method. A new dynamic phasor (DP) model of an MMC with an extended frequency range for direct interfacing with an electromagnetic transient (EMT) simulator was presented in [24]. In reference [25], a method of MMC modeling and design based on parametric and model-form uncertainty quantification is proposed, which can establish confidence in modeling and simulation in the presence of manufacturing variability and modeling errors, and may eliminate the need for heuristic safety factors. However, the high-efficiency calculation of large-scale SMs is not involved. The internal dynamics of the MMC are modeled considering the dominant harmonic components of each variable. However, the improved simulation models introduced above are based on Power Systems Computer Aided Design/Electromagnetic Transients including DC(PSCAD/EMTDC) for electromagnetic transient simulation. It is inconvenient to set variable parameters or change the topology of the whole circuit. It cannot be satisfied by loop calculation to compare the changes of parameters.

In this paper, a universal mathematical model for MMC is proposed which can reflect the steady-state and dynamic process of MMC. The model is a detailed numerical model, including the capacitor voltage and switch function of SMs. It can be implemented by a computer for any number of SMs. The whole model is parametric programmed; by setting one or several parameters, the desired results can be quickly obtained. Compared with the simulation model, this algorithm can easily modify the circuit parameters by setting cycle statements, and automatically carry out repeated simulation and multi-state simulation, so it is convenient to observe the operation characteristics of the system under different parameter values. Using the proposed MMC, the output voltage and the current THDs of MMC have been analyzed under different parameters (such as module number, capacitor voltage, arm inductance). In addition, the change in the capacitance voltage has been studied as one capacitance value decayed.

Without losing generality, in this paper the research is based on MATLAB/Simulink because of its powerful numerical calculation ability and rich processing module (such as Pulse Width Modulation (PWM) generator, and various transformation and comparison modules), especially the demonstration of control strategy. Compared with Simulink, PLECS is more professional, but does not have as many toolboxes as Simulink. The LTSpice installation package is small, easy to operate, and fast, but most of the support is for the ADI company's own chip model. PSIM has the advantages of simple operation and fast simulation speed, and supports mainstream simulation mode analysis. However, due to the use of ideal switches, the simulation accuracy is limited.

This paper is organized as follows. Section 2 introduces the MMC topology, operation, and mathematical model. The algorithm of the universal mathematical model (UMM) is explained in Section 3. The correctness of the UMM is demonstrated in Section 4. A performance analysis under different conditions is shown in Section 5. The conclusions of the study are presented in Section 4.

2. Topology and Mathematical Model of the MMC

Topology and Principles of Operation

The circuit structure of a three-phase MMC is shown in Figure 1. The single-phase consists of an upper and a lower arm. Each arm is composed of N sub-modules (SM), and an inductor and equivalent resistance are connected in series. Each individual SM contains a capacitor and two complementary insulated gate bipolar transistor modules (i.e., $S_{jm,n}$ and $S'_{jm,n}$). In this paper, the subscript $j = a, b, c$ means three-phase; $m = u, l$, where u represents the upper arm and l represents a lower arm; $n = 1, 2, 3, \ldots, N$ represents the number of sub-modules. The rest of the symbols are as follows: L (the arm inductance), R (the arm equivalent resistance), u_{jm} (the arm voltage), i_{jm} (the arm current), u_j (the AC side voltage), i_{oj} (the output current), i_{cj} (the circulating current), L_{oj} (the load inductance), R_{oj} (the load resistance), U_{dc} (the DC source voltage), $u_{jm,n}$ (the capacitor voltage).

$$\frac{du_{jm,n}}{dt} = \frac{S_{jm,n}i_{jm}}{C} \tag{1}$$

where $S_{jm,n}$ is the switch function of the nth SM in the m arm of phase-j, and its value is 1 or 0; C is the SM capacitance.

The relationship between the arm voltage and the capacitor voltage in phase-j and the switching function is:

$$u_{jm} = \sum_{n=1}^{N} S_{jm,n}u_{jm,n} \tag{2}$$

Considering a fictitious midpoint in the DC side of Figure 1 and using Kirchhoff's circuit laws, the following mathematical equations that govern the dynamic behavior of the MMC in phase-j can be obtained:

$$u_{ju} + Ri_{ju} + L\frac{di_{ju}}{dt} + R_{oj}i_{oj} + L_{oj}\frac{di_{oj}}{dt} + u_j - \frac{U_{dc}}{2} = 0 \tag{3}$$

$$u_{jl} + Ri_{jl} + L\frac{di_{jl}}{dt} - R_{oj}i_{oj} - L_{oj}\frac{di_{oj}}{dt} - u_j - \frac{U_{dc}}{2} = 0 \tag{4}$$

$$i_{oj} = i_{ju} - i_{jl} \tag{5}$$

The arm currents can be expressed as:

$$i_{ju} = \frac{i_{oj}}{2} + i_{cj}, \; i_{jl} = -\frac{i_{oj}}{2} + i_{cj} \tag{6}$$

where i_{cj} is the circulating currents flowing through phase-j of the MMC and can be calculated by Equation (7):

$$i_{cj} = \frac{i_{ju} + i_{jl}}{2} \tag{7}$$

Substitute (3) and (5) into (4), and the dynamics of phase-j AC-side currents can be obtained as:

$$\frac{di_{oj}}{dt} = -\frac{R + 2R_{oj}}{L + 2L_{oj}}i_{oj} - \frac{1}{L + 2L_{oj}}u_{ju} + \frac{1}{L + 2L_{oj}}u_{jl} - \frac{2}{L + 2L_{oj}}u_j \tag{8}$$

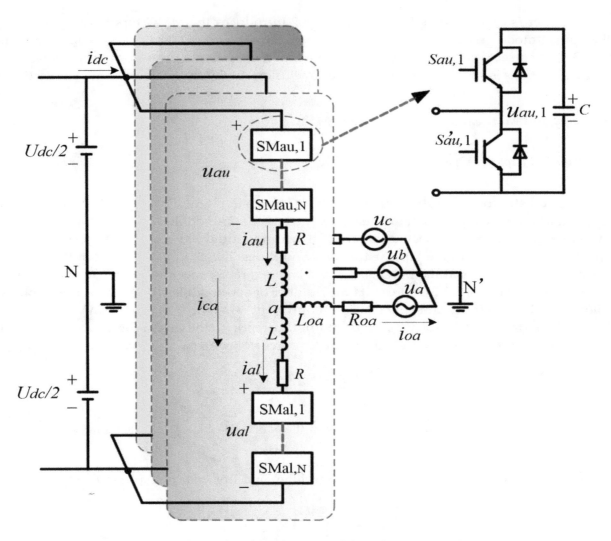

Figure 1. Structure of a three-phase MMC-based inverter and its SM.

Similarly, the dynamic behavior of the circulating current in phase-j can be obtained by substituting (3) and (7) into (4):

$$\frac{di_{cj}}{dt} = -\frac{R}{L}i_{cj} - \frac{1}{2L}u_{ju} - \frac{1}{2L}u_{jl} + \frac{1}{2L}U_{dc} \tag{9}$$

Based on (1), (2), (8), and (9), the state-space equation of the MMC in phase-j can be described as:

$$\dot{x}(t) = A(t)x(t) + Dd(t) \tag{10}$$

where $x = [i_{oj}, i_{cj}, u_{ju,1}, \ldots, u_{ju,N}, u_{jl,1}, \ldots, u_{jl,N}]^T \in \mathbb{R}^{2N+2}$ is the state vector; $d = [u_j, U_{dc}]^T$ is a perturbation vector; the desired value (u_j^*) of u_j is presented in Equation (11); $A \in \mathbb{R}^{(2N+2)\times(2N+2)}$ is a time-varying structure state matrix, presented in (12); and $D \in \mathbb{R}^{(2N+2)\times 2}$ is the perturbation coefficient matrix, presented in (20).

$$u_j^* = \sqrt{2}U\sin(2\pi ft + \varphi_j) \tag{11}$$

where U is the voltage effective value (RMS) on the AC side, f is the AC system frequency, and φ_j is the initial phase angle in phase-j.

$$A(t) = \begin{bmatrix} A_1 & A_2(t) \\ A_3(t) & 0 \end{bmatrix} \in \mathbb{R}^{(2N+2)\times(2N+2)} \tag{12}$$

where $A_1 \in \mathbb{R}^{2\times 2}$ is a constant matrix in (13), and $A_2 \in \mathbb{R}^{2\times 2N}$ and $A_3 \in \mathbb{R}^{2N\times 2}$ are time-varying matrixes in (14) and (18), respectively.

$$A_1 = \begin{bmatrix} -\frac{R+2R_{oj}}{L+2L_{oj}} & 0 \\ 0 & -\frac{R}{L} \end{bmatrix} \tag{13}$$

$$A_2(t) = A_2' diag(\boldsymbol{u}(t)) \tag{14}$$

where $A_2' = \begin{bmatrix} A_{21}' & A_{22}' \end{bmatrix} \in \mathbb{R}^{2\times 2N}$; $A_{21}' \in \mathbb{R}^{2\times N}$ is presented in (15); $A_{22}' \in \mathbb{R}^{2\times N}$ is presented in (16); and $\boldsymbol{u}(t)$ is the input control vector, presented in (17).

$$A_{21}' = \begin{bmatrix} \frac{-1}{L+2L_{oj}} & \frac{-1}{L+2L_{oj}} & \cdots & \frac{-1}{L+2L_{oj}} \\ -\frac{1}{2L} & -\frac{1}{2L} & \cdots & -\frac{1}{2L} \end{bmatrix} \tag{15}$$

$$A_{22}' = \begin{bmatrix} \frac{1}{L+2L_{oj}} & \frac{1}{L+2L_{oj}} & \cdots & \frac{1}{L+2L_{oj}} \\ -\frac{1}{2L} & -\frac{1}{2L} & \cdots & -\frac{1}{2L} \end{bmatrix} \tag{16}$$

$$\boldsymbol{u}(t) = \begin{bmatrix} S_{ju,1}, S_{ju,2}, \cdots, S_{ju,N}, S_{jl,1}, S_{jl,2}, \cdots, S_{jl,N} \end{bmatrix} \tag{17}$$

$$A_3(t) = diag(\boldsymbol{u}(t))A_3' \tag{18}$$

where $A_3' \in \mathbb{R}^{2N\times 2}$ is as follows:

$$A_3' = \begin{bmatrix} \frac{1}{2C_{ju,1}} & \cdots & \frac{1}{2C_{ju,N}} & \frac{-1}{2C_{jl,1}} & \cdots & \frac{-1}{2C_{jl,N}} \\ \frac{1}{C_{ju,1}} & \cdots & \frac{1}{C_{ju,N}} & \frac{1}{C_{jl,1}} & \cdots & \frac{1}{C_{jl,N}} \end{bmatrix}^T \tag{19}$$

$$D = \begin{bmatrix} D_1 \\ 0 \end{bmatrix} \in \mathbb{R}^{(2N+2)\times 2} \tag{20}$$

where D_1 is as follows:

$$D_1 = \begin{bmatrix} \frac{-2}{L+2L_{oj}} & 0 \\ 0 & \frac{1}{2L} \end{bmatrix} \tag{21}$$

The output voltage of the MMC, such as phase-j, is defined as the voltage difference from point j to N.

$$v_{jN} = R_{oj}i_{oj} + L_{oj}\frac{di_{oj}}{dt} + u_j \tag{22}$$

In (10), the control of the system is to adjust the structure of matrix A so that $u_j \to u_j^*$, $U_{dc} \to U_{dc}^*$ (or the active power and reactive power are close to their desired values). As the focus of the paper is the universal mathematical model of the controlled object, the control method is shown in reference [26], and will not be detailed here.

From Equation (8), we can get the formula of u_j as follows (in active inverter, $u_j \neq 0$, $R_{oj} = 0$; in passive inverter, $u_j = 0$, $R_{oj} \neq 0$):

$$u_j = -\frac{1}{2}(L + 2L_{oj})\frac{di_{oj}}{dt} - \frac{R}{2}i_{oj} - \frac{1}{2}u_{ju} + \frac{1}{2}u_{jl} \tag{23}$$

3. Algorithm of the Universal Mathematical Model

Equation (10) shows that the MMC is a nonlinear multi input system where the nonlinearity consists of the products between the states and inputs. The direct solution is difficult to find; as a result, the discrete sampling method will be used. The algorithm of the universal model is shown in Figure 2.

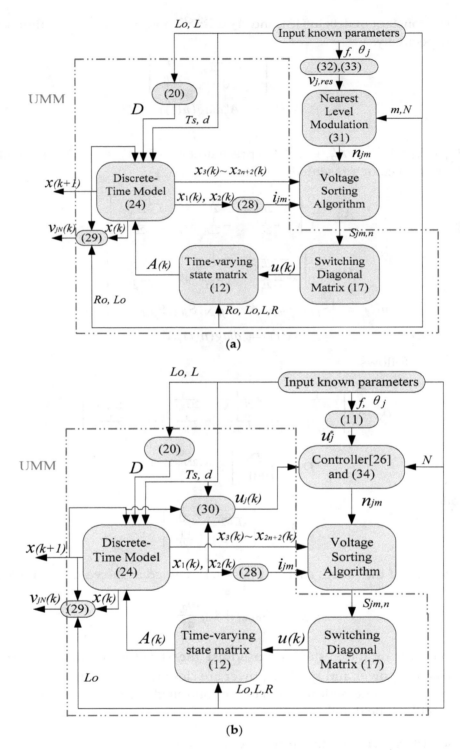

Figure 2. Block diagram of the universal mathematical model algorithm for MMC. (a) Open-loop control system, (b) closed-loop control system.

3.1. Discrete-Time Model of the MMC

It is assumed that the switching of equations occurs at the sampling points. Based on (10) and assuming a sampling time of T_s, the discrete-time model of the MMC, based on a forward Euler approximation, is obtained as:

$$x(k+1) = (I + T_s A(k))x(k) + T_s D d(k) \tag{24}$$

The output current, $i_{oj}(k)$, can be calculated as:

$$i_{oj}(k) = x_1(k) \tag{25}$$

The circulating current, $i_{cj}(k)$, is obtained by Equation (26):

$$i_{cj}(k) = x_2(k) \tag{26}$$

The capacitance voltage, $u_{jm,n}(k)$, is obtained by the following Equation:

$$u_{jm,n}(k) : x_3(k) \sim x_{2N+2}(k) \tag{27}$$

Based on Equations (6), (25), and (26), the arm currents can be calculated as:

$$i_{ju}(k) = \frac{x_1(k)}{2} + x_2(k), \ i_{jl}(k) = -\frac{x_1(k)}{2} + x_2(k) \tag{28}$$

Based on Equation (22), the output voltage of the MMC can be expressed as:

$$v_{jN}(k) = R_{oj}x_1(k) + L_{oj}\frac{x_1(k+1) - x_1(k)}{T_s} + u_j(k) \tag{29}$$

Based on Equation (23), the AC voltage of the MMC can be expressed as:

$$u_j(k) = -\frac{1}{2}(L + 2L_{oj})\frac{x_1(k+1) - x_1(k)}{T_s} - \frac{R}{2}x_1(k) - \frac{1}{2}u_{ju}(k) + \frac{1}{2}u_{jl}(k) \tag{30}$$

3.2. Nearest Level Modulation

The nearest level modulation (NLM), also known as the round method, is an approach that uses the nearest voltage level to estimate the desired output voltage. The three phases are controlled independently. Given a normalized voltage reference $v_{j,res}$, the nearest output voltage level n_{jm} can be determined by:

$$\begin{cases} n_{ju} = \frac{N}{2} - \text{round}\left(\frac{mNv_{j,res}}{2}\right) \\ n_{jl} = \frac{N}{2} + \text{round}\left(\frac{mNv_{j,res}}{2}\right) \end{cases} \tag{31}$$

where m is the modulation coefficient, and $v_{j,res}$ is defined as:

$$v_{j,res} = \sin\left(2\pi ft + \theta_j\right) \tag{32}$$

where θ_j is the initial phase angle in phase-j. Normalized voltage references for the three phases can be presented as:

$$\begin{cases} v_{a,res} = \sin(2\pi ft) \\ v_{b,res} = \sin\left(2\pi ft - \frac{2\pi}{3}\right) \\ v_{c,res} = \sin\left(2\pi ft + \frac{2\pi}{3}\right) \end{cases} \tag{33}$$

3.3. Control Systems

In this paper, the control strategy applied in [26] is utilized to obtain the optimal value u_{ju}^* and u_{jl}^* through the tracking control of the AC voltage or DC voltage. Of course, not limited to this control strategy, other controls (such as the traditional PI control) are also applicable.

$$
\begin{cases}
n_{ju} = \text{round}\left(\dfrac{u_{ju}^*}{u_C}\right) \\[2ex]
n_{jl} = \text{round}\left(\dfrac{u_{jl}^*}{u_C}\right)
\end{cases}
\tag{34}
$$

where u_C is the capacitor-rated voltage, $u_C = U_{dc}^*/N$.

The MMC control diagram is shown in Figure 3, and a more detailed closed-loop control is shown in Figure 2b. Since this study focuses on the universality of the model controlled (UMM), only the system is considered as an open-loop system (in Figure 2a) in the early stage of the design, and the control system is designed after the system-controlled parameters are fixed.

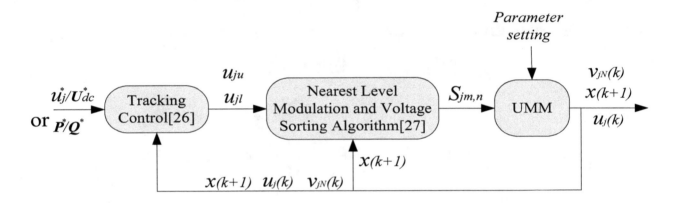

Figure 3. MMC control diagram.

3.4. Voltage Sorting Algorithm

In this paper, the voltage sorting algorithm applied in [27] is utilized to equalize all the capacitor voltages of the MMC $u_{jm,n}$. The algorithm reads the insertion indices n_{ju} and n_{jl} and determines which SMs are connected or bypassed in each arm of the MMC according to the plus-minus of the arm current i_{jm}. For example, if $i_{jm}(k) > 0$, the algorithm connects n_{jm} SMs with the lowest voltages in the corresponding arm and bypasses all the others. Conversely, if $i_{jm}(k) < 0$, the algorithm connects n_{jm} SMs with the highest voltages and bypasses the others. Therefore, the switching signals $S_{jm,n}$ to be applied in the sampling time k can be obtained. Finally, $u(k)$ is obtained based on (17).

4. Verification of Universal Mathematical Model

To evaluate the performance of the proposed UMM, a comparison between the results from the UMM and the nonlinear time-domain simulation model has been conducted. The nonlinear time-domain simulation model (SIM) is implemented in MATLAB/Simulink, and the UMM is performed using an m-file in MATLAB. The initial value is set as $x(0) = [0, 0, u_C, u_C, \ldots, u_C]^T \in \mathbb{R}^{2N+2}$. The comparison was conducted for a single-phase converter with 20 submodules. The main parameters of the MMC in the simulation are listed in Table 1. All the simulations were conducted using a Microsoft Windows 10 operating system with a 2.7 GHz Intel® core ™ i7-7500U processor and 16 GB of RAM. The test results are given in the following sub-sections.

Table 1. Main parameters of the MMC (phase-a).

Parameters	Value
AC system frequency f (Hz)	50
DC voltage U_{dc} (kV)	60
Initial capacitor voltage u_C (V)	U_{dc}/N
SM number in each arm (N)	20 (variable)
SM capacitance C (mF)	40 (variable)
Arm inductance L (mH)	3 (variable)
Arm equivalent resistance R (Ω)	0.5
Load inductance L_{oa} (mH)	400
Load resistance R_{oa} (Ω)	500
Sampling time T_s (μs)	50
AC side voltage v_a RMS (V)	0 (passive network)

4.1. Accuracy Analysis

The waveforms calculated by the UMM were compared with those generated from the SIM to evaluate the accuracy of the proposed approach. The results are shown in Figures 4 and 5. In these figures, the solid line "I" represents the result of UMM, the dashed line "II" represents the result of SIM.

4.1.1. Dynamic Simulation under Normal Operation

Figure 4 displays the out current, i_{oa}, and output voltage, v_{aN}, versus time. It shows that within a 0.2 s time range, the results calculated via UMM are favorable compared with the result obtained from SIM. According to Table 2, the root mean square errors are within a reasonable range. These results demonstrate the accuracy of the UMM method at steady state.

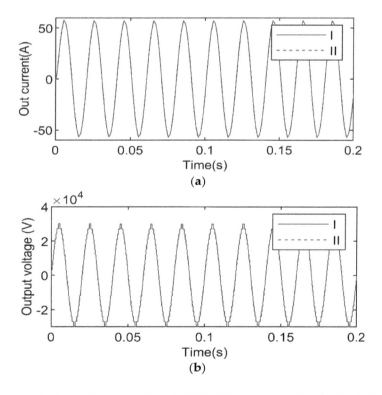

Figure 4. Output current/voltage diagram of the MMC. "I" represents the result of UMM, "II" represents the result of SIM. (**a**) current i_{oa}, (**b**) voltage v_{aN}.

For example, Figure 5a shows the transient behaviors of the circulating current. Figure 5b shows a trend of capacitor voltages for the upper and lower arm from transient to steady state. The upper arm current waveform is shown in Figure 5c. Before 0.06 s, the system was at a transient state, and the UMM and SIM results had shown similar behavior with a slight amplitude difference. After the transient state, the system tended to be stable and the two results were completely coincident.

The root mean square errors are shown in Table 2. From the numerical point of view, except for v_{aN} the errors of other variables are small. In the meantime, the value of v_{aN} is only less than 3/10,000 relative to its RMS (21,216 V).

In short, the perfect coincidence of the above-mentioned various dynamic waveforms (output current i_{oa}/voltage v_{aN}, circulating current i_{ca}, capacitive voltage $u_{am,n}$, and upper arm current i_{au}) has proved the accuracy and correctness of the proposed algorithm UMM.

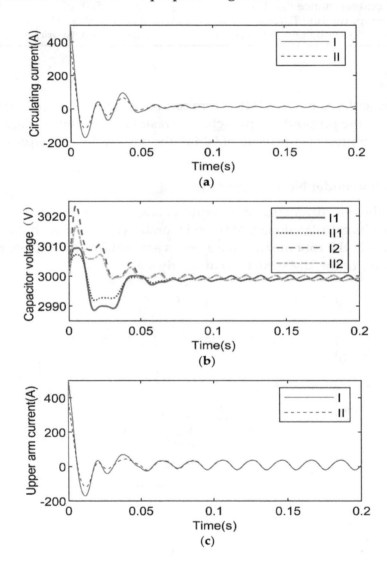

Figure 5. Circulating current/capacitor voltage diagram of the MMC. "I" represents the result of UMM, "II" represents the result of SIM. (**a**) circulating current i_{ca}, (**b**) voltage v_{aN}. "1" represents the upper arm, "2" represents the lower arm, (**c**) upper arm current i_{au}.

Table 2. Root mean square error of UMM and SIM.

Parameters	Root Mean Square Error	Parameters	Root Mean Square Error
i_{oa}	0.0061 A	i_{au}	0.0638 A
v_{aN}	6.4867 V	$v_{au,1}$	0.2855 V
i_{ca}	0.0668 A	$v_{al,1}$	0.6646 V

4.1.2. Dynamic Simulation of SM Open Circuit Fault

The open circuit fault of SM was assumed to study the effect of the internal fault of sub module on the MMC system. The time interval of failure is given as [3, 3.04]—i.e., two cycles.

In Figure 6, it can be seen that in the transient process of fault, the waveforms of UMM and that of SIM also match well.

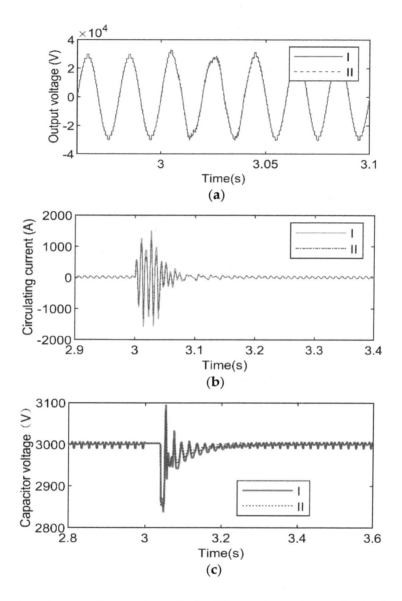

Figure 6. The waveform in case of open circuit fault. "I" represents the result of UMM, "II" represents the result of SIM. (**a**) output voltage, (**b**) circulating current, (**c**) capacitor voltage waveform in the case of open circuit fault.

4.2. Computational Efficiency

A 5 s period was tested with a 50 μs simulation time step. In this paper, we considered three groups of data (number of submodules, $N = 8, 12, 20$) for calculation, and the calculation results are shown in Figure 7. For example, when $N = 20$, the UMM took 7.1 s to get the result, while SIM consumed 62.2 s. The computational efficiency of UMM is significantly improved, which is about 8.7 times faster than that of SIM. However, the computational efficiency of the average value model (AVM) lies in the middle of the three.

Another advantage of UMM over SIM is that the SIM system took a large amount of time to build a system with many SMs.

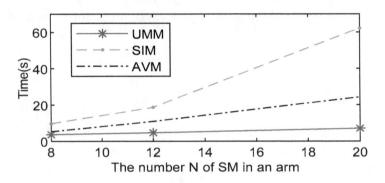

Figure 7. CPU times for UMM and SIM.

Under different sampling times (10, 30, 50 μs), the Central Processing Unit (CPU) times for UMM have been shown in Figure 8 as a 0.1 s period. It can be seen in Figure 8 that at $N < 150$, the three sampling times have little difference. However, with the increase in N, the time gap among them becomes larger. Especially when N is 404, the CPU times of the three were 162.8, 44.0, and 25.5 s, respectively. However, in fact, only the sampling time 50 μs for the UMM system met the accuracy requirements.

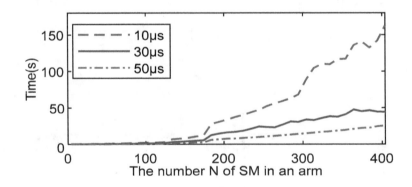

Figure 8. CPU times for UMM under different sampling times.

5. Application Based on UMM

Section 4 had proved the accuracy and efficiency of UMM. However, in this section, UMM will be used to study the MMC system, which is also the biggest difference from most existing models. The quality of the output voltage and current is a criterion to judge the performance of an inverter. For a long time, it was believed that the output current and voltage quality would be improved with an increase in the SM quantity, but there is no definite conclusion and mathematical evidence for this.

It is time-consuming to build simulation models, so it is impossible to simulate and analyze a large number of SMs. The utilization of UMM can facilitate this analysis. In this section, the influence of the number N of SMs on the output voltage and current THD under different conditions will be investigated. The dynamic performance of the MMC system was analyzed on the effects of different Ns.

5.1. Output Voltage/Current Harmonic Performance under Different N Values

The calculation parameters are given with Table 1, except for the parameter N, where N is an independent variable from 0 to 400. The simulation results are shown in Figure 9.

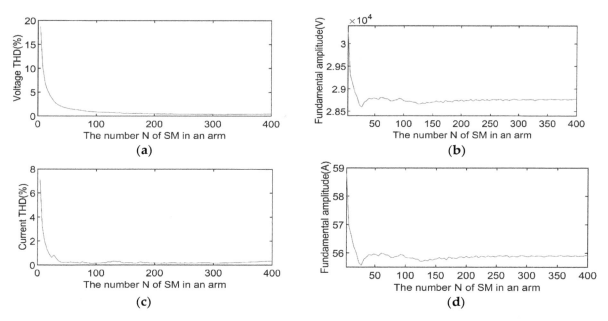

Figure 9. The oscillogram varying with the number of SM. (**a**) Output voltage THD, (**b**) output voltage fundamental amplitude, (**c**) output current THD, (**d**) output current fundamental amplitude.

In Figure 9, during the period when N was increased from 4 to 50, both the THD and fundamental amplitude decreased rapidly. After that, as N increased, the change rate tends to reduce. The optimal value was at (308, 0.352%) in Figure 9a, and (92, 0.1335%) in Figure 9c. The fundamental amplitude reached the minimum value at $N = 28$, then increased slightly in Figure 9b,d. After $N > 50$, it reached a stable state with small fluctuations.

In summary, the results show that the relationship between the quality of output voltage (or current) and the number N of SMs is not proportional. It has an optimal solution; using UMM, it is possible to find the optimal solution. In practice, in the design of N should be also considered the cost of hardware, the voltage grade, and the complexity of control.

5.2. Output Voltage/Current Harmonic Performance under Different SM Capacitance C and N Values

The influence of different SM capacitances, C, is studied in this section. The results are shown in Figure 10.

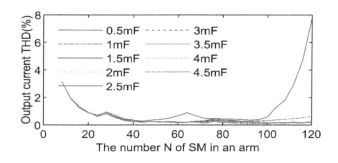

Figure 10. The THD diagram of the output current varying with C and N. The C value is shown in the legend.

Figure 10 displays the THD value of the output current and voltage changing with a different number N of SMs and SM capacitance C, respectively. As illustrated, when the N is small, less than 50, the THD basically coincides with different C values. After $N > 100$, the THD of the curve of $C = 0.5$ mF rises rapidly and become unstable. This phenomenon indicates that, along with the increase in the SMs, the capacity should be increased accordingly to keep the system stable.

According to [28], the voltage fluctuation rate of the SM capacitor can be calculated in the following equation:

$$\varepsilon = \frac{1}{3} \frac{S_{vN}}{N\omega C u_C^2} \tag{35}$$

where S_{vN} is the MMC nominal power, and $\omega = 2\pi f$.

Substitute $u_C = U_{dc}/N$ into (32), then:

$$\varepsilon = \frac{NS_{vN}}{3\omega C U_{dc}^2} \tag{36}$$

Generally, S_{vN}, ω, and U_{dc} are constants. As a result, ε is directly proportional to N and inversely proportional to C. When N increases and C is a fixed value, the system will diverge and become unstable after ε exceeds the allowable value.

5.3. Output Voltage/Current Harmonic Performance under Different Arm Inductance L and N Values

The influence of different SM inductances, L, is studied in this section. The simulation results are shown in Figure 11.

In Figure 11, when N is less than 110, the THD of the output voltage is basically the same for all the inductance values; after that, the THD decreases slowly as N increases, but the larger the arm inductance value, the higher the THD. It shows a small mutation near $N = 140$.

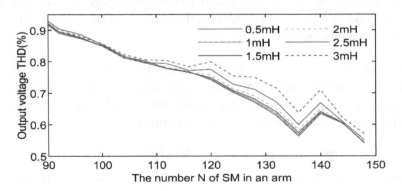

Figure 11. The THD diagram of the output voltage varying with L and N. The L value is shown in the legend.

Figure 12 shows the THD value of the output voltage changing with N and L. When N is less than 40, the output current THD is around the same value. The inductance value has different effects on THD in different areas of N. For example, the areas with the smallest THD are as follows: $L = 1$ mF, $N \in [44, 56]$; $L = 1.5$ mF, $N \in [56, 68] \cap [100, 116]$; $L = 3$ mF in the interval $[68, 80]$, $L = 0.5$ mF, $N \in [68, 100] \cap [116, 144]$ in the interval $[68, 100]$. After N increased to greater than 96, the THD increases significantly with the increase in L.

From the general trend, as N becomes larger, the inductance value should become smaller. However, if the inductance value is too small, it will lose the effect of suppressing the arm current and fault tolerance. Therefore, the design should be selected based on the actual situation.

5.4. Dynamic Performance under SM Capacitance Decay Fault

In this test, the capacitance of the SM_{au1} was attenuated by 0.6 times. Shown as Figure 13, the failure occurred at 3 s and continued until the end. From the SM voltage waveform, the difference between the waveforms before and after 3 s is obvious. This means that, in addition to the IGBT open circuit fault mentioned above, the proposed model can also study the parameter fault.

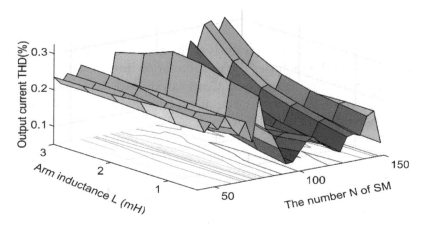

Figure 12. The THD diagram of the output current varying with *L* and *N*.

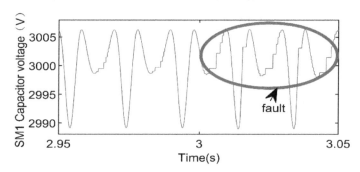

Figure 13. The SM1 capacitor voltage diagram under capacitance decay fault.

5.5. Object-Oriented Parametric Design

Through object-oriented programming in Figure 14, the program can realize human-computer interaction and reduce the workload of designers, so as to realize the universality, versatility, ease of use, and operation of MMC. Of course, this is only an example. See reference [28] for the calculation of relevant parameters, which will not be discussed here.

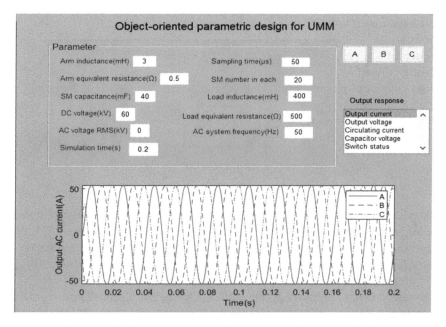

Figure 14. Design and verification interface for MMC.

6. Conclusions

A general mathematical model has been derived in detail from the circuit structure of MMC in this paper. The mathematical model includes the non-linear or linear characteristics of switching state, capacitance voltage, inductance, and resistance. Compared with the traditional MATLAB/Simulink (21-level MMC), the output voltage/current, circulating current, and capacitor voltage coincide, and the accuracy of the proposed model is fully proved. Additionally, this model is 8.7 ($N = 20$) times faster than the traditional simulation method. By changing the arm induction, SM capacitance, and the number of SMs to study the impact of each parameter on the output current/voltage THD, the optimal number of SMs has been found. In addition, the proposed model can also analyze the structure and parameter faults of MMC. The UMM proposed lays a solid foundation for further research into the dynamic performances of MMC with a large number of SMs.

The UMM mainly analyzes the static/dynamic system by discretizing the state equation of MMC. In essence, it belongs to the analytical analysis mathematical model. Its iterative calculation speed is much faster than that of the simulation model. At the same time, it also reduces the modeling time of the large-scale component modules and improves the universality. This is the reason that although the simulation model compared is only based on MATLAB/Simulink, it is not lost generality.

Author Contributions: Conceptualization, M.L. and Z.L.; methodology, M.L.; software, M.L.; validation, M.L.; resources, X.Y.; data curation, X.Y.; writing—original draft preparation, M.L.; writing—review and editing, M.L.; project administration, X.Y.; funding acquisition, Z.L. All authors have read and agreed to the published version of the manuscript.

Abbreviations

MMC	modular multilevel converter
THD	total harmonic distortion
UMM	universal mathematical model
SIM	simulation model
NLM	nearest level modulation
TDM	traditional detailed model
AM	accelerated model
DP	dynamic phasor
HVDC	High-voltage direct voltage
SM	sub-module
DC	direct current
AC	alternating current
IGBT	insulated gate bipolar transistor
AVM	average value model
EAM	enhanced accelerated model
EMT	electromagnetic transient

References

1. Raju, M.N.; Sreedevi, J.; Mandi, R.P.; Meera, K.S. Modular multilevel converters technology: A comprehensive study on its topologies, modelling, control and applications. *IET Power Electron.* **2019**, *12*, 149–169. [CrossRef]
2. Kouro, S.; Malinowski, M.; Gopakumar, K.; Pou, J.; Franquelo, L.G.; Wu, B.; Rodriguez, J.; Perez, M.A.; Leon, J.I. Recent advances and industrial applications of multilevel converters. *IEEE Trans. Ind. Electron.* **2010**, *57*, 2553–2580. [CrossRef]
3. Debnath, S.; Member, S.; Qin, J.; Member, S.; Bahrani, B. Operation, Control, and Applications of the Modular Multilevel Converter: A Review. *IEEE Trans. Power Electron.* **2015**, *30*, 37–53. [CrossRef]
4. Nguyen, T.H.; Al Hosani, K.; El Moursi, M.S.; Blaabjerg, F. An Overview of Modular Multilevel Converters

in HVDC Transmission Systems with STATCOM Operation during Pole-to-Pole DC Short Circuits. *IEEE Trans. Power Electron.* **2019**, *34*, 4137–4160. [CrossRef]

5. Picas, R.; Zaragoza, J.; Pou, J.; Ceballos, S.; Konstantinou, G.; Capella, G.J. Study and comparison of discontinuous modulation for modular multilevel converters in motor drive applications. *IEEE Trans. Ind. Electron.* **2019**, *66*, 2376–2386. [CrossRef]

6. Li, B.; Zhou, S.; Xu, D.; Yang, R.; Xu, D.; Buccella, C.; Cecati, C. An Improved Circulating Current Injection Method for Modular Multilevel Converters in Variable-Speed Drives. *IEEE Trans. Ind. Electron.* **2016**, *63*, 7215–7225. [CrossRef]

7. Perez, M.A.; Bernet, S.; Rodriguez, J.; Kouro, S.; Lizana, R. Circuit topologies, modeling, control schemes, and applications of modular multilevel converters. *IEEE Trans. Power Electron.* **2015**, *30*, 4–17. [CrossRef]

8. Li, J.; Konstantinou, G.; Member, S.; Wickramasinghe, H.R.; Member, S.; Pou, J. Operation and Control Methods of Modular Multilevel Converters in Unbalanced AC Grids: A Review. *IEEE J. Emerg. Sel. Top. Power Electron.* **2019**, *7*, 1258–1271. [CrossRef]

9. Ilves, K.; Antonopoulos, A.; Norrga, S.; Nee, H.-P. Steady-state analysis of interaction between harmonic components of arm and line quantities of modular multilevel converters. *IEEE Trans. Power Electron.* **2012**, *27*, 57–68. [CrossRef]

10. Vatani, M.; Member, S.; Hovd, M.; Member, S. Control of the Modular Multilevel Converter Based on a Discrete-Time Bilinear Model Using the Sum of Squares Decomposition Method. *IEEE Trans. Power Del.* **2015**, *30*, 2179–2188.

11. Dekka, A.; Wu, B.; Fuentes, R.; Perez, M.; Zargari, N. Evolution of Topologies, Modeling, Control Schemes, and Applications of Modular Multilevel Converters. *IEEE J. Emerg. Sel. Top. Power Electron.* **2017**, *5*, 1631–1656. [CrossRef]

12. Vatani, M.; Member, S.; Bahrani, B. Indirect Finite Control Set Model Predictive Control of Modular Multilevel Converters. *IEEE Trans. Smart Grid* **2015**, *6*, 1520–1529. [CrossRef]

13. Wang, J.; Liang, J.; Gao, F.; Xiaoming, D.; Wang, C.; Zhao, B. A Closed-Loop Time-Domain Analysis Method for Modular Multilevel Converter. *IEEE Trans. Power Electron.* **2017**, *32*, 7494–7508. [CrossRef]

14. Saad, H.; Peralta, J.; Dennetiere, S.; Mahseredjian, J.; Jatskevich, J.; Martinez, J.A.; Davoudi, A.; Saeedifard, M.; Sood, V.; Wang, X. Dynamic averaged and simplified models for MMC based HVDC transmission systems. *IEEE Trans. Power Del.* **2013**, *28*, 1723–1730.

15. Leon, A.E.; Amodeo, S.J. Modeling, control, and reduced-order representation of modular multilevel converters. *Electr. Power Syst. Res.* **2018**, *163*, 196–210. [CrossRef]

16. Belhaouane, M.M.; Ayari, M.; Guillaud, X.; Braiek, N.B. Robust Control Design of MMC-HVDC Systems Using Multivariable Optimal Guaranteed. *IEEE Trans. Ind. Appl.* **2019**, *55*, 2952–2963. [CrossRef]

17. Xu, J.; Gole, A.M.; Zhao, C. The use of averaged-value model of modular multilevel converter in DC grid. *IEEE Trans. Power Del.* **2015**, *30*, 519–528.

18. Meng, X.; Han, J.; Bieber, L.; Wang, L.; Li, W.; Belanger, J. A Universal Blocking-Module-Based Average Value Model of Modular Multilevel Converters with Different Types of Submodules. *IEEE Trans. Energy Convers.* **2020**, *35*, 53–66. [CrossRef]

19. Lyu, J.; Zhang, X.; Cai, X.; Molinas, M. Harmonic State-Space Based Small-Signal Impedance Modeling of a Modular Multilevel Converter with Consideration of Internal Harmonic Dynamics. *IEEE Trans. Power Electron.* **2019**, *34*, 2134–2148. [CrossRef]

20. Freitas, C.M.; Watanabe, E.H.; Monterio, L.F.C. A linearized small-signal Thévenin-equivalent model of a voltage-controlled modular multilevel converter. *Electr. Power Syst. Res.* **2020**, *182*, 106231–106241. [CrossRef]

21. Gnanarathna, U.N.; Gole, A.M.; Jayasinghe, R.P. Efficient modeling of modular multilevel HVDC converters (MMC) on electromagnetic transient simulation programs. *IEEE Trans. Power Del.* **2011**, *26*, 316–324.

22. Xu, J.; Zhao, C.; Liu, W.; Guo, C. Accelerated model of modular multilevel converters in PSCAD/EMTDC. *IEEE Trans. Power Del.* **2013**, *28*, 129–136.

23. Beddard, A.; Barnes, M.; Preece, R. Comparison of Detailed Modeling Techniques for MMC Employed on VSC-HVDC Schemes. *IEEE Trans. Power Deliv.* **2015**, *30*, 579–589.

24. Rupasinghe, J.; Member, S.; Filizadeh, S.; Member, S. A Dynamic Phasor Model of an MMC with Extended Frequency Range for EMT Simulations. *IEEE J. Emerg. Sel. Top. Power Electron.* **2019**, *7*, 30–40. [CrossRef]

25. Rashidi, N.; Burgos, R.; Roy, C.; Boroyevich, D. On the Modeling and Design of Modular Multilevel

Converters with Parametric and Model-Form Uncertainty Quantification. *IEEE Trans. Power Electron.* **2020**, *35*, 10168–10179. [CrossRef]

26.   Liu, M.; Li, Z.; Yang, X. Tracking Control of Modular Multilevel Converter Based on Linear Matrix Inequality without Coordinate Transformation. *Energies* **2020**, *13*, 1978. [CrossRef]

27.   Gutierrez, B. Modular Multilevel Converters (MMCs) Controlled by Model Predictive Control with Reduced Calculation Burden. *IEEE Trans. Power Electron.* **2018**, *33*, 9176–9187. [CrossRef]

28.   Zheng, X. *HVDC System*, 2nd ed.; China Machine Press: Beijing, China, 2016; pp. 69–71.

Permissions

The contributors of this book come from diverse backgrounds, making this book a truly international effort. This book will bring forth new frontiers with its revolutionizing research information and detailed analysis of the nascent developments around the world.

We would like to thank all the contributing authors for lending their expertise to make the book truly unique. They have played a crucial role in the development of this book. Without their invaluable contributions this book wouldn't have been possible. They have made vital efforts to compile up to date information on the varied aspects of this subject to make this book a valuable addition to the collection of many professionals and students.

This book was conceptualized with the vision of imparting up-to-date information and advanced data in this field. To ensure the same, a matchless editorial board was set up. Every individual on the board went through rigorous rounds of assessment to prove their worth. After which they invested a large part of their time researching and compiling the most relevant data for our readers.

The editorial board has been involved in producing this book since its inception. They have spent rigorous hours researching and exploring the diverse topics which have resulted in the successful publishing of this book. They have passed on their knowledge of decades through this book. To expedite this challenging task, the publisher supported the team at every step. A small team of assistant editors was also appointed to further simplify the editing procedure and attain best results for the readers.

Apart from the editorial board, the designing team has also invested a significant amount of their time in understanding the subject and creating the most relevant covers. They scrutinized every image to scout for the most suitable representation of the subject and create an appropriate cover for the book.

The publishing team has been an ardent support to the editorial, designing and production team. Their endless efforts to recruit the best for this project, has resulted in the accomplishment of this book. They are a veteran in the field of academics and their pool of knowledge is as vast as their experience in printing. Their expertise and guidance has proved useful at every step. Their uncompromising quality standards have made this book an exceptional effort. Their encouragement from time to time has been an inspiration for everyone.

The publisher and the editorial board hope that this book will prove to be a valuable piece of knowledge for researchers, students, practitioners and scholars across the globe.

List of Contributors

Stefania Cuoghi, Lorenzo Ntogramatzidis and Fabrizio Padula
School of Electrical Engineering, Computing and Mathematical Sciences, Curtin University, Bentley 6102, Western Australia, Australia

Gabriele Grandi
Department of Electrical, Electronic, and Information Engineering, University of Bologna, 40136 Bologna, Italy

Yai Wang and Zhigang Liu
School of Electrical Engineering, Southwest Jiaotong University, Chengdu 610031, China

S. M. Rakiul Islam, Sung-Yeul Park and Sung-Min Park
Electrical and Computer Engineering Department, University of Connecticut, Storrs, CT 06269, USA

Shaobo Zheng and Song Han
Computer Science Engineering Department, University of Connecticut, Storrs, CT 06269, USA

Sung-Min Park
Electrical and Electronic Engineering Department, Hongik University, Sejong-si 30016, Korea

Maria R. Rogina, Alberto Rodriguez and Diego G. Lamar
Electrical, Electronic, Computers and Systems Engineering Department, University of Oviedo, 33204 Gijón, Spain

Jaume Roig, German Gomez and Piet Vanmeerbeek
On Semiconductor, 9700 Oudenaarde, Belgium

Jiang You and Mengyan Liao
College of Automation, Harbin Engineering University, Harbin 150001, China

Hailong Chen
College of Shipbuilding Engineering, Harbin Engineering University, Harbin 150001, China

Negareh Ghasemi
School of Information Technology and Electrical Engineering, The University of Queensland, Brisbane 4072, Australia

Mahinda Vilathgamuwa
School of Electrical Engineering and Computer Science, Queensland University of Technology, Brisbane 4000, Australia

Jorge Garcia and Pablo Garcia
LEMUR Group, Department of Electrical Engineering, University of Oviedo, 33204 Gijon, Spain

Fabio Giulii Capponi and Giulio De Donato
Department of Astronautical, Electrical and Energy Engineering, University of Roma "La Sapienza", 00184 Roma, Italy

Lan Li, Hao Wang, Lun Chai and Bing Li
College of Electrical and Power Engineering, Taiyuan University of Technology, Shanxi 030024, China

Xiangping Chen
Electrical Engineering School, Guizhou University, Guiyang 550025, China
Faculty of Engineering, Cardiff University, Cardiff CF24 3AA, UK

Abid Ali Shah Bukhari
School of Engineering and Applied Science, Aston University, Birmingham B4 7ET, UK

Goh Teck Chiang and Takahide Sugiyama
Toyota Central R&D Labs Inc., Nagakute City 480-1192, Japan

Pawel Szczesniak
Institute of Electrical Eng., University of Zielona Góra, 65-516 Zielona Góra, Poland

En-Chih Chang and Chun-An Cheng
Department of Electrical Engineering, I-Shou University, No.1, Sec. 1, Syuecheng Rd., Dashu District, Kaohsiung City 84001, Taiwan

Lung-Sheng Yang
Department of Electrical Engineering, Far East University, No.49, Zhonghua Rd., Xinshi Dist., Tainan City 74448, Taiwan

G. Kiran Kumar and D. Elangovan
School of Electrical Engineering, VIT Vellore, Tamil Nadu 632014, India

E. Parimalasundar
Department of Electrical and Electronics Engineering, Sree Vidyanikethan Engineering College, Tirupati 517102, India

P. Sanjeevikumar and Jens Bo Holm-Nielsen
Center for Bioenergy and Green Engineering, Department of Energy Technology, Aalborg University, 6700 Esbjerg, Denmark

Francesco Lannuzzo
Department of Energy Technology, Aalborg University, 9220 Aalborg, Denmark

Mahmoud Nassary, Mohamed Orabi, Emad M. Ahmed and El-Sayed Hasaneen
 APAERC, Faculty of Engineering, Aswan University, 81542, Aswan, Egypt

Manuel Arias
Departamento de Ingeniería Eléctrica, Electrónica, de Computadores y Sistemas University of Oviedo, 33204 Gijón, Spain

Ming Liu
College of Electrical Engineering, Guizhou University, Guiyang 550025, China
Department of Mechanical Engineering, Guizhou College of Electronic Science and Technology, Guian 550003, China

Zetao Li
College of Electrical Engineering, Guizhou University, Guiyang 550025, China
Guizhou Provincial Key Laboratory of Internet Intelligent Manufacturing, Guiyang 550025, China

Xiaoliu Yang
College of Electrical Engineering, Guizhou University, Guiyang 550025, China

Index

Printed in the USA
CPSIA information can be obtained
at www.ICGtesting.com
JSHW051407091023
49903JS00006B/324

9 781639 897414